Quantity	Units/Conversion
General	
Acceleration	1 in/s^2 = 0.0254 m/s^2
Area	1 in^2 = 645.16 mm^2
Density (i)	1 lbm/in^3 = 27679.905 kg/m^3
(ii)	1 slug/ft^3 = 515.379 kg/m^3
Force	1 lb = 4.448 N (N = Newton)
Frequency	Hz (hertz = cycle/s)
Length	1 in = 0 .0254 m; 1 ft = 0.3048 m
Mass (i)	1 lbm = 0.45359 kg
(ii)	1 slug = 14.594 kg
Moment	1 in.-lb = 0.1130 N $\cdot$ m
Moment of inertia (area)	1 in^4 = 416231.4 mm^4
Moment of inertia (mass) (i)	1 lbm-in^2 = 2.9264E $-$ 4 kg $\cdot$ m^2
(ii)	1 slug-in^2 = 0.009415 kg $\cdot$ m^2
Power (i)	1 in.-lb/s = 0.1130 W (watt = J/s)
(ii)	1 hp = 0.746 kW (1 hp = 550 ft-lb)
Pressure	1 psi = 6894.8 Pa (psi = pounds/in^2; Pa = N/m^2
Stiffness	1 lb/in. = 175.1 N/m
Stress (i)	1 psi = 6894.8 Pa
(ii)	1 ksi = 6.8948 MPa; 1 MPa = 145.04 psi
	(ksi = 1000 psi; Mpa = 10^6 Pa)
Time	s (second)
Velocity	1 in./s = 0.0254 m/s
Volume	1 in.^{3} = 16.3871E $-$ 6 m^3
Work, energy	1 in.-lb = 0.1130 J (joule = N $\cdot$ m)
Heat Transfer	
Convection coefficient	1 Btu/h.ft^2.°F = 5.6783 W/m^2.°C
Heat	1 Btu = 1055.06 J (1 Btu = 778.17 ft-lb)
Heat flux	1 Btu/h.ft^2 = 3.1546 W/m^2
Specific heat	1 Btu/°F = 1899.108 J/°C
Temperature (i)	T °F = [(9/5)T + 32]°C
(ii)	T °K = T °C + 273.15 (K = kelvin)
Thermal conductivity	1 Btu/h.ft.°F = 1.7307 W/m.°C
Fluid Flow	
Absolute viscosity	1 lb.s/ft^2 = 478.803 P (poise = g/cm $\cdot$ s)
Kinematic viscosity	1 ft^2/s = 929.03 St (stoke = cm^2/s)
Electric and Magnetic Fields	
Capicitance	F (farad)
Charge	C (coulomb)
Electric charge density	C/m^3
Electric potential	V (volt)
Inductance	H (henry)
Permeability	H/m
Permittivity	F/m
Scalar magnetic potential	A (ampere)

Introduction
to Finite Elements
in Engineering

Introduction to Finite Elements in Engineering

TIRUPATHI R. CHANDRUPATLA, Ph.D., P.E.

GMI Engineering & Management Institute
Flint, Michigan

ASHOK D. BELEGUNDU, Ph.D.

The Pennsylvania State University
University Park, Pennsylvania

PRENTICE HALL, Englewood Cliffs, New Jersey 07632

Library of Congress Cataloging-in-Publication Data

CHANDRUPATLA, TIRUPATHI R. (date)
    Introduction to finite elements in engineering / Tirupathi R.
Chandrupatla and Ashok D. Belegundu.
        p.   cm.
    Includes bibliographical references and index.
    ISBN 0-13-483082-2
    1. Finite element method.   2. Engineering mathematics.
I. Belegundu, Ashok D.   II. Title.
TA347.F5C463   1991
620'.001'51535—dc20   90-42503
                                                          CIP

Editorial/production supervision: *Merrill Peterson*
Interior design: *Joan Stone*
Cover design: *Ben Santora*
Prepress Buyer: *Linda Behrens*
Manufacturing buyer: *Dave Dickey*

© 1991 by Prentice-Hall, Inc.
A Division of Simon & Schuster
Englewood Cliffs, New Jersey 07632

Printed in the United States of America

10 9 8 7 6 5 4 3 2

ISBN 0-13-483082-2

PRENTICE-HALL INTERNATIONAL (UK) LIMITED, *London*
PRENTICE-HALL OF AUSTRALIA PTY. LIMITED, *Sydney*
PRENTICE-HALL CANADA INC., *Toronto*
PRENTICE-HALL HISPANOAMERICANA, S.A., *Mexico*
PRENTICE-HALL OF INDIA PRIVATE LIMITED, *New Delhi*
PRENTICE-HALL OF JAPAN, INC., *Tokyo*
SIMON & SCHUSTER ASIA PTE. LTD., *Singapore*
EDITORA PRENTICE-HALL DO BRASIL, LTDA., *Rio de Janeiro*

To our parents

ABOUT THE AUTHORS

Tirupathi R. Chandrupatla is a Professor of Mechanical Engineering at GMI Engineering & Management Institute, which was formerly General Motors Institute. He received his Bachelor's degree from the Regional Engineering College, Warangal, which was affiliated with Osmania University, India. After obtaining his Master's degree in Design and Manufacturing from I.I.T., Bombay, he worked as Design Engineer with Hindustan Machine Tools, Bangalore. He then taught at the Indian Institute of Technology, Bombay. He pursued his graduate studies in the Department of Aerospace Engineering and Engineering Mechanics of the University of Texas at Austin and received his Ph.D. in 1977. He subsequently taught at the University of Kentucky before joining GMI in 1979.

Dr. Chandrupatla has broad interests, which include finite element analysis, design, and manufacturing engineering. He is a consultant to industry in these areas. Dr. Chandrupatla is a registered professional engineer and also a Certified Manufacturing Engineer. He is an active member of ASME and SME.

Ashok D. Belegundu is an Associate Professor of Mechanical Engineering at The Pennsylvania State University. He was on the faculty at GMI from 1982 through 1986. He received his Ph.D. in 1982 at the University of Iowa and his Bachelor's degree at the Indian Institute of Technology, Madras.

Dr. Belegundu's teaching and research interests include finite elements, machine design, reliability, and optimization techniques. He has published several papers on optimum design and worked on sponsored projects for government and industry. He is a member of AIAA, ASCE, ASME.

Contents

PREFACE   *xi*

1 FUNDAMENTAL CONCEPTS   **1**

1.1   Introduction   1

1.2   Historical Background   1

1.3   Outline of Presentation   2

1.4   Stresses and Equilibrium   2

1.5   Boundary Conditions   4

1.6   Strain–Displacement Relations   5

1.7   Stress–Strain Relations   6

1.8   Temperature Effects   9

1.9   Potential Energy and Equilibrium;
         The Rayleigh–Ritz Method   9

1.10   Galerkin's Method   14

1.11   Saint Venant's Principle   18

1.12   Conclusion   18

         Historical References   18

2 MATRIX ALGEBRA AND GAUSSIAN ELIMINATION   21

    2.1   Matrix Algebra   21

    2.2   Gaussian Elimination   29

3 ONE-DIMENSIONAL PROBLEMS   44

    3.1   Introduction   44

    3.2   Finite Element Modeling   45

    3.3   Coordinates and Shape Functions   48

    3.4   The Potential Energy Approach   52

    3.5   The Galerkin Approach   56

    3.6   Assembly of the Global Stiffness Matrix and Load Vector   58

    3.7   Properties of $\mathbf{K}$ 61

    3.8   The Finite Element Equations; Treatment of Boundary Conditions   63

    3.9   Quadratic Shape Functions   79

    3.10   Temperature Effects   86

4 TRUSSES   100

    4.1   Introduction   100

    4.2   Plane Trusses   101

    4.3   Three-Dimensional Trusses   113

    4.4   Assembly of Global Stiffness Matrix for the Banded and Skyline Solutions   114

5 TWO-DIMENSIONAL PROBLEMS USING CONSTANT STRAIN TRIANGLES   130

    5.1   Introduction   130

    5.2   Finite Element Modeling   131

    5.3   Constant Strain Triangle (CST)   133

    5.4   Problem Modeling and Boundary Conditions   150

6 AXISYMMETRIC SOLIDS SUBJECTED TO AXISYMMETRIC LOADING   **165**

6.1   Introduction 165

6.2   Axisymmetric Formulation 165

6.3   Finite Element Modeling: Triangular Element 168

6.4   Problem Modeling and Boundary Conditions 179

7 TWO-DIMENSIONAL ISOPARAMETRIC ELEMENTS AND NUMERICAL INTEGRATION   **194**

7.1   Introduction 194

7.2   The Four-Node Quadrilateral 194

7.3   Numerical Integration 201

7.4   Higher-Order Elements 207

8 BEAMS AND FRAMES   **223**

8.1   Introduction 223

8.2   Finite Element Formulation 226

8.3   Load Vector 230

8.4   Boundary Considerations 231

8.5   Shear Force and Bending Moment 232

8.6   Beams on Elastic Supports 234

8.7   Plane Frames 235

8.8   Some Comments 240

9 THREE-DIMENSIONAL PROBLEMS IN STRESS ANALYSIS   **251**

9.1   Introduction 251

9.2   Finite Element Formulation 252

9.3   Stress Calculations 257

9.4   Mesh Preparation 258

9.5   Hexahedral Elements and Higher-Order
       Elements 261

9.6   Problem Modeling 262

10 SCALAR FIELD PROBLEMS   273

10.1   Introduction 273

10.2   Steady-State Heat Transfer 275

10.3   Torsion 302

10.4   Potential Flow, Seepage, Electric and Magnetic
       Fields, Fluid Flow in Ducts 309

10.5   Conclusion 315

11 DYNAMIC CONSIDERATIONS   334

11.1   Introduction 334

11.2   Formulation 334

11.3   Element Mass Matrices 338

11.4   Evaluation of Eigenvalues and Eigenvectors 343

11.5   Interfacing with Previous Finite Element Programs
       and a Program for Determining Critical Speeds of
       Shafts 354

11.6   Conclusion 355

12 PREPROCESSING AND POSTPROCESSING   372

12.1   Introduction 372

12.2   Mesh Generation 373

12.3   Postprocessing 381

12.4   Conclusion 385

APPENDIX   403

Proof of $dA = \det \mathbf{J}\, d\xi d\eta$ 403

BIBLIOGRAPHY   407

INDEX   411

Preface

The text material evolved from over ten years of teaching courses on finite elements, primarily at the undergraduate level and also at the introductory graduate level. The period also involved close interaction with industry through consulting and developmental projects and continuing education courses in finite element theory and applications. Our approach has been appreciated by the students and practicing engineers who have taken our courses. This gave us the motivation to develop our course material into this book.

This book is intended as a textbook for senior year undergraduate and first year graduate students and also as a learning source for practicing engineers. We have strived to integrate finite element theory, problem formulation, and computer analysis. The isoparametric concept is introduced at the beginning while considering one-dimensional problems and is subsequently used in a consistent and unified manner for two- and three-dimensional problems. Both the energy and the Galerkin approaches are used in the development of stiffness and load matrices and governing equations. Attention is given throughout to the modeling of engineering problems, specification of boundary conditions, and handling of temperature effects. Pre- and postprocessing concepts are discussed. Deformation and stress analyses are introduced first to solidify the concepts and then a detailed formulation of heat, fluid flow, and electric and magnetic fields is presented.

Complete computer programs, which implement and parallel the theory, are included at the end of each chapter starting with Chapter 2. Listings of BASIC programs are given in the main text. The $5\frac{1}{4}$-in. diskette (formatted for IBM-compatible PCs) included with this book contains a set of BASIC programs and their FORTRAN versions stored in text form. BASIC programs are written using

BASICA/GWBASIC, which are available on most PCs. They can also be readily run using QUICKBASIC or TURBOBASIC. Data preparation for computer programs is presented through examples at the end of each chapter.

Chapter 1 gives a brief historical background and develops the fundamental concepts. Equations of equilibrium, stress–strain relations, strain–displacement relations, and the principle of potential energy are reviewed. The concept of Galerkin's method is introduced.

Properties of matrices and determinants are reviewed in Chapter 2. The Gaussian elimination method is presented, and its relationship to the solution of symmetric banded matrix equations and the skyline solution is discussed.

Chapter 3 develops the key concepts of finite element formulation by considering one-dimensional problems. The steps include development of shape functions, derivation of element stiffness, formation of global stiffness, treatment of boundary conditions, solution of equations, and stress calculations. Both the potential energy approach and Galerkin's formulations are presented. Consideration of temperature effects is also included.

Finite element formulation for plane and three-dimensional trusses is developed in Chapter 4. The assembly of global stiffness in banded and skyline forms is explained. Computer programs for both banded and skyline solutions are given.

Chapter 5 introduces the finite element formulation for two-dimensional plane stress and plane strain problems using constant strain triangle (CST) elements. Problem modeling and treatment of boundary conditions are presented in detail. Chapter 6 treats the modeling aspects of axisymmetric solids subjected to axisymmetric loading. Several real-world problems are included in this chapter. Chapter 7 introduces the concepts of isoparametric quadrilateral and higher-order elements and numerical integration using Gaussian quadrature.

Beams and frames are presented in Chapter 8 using Hermite shape functions. Chapter 9 gives the tetrahedral element for three-dimensional stress analysis.

Scalar field problems are treated in detail in Chapter 10. Galerkin formulation for steady-state heat transfer, torsion, potential flow, seepage flow, electric and magnetic fields, and fluid flow in ducts are covered. The computer programs that are included can be applied to several field problems.

Chapter 11 introduces dynamic considerations. Element mass matrices are given. Techniques for evaluation of eigenvalues (natural frequencies) and eigenvectors (mode shapes) are discussed. The programs include the inverse iteration scheme and the generalized Jacobi method.

Preprocessing and postprocessing concepts are presented in Chapter 12. This chapter includes several useful programs for mesh generation, 2-D plotting, finite element data handling, and contour plotting.

At the undergraduate level some topics may be dropped in a one-semester course without breaking the continuity of presentation. A typical one-semester undergraduate course may cover the following topics:

Chapters 1 and 2 (review)

Chapters 3–6, 8

Chapter 10 (those covering structural aspects may cover the torsion part of the chapter)

Chapters 7, 9, and 11 through projects

Topics from Chapter 12, particularly the programs, may be introduced toward the end of Chapter 5. Either the energy or the Galerkin approach may be chosen for deriving element matrices and obtaining governing equations.

We have included term projects (denoted by *) along with the homework problems. Work on these project problems should give the students an opportunity to write their own programs.

At the graduate level, Chapters 1 and 2 may be left as self-study. Material from Chapters 7, 9, and 11 can be covered in the course.

We thank Jasbir S. Arora, University of Iowa; Terence Frankland, Geneva College, Pennsylvania; and John E. Jackson, Jr., Clemson University, South Carolina, who reviewed our manuscript and gave many constructive suggestions that helped us improve some of the material presented in this book. Their comments have been a source of encouragement for us. We also had the benefit of reviews, comments, and suggestions from Walter E. Haizler of Texas A & M University, Walter Pilkey of the University of Virginia, Tom Rogge of Iowa State University, and R. L. Spiker of Rensselaer Polytechnic Institute, who reviewed some of the early chapters. We express our sincere thanks to all of them.

Dr. Chandrupatla expresses his gratitude to J. Tinsley Oden, whose teaching and encouragement have been a source of inspiration to him.

Dr. Belegundu expresses his deep appreciation and gratitude to his former advisor, Jasbir S. Arora, and to his former professor, Edward J. Haug, for their guidance and teaching.

We are grateful to the chairmen of our departments, George T. Kartsounes (GMI) and Harold R. Jacobs (Penn State), who encouraged and supported our effort. We thank Frank W. Schmidt and Fan-Bill Cheung, Penn State; S. D. Rajan, Arizona State University; and Huseyin R. Hiziroglu and Maciej Zgorzelski, GMI, for valuable suggestions on the manuscript. Encouragement received from former department chairmen at GMI, David H. Harry and Raymond E. Trent, is gratefully acknowledged.

We thank our editor, Doug Humphrey, for providing us with guidance and encouragement throughout this project. We thank Sue Girardi for doing an excellent job in the preparation of the manuscript. We thank our copy editor Virginia Dunn and production editor Merrill Peterson.

We thank our families Suhasini, Sreekanth, and Hareesh and Uma, Shoba, and Soumya, whose encouragement has been ever present.

We thank all the students who took our courses, used the programs, and made many valuable suggestions. We thank our graduate students at Penn State for their teaching assistance. We turn to our readers and users for their suggestions for future editions.

T. R. Chandrupatla
A. D. Belegundu

Introduction
to Finite Elements
in Engineering

chapter one

Fundamental Concepts

1.1 INTRODUCTION

The finite element method has become a powerful tool for the numerical solution of a wide range of engineering problems. Applications range from deformation and stress analysis of automotive, aircraft, building, and bridge structures to field analysis of heat flux, fluid flow, magnetic flux, seepage, and other flow problems. With the advances in computer technology and CAD systems, complex problems can be modeled with relative ease. Several alternative configurations can be tried out on a computer before the first prototype is built. All of this suggests that we need to keep pace with these developments by understanding the basic theory, modeling techniques, and computational aspects of the finite element method. In this method of analysis, a complex region defining a continuum is discretized into simple geometric shapes called finite elements. The material properties and the governing relationships are considered over these elements and expressed in terms of unknown values at element corners. An assembly process, duly considering the loading and constraints, results in a set of equations. Solution of these equations gives us the approximate behavior of the continuum.

1.2 HISTORICAL BACKGROUND

Basic ideas of the finite element method originated from advances in aircraft structural analysis. In 1941, Hrenikoff presented a solution of elasticity problems using the "frame work method." Courant's paper, which used piecewise polynomial interpolation over triangular subregions to model torsion problems, appeared in 1943. Turner et al. derived stiffness matrices for truss, beam, and other elements and presented their findings in 1956. The term *finite element* was first coined and used by Clough in 1960.

In the early 1960s, engineers used the method for approximate solution of problems in stress analysis, fluid flow, heat transfer, and other areas. A book by Argyris in 1955 on energy theorems and matrix methods laid a foundation for further developments in finite element studies. The first book on finite elements by Zienkiewicz and Chung was published in 1967. In the late 1960s and early 1970s, finite element analysis was applied to nonlinear problems and large deformations. Oden's book on nonlinear continua appeared in 1972.

Mathematical foundations were laid in the 1970s. New element development, convergence studies, and other related areas fall in this category.

Today, the developments in mainframe computers and availability of powerful microcomputers has brought this method within reach of students and engineers working in small industries.

1.3 OUTLINE OF PRESENTATION

In this book, we adopt the potential energy and the Galerkin approaches for the presentation of the finite element method. The area of solids and structures is where the method originated and we start our study with these ideas to solidify understanding. For this reason several early chapters deal with rods, beams, and elastic deformation problems. The same steps are used in the development of material throughout the book, so that the similarity of approach is retained in every chapter. The finite element ideas are then extended to field problems in Chapter 10. Every chapter includes a set of problems and computer programs for interaction.

We now recall some fundamental concepts needed in the development of the finite element method.

1.4 STRESSES AND EQUILIBRIUM

A three-dimensional body occupying a volume V and having a surface S is shown in Fig. 1.1. Points in the body are located by x, y, z coordinates. The boundary is constrained on some region, where displacement is specified. On part of the boundary, distributed force per unit area $\mathbf{T}$, also called traction, is applied. Under the force, the body deforms. The deformation of a point $\mathbf{x}$ ($= [x, y, z]^T$) is given by the three components of its displacement:

$$\mathbf{u} = [u, v, w]^T \tag{1.1}$$

The distributed force per unit volume, for example, the weight per unit volume, is

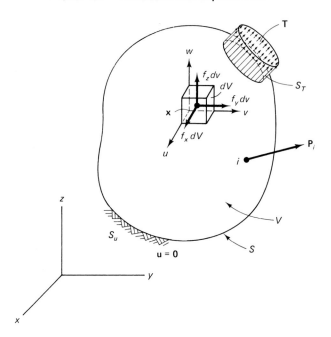

Figure 1.1   Three-dimensional body.

the vector **f** given by

$$\mathbf{f} = [f_x, f_y, f_z]^T \tag{1.2}$$

The body force acting on the elemental volume dV is shown in Fig. 1.1. The surface traction **T** may be given by its component values at points on the surface.

$$\mathbf{T} = [T_x, T_y, T_z]^T \tag{1.3}$$

Examples of traction are distributed contact force and action of pressure. A load **P** acting at a point i is represented by its three components

$$\mathbf{P}_i = [P_x, P_y, P_z]_i^T \tag{1.4}$$

The stresses acting on the elemental volume dV are shown in Fig. 1.2. When the volume dV shrinks to a point, the stress tensor is represented by placing its components in a (3×3) symmetric matrix. However, we represent stress by the six independent components as follows:

$$\boldsymbol{\sigma} = [\sigma_x, \sigma_y, \sigma_z, \tau_{yz}, \tau_{xz}, \tau_{xy}]^T \tag{1.5}$$

where σ_x, σ_y, σ_z are normal stresses and τ_{yz}, τ_{xz}, τ_{xy}, are shear stresses. Let us consider equilibrium of the elemental volume shown in Fig. 1.2. First we get forces on faces by multiplying the stresses by the corresponding areas. Writing $\Sigma F_x = 0$, $\Sigma F_y = 0$ and $\Sigma F_z = 0$, and recognizing $dV = dx\ dy\ dz$, we get the equilibrium equations:

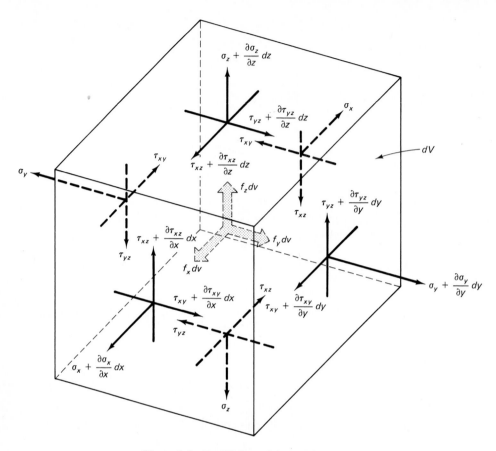

Figure 1.2 Equilibrium of elemental volume.

$$\frac{\partial \sigma_x}{\partial x} + \frac{\partial \tau_{xy}}{\partial y} + \frac{\partial \tau_{xz}}{\partial z} + f_x = 0$$

$$\frac{\partial \tau_{xy}}{\partial x} + \frac{\partial \sigma_y}{\partial y} + \frac{\partial \tau_{yz}}{\partial z} + f_y = 0 \qquad (1.6)$$

$$\frac{\partial \tau_{xz}}{\partial y} + \frac{\partial \tau_{yz}}{\partial y} + \frac{\partial \sigma_z}{\partial z} + f_z = 0$$

1.5 BOUNDARY CONDITIONS

Referring to Fig. 1.1, we find that there are displacement boundary conditions and surface loading conditions. If **u** is specified on part of the boundary denoted by S_u, we have

$$\mathbf{u} = \mathbf{0} \quad \text{on} \quad S_u \qquad (1.7)$$

We can also consider boundary conditions such as $\mathbf{u} = \mathbf{a}$, where $\mathbf{a}$ is a given displacement.

We now consider the equilibrium of an elemental tetrahedron $ABCD$, shown in Fig. 1.3, where DA, DB, and DC are parallel to the x, y, and z axes, respectively, and area ABC, denoted by dA, lies on the surface. If $\mathbf{n} = [n_x, n_y, n_z]^T$ is the unit normal to dA, then area $BDC = n_x dA$, area $ADC = n_y dA$, and area $ADB = n_z dA$. Consideration of equilibrium along the three axes directions gives

$$\sigma_x n_x + \tau_{xy} n_y + \tau_{xz} n_z = T_x$$
$$\tau_{xy} n_x + \sigma_y n_y + \tau_{yz} n_z = T_y \qquad (1.8)$$
$$\tau_{xz} n_x + \tau_{yz} n_y + \sigma_z n_z = T_z$$

These conditions must be satisfied on the boundary, S_T, where the tractions are applied. In this description, the point loads must be treated as loads distributed over small but finite areas.

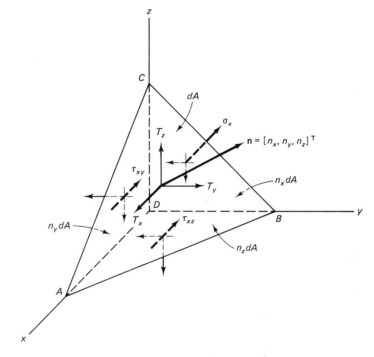

Figure 1.3   An elemental volume at surface.

1.6 STRAIN–DISPLACEMENT RELATIONS

We represent the strains in a vector form that corresponds to the stresses in Eq. 1.5,

$$\boldsymbol{\epsilon} = [\epsilon_x, \epsilon_y, \epsilon_z, \gamma_{yz}, \gamma_{xz}, \gamma_{xy}]^T \qquad (1.9)$$

where ϵ_x, ϵ_y, and ϵ_z are normal strains and γ_{yz}, γ_{xz}, and γ_{xy} are the engineering shear strains.

Figure 1.4 gives the deformation of the *dx-dy* face for small deformations, which we consider here. Also considering other faces, we can write

$$\boldsymbol{\epsilon} = \left[\frac{\partial u}{\partial x}, \frac{\partial v}{\partial y}, \frac{\partial w}{\partial z}, \frac{\partial v}{\partial z} + \frac{\partial w}{\partial y}, \frac{\partial u}{\partial z} + \frac{\partial w}{\partial x}, \frac{\partial u}{\partial y} + \frac{\partial v}{\partial x} \right]^{\mathrm{T}} \quad (1.10)$$

The strain relations above hold for small deformations.

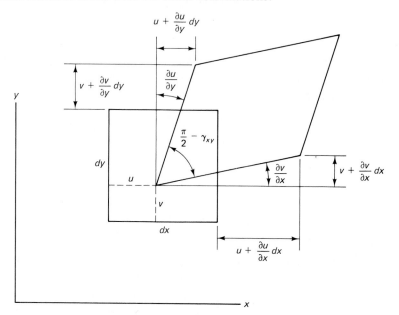

Figure 1.4   Deformed elemental surface.

1.7 STRESS–STRAIN RELATIONS

For linear elastic materials, the stress–strain relations come from the generalized Hooke's law. For isotropic materials, the two material properties are Young's modulus (or modulus of elasticity) E and Poisson's ratio ν. Considering an elemental cube inside the body, Hooke's law gives

$$\epsilon_x = \frac{\sigma_x}{E} - \nu \frac{\sigma_y}{E} - \nu \frac{\sigma_z}{E}$$

$$\epsilon_y = -\nu \frac{\sigma_x}{E} + \frac{\sigma_y}{E} - \nu \frac{\sigma_z}{E}$$

$$\epsilon_z = -\nu \frac{\sigma_x}{E} - \nu \frac{\sigma_y}{E} + \frac{\sigma_z}{E} \quad (1.11)$$

$$\gamma_{yz} = \frac{\tau_{yz}}{G}$$

$$\gamma_{xz} = \frac{\tau_{xz}}{G}$$

$$\gamma_{xy} = \frac{\tau_{xy}}{G}$$

The shear modulus (or modulus of rigidity), G, is given by

$$G = \frac{E}{2(1 + \nu)} \tag{1.12}$$

From Hooke's law relationships (Eq. 1.11), note that

$$\epsilon_x + \epsilon_y + \epsilon_z = \frac{(1 - 2\nu)}{E}(\sigma_x + \sigma_y + \sigma_z) \tag{1.13}$$

Substituting for $(\sigma_y + \sigma_z)$ and so on into Eq. 1.11, we get the inverse relations

$$\boldsymbol{\sigma} = \mathbf{D}\boldsymbol{\epsilon} \tag{1.14}$$

$\mathbf{D}$ is the symmetric (6×6) material matrix given by

$$\mathbf{D} = \frac{E}{(1 + \nu)(1 - 2\nu)} \begin{bmatrix} 1 - \nu & \nu & \nu & 0 & 0 & 0 \\ \nu & 1 - \nu & \nu & 0 & 0 & 0 \\ \nu & \nu & 1 - \nu & 0 & 0 & 0 \\ 0 & 0 & 0 & 0.5 - \nu & 0 & 0 \\ 0 & 0 & 0 & 0 & 0.5 - \nu & 0 \\ 0 & 0 & 0 & 0 & 0 & 0.5 - \nu \end{bmatrix} \tag{1.15}$$

Special Cases

One dimension.   In one dimension, we have normal stress σ along x and the corresponding normal strain ϵ. Stress–strain relations (Eq. 1.14) are simply

$$\sigma = E\epsilon \tag{1.16}$$

Two dimensions.   In two dimensions, the problems are modeled as plane stress and plane strain.

Plane Stress.   A thin planar body subjected to in-plane loading on its edge surface is said to be in plane stress. A ring press fitted on a shaft, Fig. 1.5a, is an example. Here stresses σ_z, τ_{xz}, and τ_{yz} are set as zero. The Hooke's law relations (Eq. 1.11) then give us

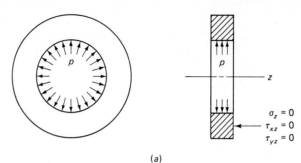

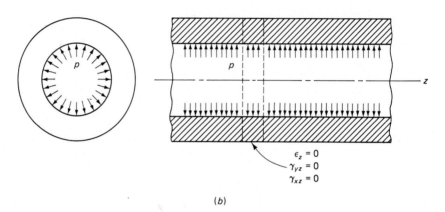

Figure 1.5   (a) Plane stress and (b) plane strain.

$$\epsilon_x = \frac{\sigma_x}{E} - \nu\frac{\sigma_y}{E}$$

$$\epsilon_y = -\nu\frac{\sigma_x}{E} + \frac{\sigma_y}{E}$$

$$\gamma_{xy} = \frac{2(1+\nu)}{E}\tau_{xy}$$

(1.17)

$$\epsilon_z = -\frac{\nu}{E}(\sigma_x + \sigma_y)$$

The inverse relations are given by

$$\begin{Bmatrix} \sigma_x \\ \sigma_y \\ \tau_{xy} \end{Bmatrix} = \frac{E}{1-\nu^2}\begin{bmatrix} 1 & \nu & 0 \\ \nu & 1 & 0 \\ 0 & 0 & \frac{1-\nu}{2} \end{bmatrix}\begin{Bmatrix} \epsilon_x \\ \epsilon_y \\ \gamma_{xy} \end{Bmatrix}$$

(1.18)

which is used as $\boldsymbol{\sigma} = \mathbf{D}\boldsymbol{\epsilon}$

Plane Strain.   If a long body of uniform cross section is subjected to transverse loading along its length, a small thickness in the loaded area, as shown in Fig. 1.5*b*, can be treated as subjected to plane strain. Here ϵ_z, γ_{xz}, γ_{yz} are taken as zero. Stress σ_z may not be zero in this case. The stress–strain relations can be obtained directly from Eqs. 1.14 and 1.15:

$$\begin{Bmatrix} \sigma_x \\ \sigma_y \\ \tau_{xy} \end{Bmatrix} = \frac{E}{(1+\nu)(1-2\nu)} \begin{bmatrix} 1-\nu & \nu & 0 \\ \nu & 1-\nu & 0 \\ 0 & 0 & \frac{1}{2}-\nu \end{bmatrix} \begin{Bmatrix} \epsilon_x \\ \epsilon_y \\ \gamma_{xy} \end{Bmatrix} \quad (1.19)$$

D here is a (3×3) matrix, which relates three stresses and three strains.

Anisotropic bodies, with uniform orientation, can be considered by using the appropriate **D** matrix for the material.

1.8 TEMPERATURE EFFECTS

If the temperature rise $\Delta T(x, y, z)$ with respect to the original state is known, then the associated deformation can be considered easily. For isotropic materials, the temperature rise ΔT results in a uniform strain, which depends on the coefficient of linear expansion α of the material. α, which represents the change in length per unit temperature rise, is assumed to be a constant within the range of variation of the temperature. Also this strain does not cause any stresses when the body is free to deform. The temperature strain is represented as an initial strain:

$$\epsilon_0 = [\alpha \, \Delta T, \, \alpha \, \Delta T, \, \alpha \, \Delta T, \, 0, \, 0, \, 0]^T \quad (1.20)$$

The stress–strain relations then become

$$\sigma = D(\epsilon - \epsilon_0) \quad (1.21)$$

In **plane stress**, we have

$$\epsilon_0 = [\alpha \Delta T, \, \alpha \Delta T, \, 0]^T \quad (1.22)$$

In **plane strain**, the constraint that $\epsilon_z = 0$ results in a different ϵ_0,

$$\epsilon_0 = (1 + \nu)[\alpha \, \Delta T, \, \alpha \, \Delta T, \, 0]^T \quad (1.23)$$

For plane stress and plane strain, note that $\sigma = [\sigma_x, \sigma_y, \tau_{xy}]^T$ and $\epsilon = [\epsilon_x, \epsilon_y, \gamma_{xy}]^T$, and that **D** matrices are as given in Eqs. 1.18 and 1.19, respectively.

1.9 POTENTIAL ENERGY AND EQUILIBRIUM; THE RAYLEIGH–RITZ METHOD

In mechanics of solids, our problem is to determine the displacement **u** of the body shown in Fig. 1.1, satisfying the equilibrium equations 1.6. Note that stresses are related to strains, which, in turn, are related to displacements. This leads to requiring solution of second-order partial differential equations. Solution of this set of equations is generally referred to as an *exact* solution. Such exact solutions are available for simple geometries and loading conditions, and one may refer to publications in theory of elasticity. For problems of complex geometries and general boundary and loading conditions, obtaining such solutions is an almost impossible task. Approximate solution methods usually employ potential energy or variational methods, which place less stringent conditions on the functions.

Potential Energy, Π.

The total potential energy Π of an elastic body, is defined as the sum of total strain energy (U) and the work potential:

$$\Pi = \text{Strain energy} + \text{Work potential} \qquad (1.24)$$
$$(U) \qquad\qquad (\text{WP})$$

For linear elastic materials, the strain energy per unit volume in the body is $\frac{1}{2}\boldsymbol{\sigma}^T\boldsymbol{\epsilon}$. For the elastic body shown in Fig. 1.1, the total strain energy U is given by

$$U = \frac{1}{2}\int_V \boldsymbol{\sigma}^T\boldsymbol{\epsilon}\, dV \qquad (1.25)$$

The work potential WP is given by

$$\text{WP} = -\int_V \mathbf{u}^T\mathbf{f}\, dV - \int_S \mathbf{u}^T\mathbf{T}\, dS - \sum_i \mathbf{u}_i^T\mathbf{P}_i \qquad (1.26)$$

The total potential for the general elastic body shown in Fig. 1.1 is

$$\Pi = \frac{1}{2}\int_V \boldsymbol{\sigma}^T\boldsymbol{\epsilon}\, dV - \int_V \mathbf{u}^T\mathbf{f}\, dV - \int_S \mathbf{u}^T\mathbf{T}\, dS - \sum_i \mathbf{u}_i^T\mathbf{P}_i \qquad (1.27)$$

We consider here conservative systems, where the work potential is independent of the path taken. In other words, if the system is displaced from a given configuration and brought back to this state, the forces do zero work regardless of the path. The potential energy principle is now stated as follows:

Principle of Minimum Potential Energy

For conservative systems, of all the kinematically admissible displacement fields, those corresponding to equilibrium extremize the total potential energy. If the extremum condition is a minimum, the equilibrium state is stable.

Kinematically admissible displacements are those that satisfy the single-valued nature of displacements (compatibility) and the boundary conditions. In problems where displacements are the unknowns, which is the approach in this book, compatibility is automatically satisfied.

To illustrate the ideas, let us consider an example of a discrete connected system.

Example 1.1

Figure E1.1a shows a system of springs. The total potential energy is given by

$$\Pi = \tfrac{1}{2}k_1\delta_1^2 + \tfrac{1}{2}k_2\delta_2^2 + \tfrac{1}{2}k_3\delta_3^2 + \tfrac{1}{2}k_4\delta_4^2 - F_1q_1 - F_3q_3$$

where δ_1, δ_2, δ_3, and δ_4 are extensions of the four springs. Since $\delta_1 = q_1 - q_2$, $\delta_2 = q_2$, $\delta_3 = q_3 - q_2$, and $\delta_4 = -q_3$, we have

$$\Pi = \tfrac{1}{2}k_1(q_1 - q_2)^2 + \tfrac{1}{2}k_2q_2^2 + \tfrac{1}{2}k_3(q_3 - q_2)^2 + \tfrac{1}{2}k_4q_3^2 - F_1q_1 - F_3q_3$$

where q_1, q_2, and q_3 are the displacements of nodes 1, 2, and 3, respectively.

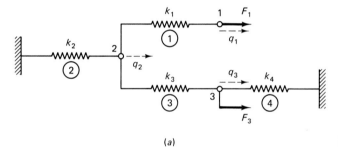

(a)   **Figure E1.1**

For equilibrium of this three degrees of freedom system, we need to minimize Π with respect to q_1, q_2, and q_3. The three equations are given by

$$\frac{\partial \Pi}{\partial q_i} = 0 \qquad i = 1, 2, 3 \tag{1.28}$$

which are

$$\frac{\partial \Pi}{\partial q_1} = k_1(q_1 - q_2) - F_1 = 0$$

$$\frac{\partial \Pi}{\partial q_2} = -k_1(q_1 - q_2) + k_2q_2 - k_3(q_3 - q_2) = 0$$

$$\frac{\partial \Pi}{\partial q_3} = k_3(q_3 - q_2) + k_4q_3 - F_3 = 0$$

These equilibrium equations can be put in the form of $\mathbf{Kq} = \mathbf{F}$ as follows:

$$\begin{bmatrix} k_1 & -k_1 & 0 \\ -k_1 & k_1 + k_2 + k_3 & -k_3 \\ 0 & -k_3 & k_3 + k_4 \end{bmatrix} \begin{Bmatrix} q_1 \\ q_2 \\ q_3 \end{Bmatrix} = \begin{Bmatrix} F_1 \\ 0 \\ F_3 \end{Bmatrix} \tag{1.29}$$

If, on the other hand, we proceed to write the equilibrium of the system by considering the equilibrium of each separate node, as shown in Fig. E1.1b, we can write

$$k_1 \delta_1 = F_1$$

$$k_2 \delta_2 - k_1 \delta_1 - k_3 \delta_3 = 0$$

$$k_3 \delta_3 - k_4 \delta_4 = F_3$$

which is precisely the set of equations represented in Eq. 1.29.

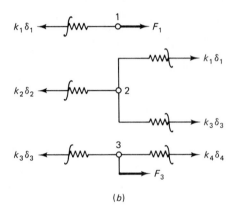

(b)   **Figure E1.1b**

We see clearly that the set of equations 1.29 is obtained in a routine manner using the potential energy approach, without any reference to the free body diagrams. This makes the potential energy approach attractive for large and complex problems. ■

Rayleigh-Ritz Method

For continua, the total potential energy Π in Eq. 1.27 can be used for finding an approximate solution. The Rayleigh–Ritz method involves the construction of an assumed displacement field, say

$$u = \sum a_i \phi_i(x, y, z) \qquad i = 1 \text{ to } \ell$$

$$v = \sum a_j \phi_j(x, y, z) \qquad j = \ell + 1 \text{ to } m \qquad (1.30)$$

$$w = \sum a_k \phi_k(x, y, z) \qquad k = m + 1 \text{ to } n$$

$$n > m > \ell$$

The functions ϕ_i are usually taken as polynomials. Displacements u, v, w must be **kinematically admissible**. That is, u, v, w must satisfy specified boundary conditions. Introducing stress–strain and strain–displacement relations, and substituting Eq. 1.30 into Eq. 1.27 gives

$$\Pi = \Pi(a_1, a_2, \ldots, a_r) \tag{1.31}$$

where r = number of independent unknowns. Now, the extremum with respect to a_i, $(i = 1$ to $r)$ yields the set of r equations

$$\frac{\partial \Pi}{\partial a_i} = 0 \qquad i = 1, 2, \ldots, r \tag{1.32}$$

Example 1.2

The potential energy for the linear elastic one-dimensional rod (Fig. E1.2), with body force neglected, is

$$\Pi = \frac{1}{2} \int_0^L EA \left(\frac{du}{dx}\right)^2 dx - 2u_1$$

where $u_1 = u(x = 1)$.

Let us consider a polynomial function

$$u = a_1 + a_2x + a_3x^2$$

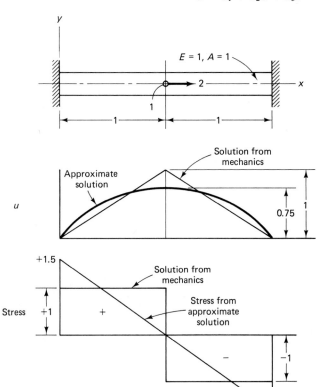

Figure E1.2

This must satisfy $u = 0$ at $x = 0$ and $u = 0$ at $x = 2$. Thus,

$$0 = a_1$$

$$0 = a_1 + 2a_2 + 4a_3$$

Hence,

$$a_2 = -2a_3$$

$$u = a_3(-2x + x^2) \qquad u_1 = -a_3$$

Then, $du/dx = 2a_3(-1 + x)$ and

$$\Pi = \frac{1}{2} \int_0^2 4a_3^2(-1 + x)^2 \, dx - 2(-a_3)$$

$$= 2a_3^2 \int_0^2 (1 - 2x + x^2) \, dx + 2a_3$$

$$= 2a_3^2 \left(\tfrac{2}{3}\right) + 2a_3$$

We set $\partial\Pi/\partial a_3 = 4a_3 \left(\tfrac{2}{3}\right) + 2 = 0$, resulting in

$$a_3 = -0.75 \qquad u_1 = -a_3 = 0.75$$

The stress in the bar is given by

$$\sigma = E\frac{du}{dx} = 1.5 \, (1 - x) \qquad\qquad \blacksquare$$

We note here that an exact solution is obtained if piecewise polynomial interpolation is used in the construction of u.

The finite element method provides a systematic way of constructing the basis functions ϕ_i used in Eq. 1.30.

1.10 GALERKIN'S METHOD

Galerkin's method uses the set of governing equations in the development of an integral form. It is usually presented as one of the weighted residual methods. For our discussion, let us consider a general representation of a governing equation on a region V:

$$Lu = P \qquad\qquad (1.33)$$

For the one-dimensional rod considered in Example 1.2, the governing equation is the differential equation

$$\frac{d}{dx}\left(EA \, \frac{du}{dx}\right) = 0$$

We may consider L as the operator

$$\frac{d}{dx} EA \frac{d}{dx} (\ )$$

operating on u.

The exact solution needs to satisfy (1.33) at every point x. If we seek an approximate solution $\tilde{u}$, it introduces an error $\epsilon(x)$, called the *residual*:

$$\epsilon(x) = L\tilde{u} - P \tag{1.34}$$

The approximate methods revolve around setting the residual relative to a weighting function W_i, to zero.

$$\int_V W_i(L\tilde{u} - P)\, dV = 0 \qquad i = 1 \text{ to } n \tag{1.35}$$

The choice of the weighting function W_i leads to various approximation methods. *In the Galerkin method, the weighting functions W_i are chosen from the basis functions used for constructing $\tilde{u}$.* Let $\tilde{u}$ be represented by

$$\tilde{u} = \sum_{i=1}^{n} Q_i G_i \tag{1.36}$$

where G_i, $i = 1$ to n, are basis functions (usually polynomials of x, y, z). Here, we choose the weighting functions to be *a linear combination of the basis functions G_i.* Specifically, consider an arbitrary function ϕ given by

$$\phi = \sum_{i=1}^{n} \phi_i G_i \tag{1.37}$$

where the coefficients ϕ_i are arbitrary, except for requiring that ϕ satisfy homogeneous (zero) boundary conditions where $\tilde{u}$ is prescribed. The fact that ϕ above is constructed in a similar manner as $\tilde{u}$ in Eq. 1.36 leads to simplified derivations in later chapters.

Galerkin's method can be stated as follows:

Choose basis functions G_i. Determine the coefficients Q_i in $\tilde{u} = \sum\limits_{i=1}^{n} Q_i G_i$ such that

$$\int_V \phi(L\tilde{u} - P)\, dV = 0 \tag{1.38}$$

for every ϕ of the type $\phi = \sum_{i=1}^{n} \phi_i G_i$, where coefficients ϕ_i are arbitrary except for requiring that ϕ satisfy homogeneous (zero) boundary conditions. The solution of the resulting equations for Q_i then yields the approximate solution $\tilde{u}$.

Usually, in the treatment of Eq. 1.38 an integration by parts is involved. The order of the derivatives is reduced and the natural boundary conditions, such as surface force conditions, are introduced.

Galerkin's method in elasticity.   Let us turn our attention to the equilibrium equations 1.6 in elasticity. Galerkin's method requires

$$\int_V \left[\left(\frac{\partial \sigma_x}{\partial x} + \frac{\partial \tau_{xy}}{\partial y} + \frac{\partial \tau_{xz}}{\partial z} + f_x \right) \phi_x + \left(\frac{\partial \tau_{xy}}{\partial x} + \frac{\partial \sigma_y}{\partial y} + \frac{\partial \tau_{yz}}{\partial z} + f_y \right) \phi_y \right.$$

$$\left. + \left(\frac{\partial \tau_{xz}}{\partial x} + \frac{\partial \tau_{yz}}{\partial y} + \frac{\partial \sigma_z}{\partial z} + f_z \right) \phi_z \right] dV = 0 \qquad (1.39)$$

where

$$\boldsymbol{\phi} = [\phi_x, \ \phi_y, \ \phi_z]^T$$

is an arbitrary displacement consistent with the boundary conditions of **u**. If $\mathbf{n} = [n_x, n_y, n_z]^T$ is a unit normal at a point **x** on the surface, the integration by parts formula is

$$\int_V \frac{\partial \alpha}{\partial x} \theta \ dV = - \int_V \alpha \frac{\partial \theta}{\partial x} \ dV + \int_S n_x \alpha \theta \ dS \qquad (1.40)$$

where α and θ are functions of (x, y, z). For multidimensional problems, Eq. 1.40 is usually referred to as the Green–Gauss theorem or the divergence theorem. Using this formula, integrating Eq. 1.39 by parts, and rearranging terms, we get

$$- \int_V \boldsymbol{\sigma}^T \boldsymbol{\epsilon}(\phi) \ dV + \int_V \boldsymbol{\phi}^T \mathbf{f} \ dV + \int_S [(n_x \sigma_x + n_y \tau_{xy} + n_z \tau_{xz}) \phi_x$$

$$+ (n_x \tau_{xy} + n_y \sigma_y + n_z \tau_{yz}) \phi_y + (n_x \tau_{xz} + n_y \tau_{yz} + n_z \sigma_z) \phi_z] \ dS = 0 \qquad (1.41)$$

where

$$\boldsymbol{\epsilon}(\phi) = \left[\frac{\partial \phi_x}{\partial x}, \frac{\partial \phi_y}{\partial y}, \frac{\partial \phi_z}{\partial z}, \frac{\partial \phi_y}{\partial z} + \frac{\partial \phi_z}{\partial y}, \frac{\partial \phi_x}{\partial z} + \frac{\partial \phi_z}{\partial x}, \frac{\partial \phi_x}{\partial y} + \frac{\partial \phi_y}{\partial x} \right]^T \qquad (1.42)$$

is the strain corresponding to the arbitrary displacement field $\boldsymbol{\phi}$.

On the boundary, from Eq. 1.8, we have $(n_x \sigma_x + n_y \tau_{xy} + n_z \tau_{xz}) = T_x$, and so on. At point loads $(n_x \sigma_x + n_y \tau_{xy} + n_z \tau_{xz}) \ dS$ is equivalent to P_x, and so on. These are the natural boundary conditions in the problem. Thus, Eq. 1.41 yields the Galerkin's "variational form" or "weak form" for three-dimensional stress analysis:

$$\boxed{\int_V \boldsymbol{\sigma}^T \boldsymbol{\epsilon}(\phi) \ dV - \int_V \boldsymbol{\phi}^T \mathbf{f} \ dV - \int_S \boldsymbol{\phi}^T \mathbf{T} \ dS - \sum_i \boldsymbol{\phi}^T \mathbf{P} = 0} \qquad (1.43)$$

where $\boldsymbol{\phi}$ is an arbitrary displacement consistent with the specified boundary conditions of **u**. We may now use Eq. 1.43 to provide us with an approximate solution.

For problems of linear elasticity, the above equation is precisely the **principle of virtual work**. ϕ is the kinematically admissible virtual displacement. The principle of virtual work may be stated as follows:

Principle of Virtual Work

A body is in equilibrium if the internal virtual work equals the external virtual work for every kinematically admissible displacement field (ϕ, $\epsilon(\phi)$).

We note that Galerkin's method and the principle of virtual work result in the same set of equations for problems of elasticity when same basis or coordinate functions are used. Galerkin's method is more general since the variational form of the type Eq. 1.43 can be developed for other governing equations defining boundary value problems. Galerkin's method works directly from the differential equation and is preferred to the Rayleigh–Ritz method for problems where a corresponding function to be minimized is not obtainable.

Example 1.3

Let us consider the problem of Example 1.2 and solve it by Galerkin's approach. The equilibrium equation is

$$\frac{d}{dx} EA \frac{du}{dx} = 0 \qquad \begin{array}{l} u = 0 \quad \text{at } x = 0 \\ u = 0 \quad \text{at } x = 2 \end{array}$$

Multiplying the differential equation above by ϕ, and integrating by parts, we get

$$\int_0^2 -EA \frac{du}{dx}\frac{d\phi}{dx}\, dx + \left(\phi\, EA \frac{du}{dx}\right)_0^1 + \left(\phi\, EA \frac{du}{dx}\right)_1^2 = 0$$

where ϕ is zero at $x = 0$ and $x = 2$. $EA\,(du/dx)$ is the tension in the rod, which takes a jump of magnitude 2 at $x = 1$ (Fig. E1.2). Thus,

$$\int_0^2 -EA \frac{du}{dx}\frac{d\phi}{dx}\, dx + 2\phi_1 = 0$$

Now, we use the same polynomial (basis) for u and ϕ. If u_1 and ϕ_1 are the values at $x = 1$, we have

$$u = (2x - x^2)u_1$$
$$\phi = (2x - x^2)\phi_1$$

Substituting these and $E = 1$, $A = 1$ in the above integral yields

$$\phi_1\left[-u_1 \int_0^2 (2 - 2x)^2\, dx + 2\right] = 0$$

$$\phi_1(- \tfrac{8}{3}u_1 + 2) = 0$$

This is to be satisfied for every ϕ_1. We get

$$u_1 = 0.75 \qquad\qquad \blacksquare$$

1.11 SAINT VENANT'S PRINCIPLE

We often have to make approximations in defining boundary conditions to represent a support–structure interface. For instance, consider a cantilever beam, free at one end and attached to a column with rivets at the other end. Questions arise as to whether the riveted joint is totally rigid or partially rigid, and as to whether each point on the cross section at the fixed end is specified to have the same boundary conditions. Saint Venant considered the effect of different approximations on the solution to the total problem. Saint Venant's principle states that as long as the different approximations are statically equivalent, the resulting solutions will be valid provided we focus on regions sufficiently far away from the support. That is, the solutions may significantly differ only within the immediate vicinity of the support.

1.12 CONCLUSION

In this chapter, we have discussed the necessary background for the finite element method. We devote the next chapter to discussing matrix algebra and techniques for solving a set of linear algebraic equations.

HISTORICAL REFERENCES

1. Hrenikoff, A., Solution of problems in elasticity by the frame work method. *Journal of Applied Mechanics, Transactions of the ASME* 8: 169–175 (1941).
2. Courant, R., Variational methods for the solution of problems of equilibrium and vibrations. *Bulletin of the American Mathematical Society* 49: 1–23 (1943).
3. Turner, M. J., R. W. Clough, H. C. Martin, and L. J. Topp, Stiffness and deflection analysis of complex structures. *Journal of Aeronautical Science* 23(9): 805–824 (1956).
4. Clough, R. W., The finite element method in plane stress analysis. *Proceedings American Society of Civil Engineers,* 2d Conference on Electronic Computation, Pittsburgh, Pennsylvania, 23: 345–378 (1960).
5. Argyris, J. H., Energy theorems and structural analysis. *Aircraft Engineering,* 26: Oct.–Nov., 1954; 27: Feb.–May, 1955.
6. Zienkiewicz, O. C., and Y. K. Cheung, *The Finite Element Method in Structural and Continuum Mechanics.* (London: McGraw-Hill, 1967).
7. Oden, J. T., *Finite Elements of Nonlinear Continua.* (New York: McGraw-Hill, 1972).

PROBLEMS

1.1. Obtain the **D** matrix given by Eq. 1.15 using the generalized Hooke's law relations (Eq. 1.11).

1.2. In a plane strain problem, we have

$$\sigma_x = 20\,000 \text{ psi}, \ \sigma_y = -10\,000 \text{ psi}$$

$$E = 30 \times 10^6 \text{ psi}, \ \nu = 0.3$$

Determine the value of the stress σ_z.

1.3. If a displacement field is described by

$$u = -x^2 + 2y^2 + 6xy$$

$$v = 3x + 6y - y^2$$

determine ϵ_x, ϵ_y, γ_{xy} at the point $x = 1$, $y = 0$.

1.4. In a solid body, the six components of the stress at a point are given by $\sigma_x = 40$ MPa, $\sigma_y = 20$ MPa, $\sigma_z = 30$ MPa, $\tau_{yz} = -30$ MPa, $\tau_{xz} = 15$ MPa, and $\tau_{xy} = 10$ MPa. Determine the normal stress at the point, on a plane for which the normal is $(n_x, n_y, n_z) = (\frac{1}{2}, \frac{1}{2}, 1/\sqrt{2})$. (*Hint:* Note that normal stress $\sigma_n = T_x n_x + T_y n_y + T_z n_z$.)

1.5. For isotropic materials, the stress–strain relations can also be expressed using Lame's constants λ and μ, as follows:

$$\sigma_x = \lambda \epsilon_v + 2\mu\epsilon_x$$

$$\sigma_y = \lambda \epsilon_v + 2\mu\epsilon_y$$

$$\sigma_z = \lambda \epsilon_v + 2\mu\epsilon_z$$

$$\tau_{yz} = \mu\gamma_{yz}, \ \tau_{xz} = \mu\gamma_{xz}, \ \tau_{xy} = \mu\gamma_{xy}$$

where ϵ_v, $\epsilon_x + \epsilon_y + \epsilon_z$. Find expressions for λ and μ in terms of E and ν.

1.6. A long rod is subjected to loading and a temperature increase of 30 °C. The total strain at a point is measured to be 1.2×10^{-5}. If $E = 200$ GPa, and $\alpha = 12 \times 10^{-6}/°C$, determine the stress at the point.

1.7. Determine the displacements of nodes of the spring system shown in Fig. P1.7.

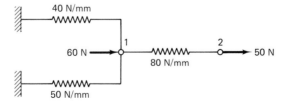

Figure P1.7

1.8. Use the Rayleigh–Ritz method to find the displacement of the midpoint of the rod shown in Fig. P1.8.

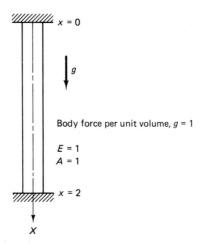

$x = 0$

g

Body force per unit volume, $g = 1$

$E = 1$
$A = 1$

$x = 2$

X

Figure P1.8

1.9. Use Galerkin's method to find the displacement at the midpoint of the rod (Figure P1.8).

1.10. Solve Example 1.2 using the potential energy approach with the polynomial $u = a_1 + a_2x + a_3x^2 + a_4x^3$.

chapter two

Matrix Algebra and Gaussian Elimination

2.1 *MATRIX ALGEBRA*

The study of matrices here is largely motivated from the need to solve systems of simultaneous equations of the form

$$a_{11}x_1 + a_{12}x_2 + \cdots + a_{1n}x_n = b_1$$

$$a_{21}x_1 + a_{22}x_2 + \cdots + a_{2n}x_n = b_2 \qquad (2.1a)$$

$$a_{n1}x_1 + a_{n2}x_2 + \cdots + a_{nn}x_n = b_n$$

where $x_1, x_2, \ldots, x_n$ are the unknowns. Equations 2.1 can be conveniently expressed in matrix form as

$$\mathbf{Ax} = \mathbf{b} \qquad (2.1b)$$

where $\mathbf{A}$ is a square matrix of dimensions $(n \times n)$, and $\mathbf{x}$ and $\mathbf{b}$ are vectors of dimension $(n \times 1)$, given as

$$
\mathbf{A} = \begin{bmatrix} a_{11} & a_{12} & \cdots & a_{1n} \\ a_{21} & a_{22} & \cdots & a_{2n} \\ \hline a_{n1} & a_{n2} & \cdots & a_{nn} \end{bmatrix}
\qquad
\mathbf{x} = \begin{Bmatrix} x_1 \\ x_2 \\ \vdots \\ x_n \end{Bmatrix}
\qquad
\mathbf{b} = \begin{Bmatrix} b_1 \\ b_2 \\ \vdots \\ b_n \end{Bmatrix}
$$

From above, we see that a matrix is simply an array of elements. The matrix **A** is also denoted as [A]. An element located at the ith row and jth column of **A** is denoted by a_{ij}.

The multiplication of two matrices, **A** and **x**, is also implicitly defined above: The dot product of the ith row of **A** with the vector **x** is equated to b_i, resulting in the ith equation of Eq. 2.1a. The multiplication operation and other operations will be discussed in detail in this chapter.

The analysis of engineering problems by the finite element method involves a sequence of matrix operations. This fact allows us to solve large-scale problems because computers are ideally suited for matrix operations. In this chapter, basic matrix operations as needed later in the text are given. The Gaussian elimination method for solving linear simultaneous equations is also discussed, and a variant of the Gaussian elimination approach, the skyline approach, is presented.

Row and Column Vectors

A matrix of dimension $(1 \times n)$ is called a **row vector**, while a matrix of dimension $(m \times 1)$ is called a **column vector**. For example,

$$\mathbf{d} = \begin{bmatrix} 1 & -1 & 2 \end{bmatrix}$$

is a (1×3) row vector, and

$$\mathbf{E} = \begin{Bmatrix} 2 \\ 2 \\ -6 \\ 0 \end{Bmatrix}$$

is a (4×1) column vector.

Addition and Subtraction

Consider two matrices **A** and **B**, both of dimension $(m \times n)$. Then, the sum **C** = **A** + **B** is defined as

$$c_{ij} = a_{ij} + b_{ij} \tag{2.2}$$

That is, the (ij)th component of **C** is obtained by adding the (ij)th component of **A** to the (ij)th component of **B**. For example,

$$\begin{bmatrix} 2 & -3 \\ -3 & 5 \end{bmatrix} + \begin{bmatrix} 2 & 1 \\ 0 & 4 \end{bmatrix} = \begin{bmatrix} 4 & -2 \\ -3 & 9 \end{bmatrix}$$

Subtraction is similarly defined.

Multiplication by a Scalar

The multiplication of a matrix $\mathbf{A}$ by a scalar c is defined as

$$c\mathbf{A} = [ca_{ij}] \tag{2.3}$$

For example, we can write

$$\begin{bmatrix} 10\,000 & 4\,500 \\ 4\,500 & -6\,000 \end{bmatrix} = 10^3 \begin{bmatrix} 10 & 4.5 \\ 4.5 & -6 \end{bmatrix}$$

Matrix Multiplication

The product of an $(m \times n)$ matrix $\mathbf{A}$ and an $(n \times p)$ matrix $\mathbf{B}$ results in an $(m \times p)$ matrix $\mathbf{C}$. That is,

$$\begin{matrix} \mathbf{A} & \mathbf{B} & = & \mathbf{C} \\ (m \times n) & (n \times p) & & (m \times p) \end{matrix} \tag{2.4}$$

The (ij)th component of $\mathbf{C}$ is obtained by taking the dot product

$$c_{ij} = (i\text{th row of } \mathbf{A}) \cdot (j\text{th column of } \mathbf{B}) \tag{2.5}$$

For example,

$$\begin{bmatrix} 2 & 1 & 3 \\ 0 & -2 & 1 \end{bmatrix} \begin{bmatrix} 1 & 4 \\ 5 & -2 \\ 0 & 3 \end{bmatrix} = \begin{bmatrix} 7 & 15 \\ -10 & 7 \end{bmatrix}$$

$$\begin{matrix} (2 \times 3) & (3 \times 2) & (2 \times 2) \end{matrix}$$

It should be noted that $\mathbf{AB} \neq \mathbf{BA}$; in fact, $\mathbf{BA}$ may not even be defined, since the number of columns of $\mathbf{B}$ may not equal the number of rows of $\mathbf{A}$.

Tranposition

If $\mathbf{A} = [A_{ij}]$, then the transpose of $\mathbf{A}$, denoted as $\mathbf{A}^T$, is given by $\mathbf{A}^T = [a_{ji}]$. Thus, the rows of $\mathbf{A}$ are the columns of $\mathbf{A}^T$. For example, if

$$\mathbf{A} = \begin{bmatrix} 1 & -5 \\ 0 & 6 \\ -2 & 3 \\ 4 & 2 \end{bmatrix}$$

then

$$\mathbf{A}^T = \begin{bmatrix} 1 & 0 & -2 & 4 \\ -5 & 6 & 3 & 2 \end{bmatrix}$$

In general, if $\mathbf{A}$ is of dimension $(m \times n)$ then $\mathbf{A}^T$ is of dimension $(n \times m)$.

The transpose of a product is given as the product of the transposes in reverse order:

$$(\mathbf{ABC})^T = \mathbf{C}^T\mathbf{B}^T\mathbf{A}^T \tag{2.6}$$

Differentiation and Integration

The components of a matrix do not have to be scalars; they may also be functions. For example,

$$\mathbf{B} = \begin{bmatrix} x + y & x^2 - xy \\ 6 + x & y \end{bmatrix}$$

In this regard, matrices may be differentiated and integrated. The derivative (or integral) of a matrix is simply the derivative (or integral) of each component of the matrix. Thus,

$$\frac{d}{dx}\mathbf{B}(x) = \left[\frac{db_{ij}(x)}{dx}\right] \tag{2.7}$$

$$\int \mathbf{B}\,dx\,dy = \left[\int\int b_{ij}\,dx\,dy\right] \tag{2.8}$$

The formula in Eq. 2.7 will now be specialized to an important case. Let $\mathbf{A}$ be an $(n \times n)$ matrix of constants, and $\mathbf{x} = [x_1, x_2, \ldots, x_n]^T$ be a column vector of n variables. Then, the derivative of $\mathbf{Ax}$ with respect to variable x_p is given by

$$\frac{d}{dx_p}(\mathbf{Ax}) = \mathbf{a}^p \tag{2.9}$$

where $\mathbf{a}^p$ is the pth column of $\mathbf{A}$. This result follows from the fact that the vector $(\mathbf{Ax})$ can be written out in full as

$$\mathbf{Ax} = \left\{\begin{array}{l} a_{11}x_1 + a_{12}x_2 + \cdots + a_{1p}x_p + \cdots + a_{1n}x_n \\ a_{21}x_1 + a_{22}x_2 + \cdots + a_{2p}x_p + \cdots + a_{2n}x_n \\ \text{------------------------} \\ \text{------------------------} \\ a_{n1}x_1 + a_{n2}x_2 + \cdots + a_{np}x_p + \cdots + a_{nn}x_n \end{array}\right\} \tag{2.10}$$

Now, we see clearly that the derivative of $\mathbf{Ax}$ with respect to x_p yields the pth column of $\mathbf{A}$ as stated in Eq. 2.9.

Square Matrix

A matrix whose number of rows equals the number of columns is called a square matrix.

Diagonal Matrix

A diagonal matrix is a square matrix with nonzero elements only along the principal diagonal. For example,

$$\mathbf{A} = \begin{bmatrix} 2 & 0 & 0 \\ 0 & 6 & 0 \\ 0 & 0 & -3 \end{bmatrix}$$

Identity Matrix

The identity (or unit) matrix is a diagonal matrix with 1's along the principal diagonal. For example,

$$\mathbf{I} = \begin{bmatrix} 1 & 0 & 0 & 0 \\ 0 & 1 & 0 & 0 \\ 0 & 0 & 1 & 0 \\ 0 & 0 & 0 & 1 \end{bmatrix}$$

If $\mathbf{I}$ is of dimension $(n \times n)$ and $\mathbf{x}$ is an $(n \times 1)$ vector, then,

$$\mathbf{Ix} = \mathbf{x}$$

Symmetric Matrix

A symmetric matrix is a square matrix whose elements satisfy

$$a_{ij} = a_{ji} \tag{2.11a}$$

or equivalently,

$$\mathbf{A} = \mathbf{A}^{\mathrm{T}} \tag{2.11b}$$

That is, elements located symmetrically with respect to the principal diagonal are equal. For example,

$$\mathbf{A} = \begin{bmatrix} 2 & 1 & 0 \\ 1 & 6 & -2 \\ 0 & -2 & 8 \end{bmatrix}$$

Upper Triangular Matrix

An upper triangular matrix is one whose elements below the principal diagonal are all zero. For example,

$$
\mathbf{U} = \begin{bmatrix} 2 & -1 & 6 & 3 \\ 0 & 14 & 8 & 0 \\ 0 & 0 & 5 & 1 \\ 0 & 0 & 0 & 3 \end{bmatrix}
$$

Determinant of a Matrix

The determinant of a square matrix $\mathbf{A}$ is a scalar quantity denoted as det $\mathbf{A}$. The determinants of a (2×2) and a (3×3) matrix are given below by the **method of cofactors**:

$$
\det \begin{bmatrix} a_{11} & a_{12} \\ a_{21} & a_{22} \end{bmatrix} = a_{11}a_{22} - a_{21}a_{12} \tag{2.12}
$$

$$
\det \begin{bmatrix} a_{11} & a_{12} & a_{13} \\ a_{21} & a_{22} & a_{23} \\ a_{31} & a_{32} & a_{33} \end{bmatrix} = a_{11}(a_{22}a_{33} - a_{32}a_{23}) \tag{2.13}
$$
$$
- a_{12}(a_{21}a_{33} - a_{31}a_{23})
$$
$$
+ a_{13}(a_{21}a_{32} - a_{31}a_{22})
$$

Matrix Inversion

Consider a square matrix $\mathbf{A}$. If det $\mathbf{A} \neq 0$, then $\mathbf{A}$ has an inverse, denoted by $\mathbf{A}^{-1}$. The inverse satisfies the relations

$$
\mathbf{A}^{-1}\mathbf{A} = \mathbf{A}\mathbf{A}^{-1} = \mathbf{I} \tag{2.14}
$$

If det $\mathbf{A} \neq 0$, then we say that $\mathbf{A}$ is **nonsingular**. If det $\mathbf{A} = 0$, then we say that $\mathbf{A}$ is **singular**, for which the inverse is not defined. The minor M_{ij} of a square matrix $\mathbf{A}$ is the determinant of the $(n - 1 \times n - 1)$ matrix obtained by eliminating the ith row and the jth column of $\mathbf{A}$. The cofactor C_{ij} of matrix $\mathbf{A}$ is given by

$$
C_{ij} = (-1)^{i+j}M_{ij}
$$

Matrix $\mathbf{C}$ with elements C_{ij} is called the cofactor matrix. The adjoint of matrix $\mathbf{A}$ is defined as

$$
\text{Adj } \mathbf{A} = \mathbf{C}^{\mathrm{T}}
$$

The inverse of a square matrix $\mathbf{A}$ is given as

$$
\mathbf{A}^{-1} = \frac{\text{adj } \mathbf{A}}{\det \mathbf{A}}
$$

For example, the inverse of a (2×2) matrix $\mathbf{A}$ is given by

$$
\begin{bmatrix} a_{11} & a_{12} \\ a_{21} & a_{22} \end{bmatrix}^{-1} = \frac{1}{\det \mathbf{A}} \begin{bmatrix} a_{22} & -a_{12} \\ -a_{21} & a_{11} \end{bmatrix}
$$

Quadratic Forms

Let A be an $(n \times n)$ matrix and x be an $(n \times 1)$ vector. Then, the scalar quantity

$$\mathbf{x}^T\mathbf{A}\mathbf{x} \tag{2.15}$$

is called a quadratic form, since upon expansion we obtain the quadratic expression

$$
\begin{aligned}
\mathbf{x}^T\mathbf{A}\mathbf{x} = \quad & x_1a_{11}x_1 + x_1a_{12}x_2 + \cdots + x_1a_{1n}x_n \\
+ \; & x_2a_{21}x_1 + x_2a_{22}x_2 + \cdots + x_2a_{2n}x_n \\
& \text{- -} \\
+ \; & x_na_{n1}x_1 + x_na_{n2}x_2 + \cdots + x_na_{nn}x_n
\end{aligned} \tag{2.16}
$$

As an example, the quantity

$$u = 3x_1^2 - 4x_1x_2 + 6x_1x_3 - x_2^2 + 5x_3^2$$

can be expressed in matrix form as

$$
u = [x_1 \; x_2 \; x_3] \begin{bmatrix} 3 & -2 & 3 \\ -2 & -1 & 0 \\ 3 & 0 & 5 \end{bmatrix} \begin{Bmatrix} x_1 \\ x_2 \\ x_3 \end{Bmatrix}
$$

$$= \mathbf{x}^T\mathbf{A}\mathbf{x}$$

Eigenvalues and Eigenvectors

Consider the eigenvalue problem

$$\mathbf{A}\mathbf{y} = \lambda\mathbf{y} \tag{2.17a}$$

where $\mathbf{A}$ is a square matrix, $(n \times n)$. We wish to find a nontrivial solution. That is, we wish to find a nonzero eigenvector $\mathbf{y}$ and the corresponding eigenvalue λ that satisfy the above equation. If we rewrite Eq. 2.17a as

$$(\mathbf{A} - \lambda\mathbf{I})\mathbf{y} = \mathbf{0} \tag{2.17b}$$

we see that a nonzero solution for $\mathbf{y}$ will occur when $\mathbf{A} - \lambda\mathbf{I}$ is a singular matrix, or

$$\det(\mathbf{A} - \lambda\mathbf{I}) = 0 \tag{2.18}$$

Equation 2.18 is called the *characteristic equation*. We can solve Eq. 2.18 for the n roots or eigenvalues $\lambda_1, \lambda_2, \ldots, \lambda_n$. For each eigenvalue λ_i, the associated eigenvector $\mathbf{y}^i$ is then obtained from Eq. 2.17b:

$$(\mathbf{A} - \lambda_i\mathbf{I})\mathbf{y}^i = \mathbf{0} \tag{2.19}$$

Note that the eigenvector $\mathbf{y}^i$ is determined only to within a multiplicative constant since $(\mathbf{A} - \lambda_i\mathbf{I})$ is a singular matrix.

Example 2.1

Consider the matrix

$$\mathbf{A} = \begin{bmatrix} 4 & -2.236 \\ -2.236 & 8 \end{bmatrix}$$

The characteristic equation is

$$\det \begin{bmatrix} 4 - \lambda & -2.236 \\ -2.236 & 8 - \lambda \end{bmatrix} = 0$$

which yields

$$(4 - \lambda)(8 - \lambda) - 5 = 0$$

Solving the above equation, we get

$$\lambda_1 = 3 \qquad \lambda_2 = 9$$

To get the eigenvector $\mathbf{y}^1 = [y_1^1, y_2^1]^T$ corresponding to the eigenvalue λ_1, we substitute $\lambda_1 = 3$ into Eq. 2.19:

$$\begin{bmatrix} (4 - 3) & -2.236 \\ -2.236 & (8 - 3) \end{bmatrix} \begin{Bmatrix} y_1^1 \\ y_2^1 \end{Bmatrix} = \begin{Bmatrix} 0 \\ 0 \end{Bmatrix}$$

Thus, the components of $\mathbf{y}^1$ satisfy the equation

$$y_1^1 - 2.236 y_2^1 = 0$$

We may now *normalize* the eigenvector, say, by making $\mathbf{y}^1$ a unit vector. This is done by setting $y_2^1 = 1$, resulting in $\mathbf{y}^1 = [2.236, 1]$. Dividing $\mathbf{y}^1$ by its length yields

$$\mathbf{y}^1 = [0.913, 0.408]^T$$

Now, $\mathbf{y}^2$ is obtained in a similar manner by substituting λ_2 into Eq. 2.19. After normalization

$$\mathbf{y}^2 = [-0.408, 0.913]^T \qquad\qquad \blacksquare$$

Eigenvalue problems in finite element analysis are of the type $\mathbf{Ay} = \lambda \mathbf{By}$. Solution techniques for these problems are discussed in Chapter 11.

Positive Definite Matrix

A symmetric matrix is said to be **positive definite** if all its eigenvalues are strictly positive (greater than zero). In the previous example, the symmetric matrix

$$\mathbf{A} = \begin{bmatrix} 4 & -2.236 \\ -2.236 & 8 \end{bmatrix}$$

had eigenvalues $\lambda_1 = 3 > 0$ and $\lambda_2 = 9 > 0$ and, hence, is positive definite. An alternative definition of a positive definite matrix is given below.

A symmetric matrix $\mathbf{A}$ of dimension $(n \times n)$ is positive definite if, for *every* nonzero vector $\mathbf{x} = [x_1, x_2, \ldots, x_n]^{\mathrm{T}}$,

$$\mathbf{x}^{\mathrm{T}}\mathbf{A}\mathbf{x} > 0 \qquad\qquad (2.20)$$

2.2 GAUSSIAN ELIMINATION

Consider a linear system of simultaneous equations in matrix form as

$$\mathbf{A}\mathbf{x} = \mathbf{b}$$

where $\mathbf{A}$ is $(n \times n)$, $\mathbf{b}$ and $\mathbf{x}$ are $(n \times 1)$. If $\det \mathbf{A} \neq 0$, then we can premultiply both sides of the above equation by $\mathbf{A}^{-1}$ to write the unique solution for $\mathbf{x}$ as $\mathbf{x} = \mathbf{A}^{-1}\mathbf{b}$. However, the explicit construction of $\mathbf{A}^{-1}$, say, by the cofactor approach, is computationally expensive and prone to round-off errors. Instead, an elimination scheme is better. The powerful Gaussian elimination approach for solving $\mathbf{A}\mathbf{x} = \mathbf{b}$ is discussed below.

Gaussian elimination is the name given to a well-known method of solving simultaneous equations by successively eliminating unknowns. We will first present the method by means of an example, followed by a general solution and algorithm. Consider the simultaneous equations

$$
\begin{aligned}
x_1 - 2x_2 + 6x_3 &= 0 \qquad &\text{I} \\
2x_1 + 2x_2 + 3x_3 &= 3 \qquad &\text{II} \qquad\qquad (2.21) \\
-x_1 + 3x_2 &= 2 \qquad &\text{III}
\end{aligned}
$$

The equations are labeled as I, II, and III. Now, we wish to eliminate x_1 from II and III. We have, from Eq. I, $x_1 = +2x_2 - 6x_3$. Substituting for x_1 into Eqs. II and III yields

$$
\begin{aligned}
x_1 - 2x_2 + 6x_3 &= 0 \qquad &\text{I} \\
0 + 6x_2 - 9x_3 &= 3 \qquad &\text{II}^{(1)} \qquad\qquad (2.22) \\
0 + x_2 + 6x_3 &= 2 \qquad &\text{III}^{(1)}
\end{aligned}
$$

It is important to realize that Eq. 2.22 can also be obtained from Eq. 2.21 by **row operations**. Specifically, in Eq. 2.21, to eliminate x_1 from II we subtract 2 times I from II, and to eliminate x_1 from III we subtract -1 times I from III. The result is Eq. 2.22 as above. Notice the zeroes below the main diagonal in column 1, representing the fact that x_1 has been eliminated from Eqs. II and III. The superscript (1) on the labels in Eqs. 2.22 denotes the fact that the equations have been modified once.

We now proceed to eliminate x_2 from III in Eqs. 2.22. For this, we subtract $\frac{1}{6}$ times II from III. The resulting system is

$$\begin{bmatrix} x_1 - 2x_2 + 6x_3 = 0 \\ 0 + 6x_2 - 9x_3 = 3 \\ 0 \quad 0 \quad \frac{15}{2}x_3 = \frac{3}{2} \end{bmatrix} \begin{Bmatrix} \text{I} \\ \text{II}^{(1)} \\ \text{III}^{(2)} \end{Bmatrix} \tag{2.23}$$

The coefficient matrix on the left side of Eqs. 2.23 is upper triangular. The solution now is virtually complete, since the last equation yields $x_3 = \frac{1}{5}$, which, upon substitution into the second equation, yields $x_2 = \frac{4}{5}$, and then $x_1 = \frac{2}{5}$ from the first equation. This process of obtaining the unknowns in reverse order is called **back-substitution**.

The above operations can be expressed more concisely in matrix form as follows. Working with the augmented matrix $[\mathbf{A}, \mathbf{b}]$, the Gaussian elimination process is

$$\begin{bmatrix} 1 & -2 & 6 & 0 \\ 2 & 2 & 3 & 3 \\ -1 & 3 & 0 & 2 \end{bmatrix} \rightarrow \begin{bmatrix} 1 & -2 & 6 & 0 \\ 0 & 6 & -9 & 3 \\ 0 & 1 & 6 & 2 \end{bmatrix} \rightarrow \begin{bmatrix} 1 & -2 & 6 & 0 \\ 0 & 6 & -9 & 3 \\ 0 & 0 & 15/2 & 3/2 \end{bmatrix} \tag{2.24}$$

which, upon back-substitution, yields

$$x_3 = \tfrac{1}{5} \qquad x_2 = \tfrac{4}{5} \qquad x_1 = \tfrac{2}{5} \tag{2.25}$$

General Algorithm for Gaussian Elimination

We discussed the Gaussian elimination process above by means of an example. This process will now be stated as an algorithm, suitable for computer implementation.

Let the original system of equations be as given in Eqs. 2.1, which can be restated as

$$\begin{bmatrix} a_{11} & a_{12} & a_{13} & \cdots & a_{1j} & \cdots & a_{1n} \\ a_{21} & a_{22} & a_{23} & \cdots & a_{2j} & \cdots & a_{2n} \\ a_{31} & a_{32} & a_{33} & \cdots & a_{3j} & \cdots & a_{3n} \\ \hline a_{i1} & a_{i2} & a_{i3} & \cdots & a_{ij} & \cdots & a_{in} \\ \hline a_{n1} & a_{n2} & a_{n3} & \cdots & a_{nj} & \cdots & a_{nn} \end{bmatrix} \begin{Bmatrix} x_1 \\ x_2 \\ x_3 \\ \vdots \\ x_i \\ \vdots \\ x_n \end{Bmatrix} = \begin{Bmatrix} b_1 \\ b_2 \\ b_3 \\ \vdots \\ b_i \\ \vdots \\ b_n \end{Bmatrix} \tag{2.26}$$

Row i (on the left), Column j (below)

Gaussian elimination is a systematic approach to successively eliminate variables x_1, $x_2, x_3, \ldots, x_{n-1}$ until only one variable, x_n, is left. This results in an upper triangular matrix with reduced coefficients and reduced right side. This process is called forward elimination. It is then easy to determine $x_n, x_{n-1}, \ldots, x_3, x_2, x_1$ successively by the process of back-substitution. Let us consider the start of step 1, with $\mathbf{A}$ and $\mathbf{b}$ written as follows:

$$
\begin{bmatrix}
a_{11} & a_{12} & a_{13} & \cdots & a_{1j} & \cdots & a_{1n} \\
a_{21} & a_{22} & a_{23} & \cdots & a_{2j} & \cdots & a_{2n} \\
\vdots & \vdots & \vdots & & \vdots & & \vdots \\
a_{i1} & a_{i2} & a_{i3} & \cdots & a_{ij} & \cdots & a_{in} \\
\hline
a_{n1} & a_{n2} & a_{n3} & \cdots & a_{nj} & \cdots & a_{nn}
\end{bmatrix}
\begin{array}{c} \text{Start of} \\ \text{step} \\ k=1 \end{array}
\begin{Bmatrix}
b_1 \\ b_2 \\ \vdots \\ b_i \\ \vdots \\ b_n
\end{Bmatrix}
\tag{2.27}
$$

The idea at step 1 is to use equation 1 (the first row) in eliminating x_1 from remaining equations. We denote the step number as a superscript set in parentheses. The reduction process at step 1 is

$$
a_{ij}^{(1)} = a_{ij} - \frac{a_{i1}}{a_{11}} \cdot a_{1j}
$$

and   (2.28)

$$
b_i^{(1)} = b_i - \frac{a_{i1}}{a_{11}} \cdot b_1
$$

We note that the ratios a_{i1}/a_{11} are simply the row multipliers that were referred to in the example discussed previously. Also, a_{11} is referred to as a *pivot*. The reduction is carried out for all the elements in the shaded area in (2.27) for which i and j range from 2 to n. The elements in rows 2 to n of the first column are zeroes since x_1 is eliminated. In the computer implementation, we need not set them to zero, but they are zeroes for our consideration. At the start of step 2, we thus have

$$
\begin{bmatrix}
a_{11} & a_{12} & a_{13} & \cdots & a_{1j} & \cdots & a_{1n} \\
0 & a_{22}^{(1)} & a_{23}^{(1)} & \cdots & a_{2j}^{(1)} & \cdots & a_{2n}^{(1)} \\
0 & a_{32}^{(1)} & a_{33}^{(1)} & \cdots & a_{3j}^{(1)} & \cdots & a_{3n}^{(1)} \\
\vdots & \vdots & \vdots & & \vdots & & \vdots \\
0 & a_{i2}^{(1)} & a_{i3}^{(1)} & \cdots & a_{ij}^{(1)} & \cdots & a_{in}^{(1)} \\
\vdots & \vdots & \vdots & & \vdots & & \vdots \\
0 & a_{n2}^{(1)} & a_{n3}^{(1)} & \cdots & a_{nj}^{(1)} & \cdots & a_{nn}^{(1)}
\end{bmatrix}
\begin{array}{c} \\ \text{Start of} \\ \text{step } k=2 \\ \\ \\ \end{array}
\begin{bmatrix}
b_1 \\ b_2^{(1)} \\ b_3^{(1)} \\ \vdots \\ b_i^{(1)} \\ \vdots \\ b_n^{(1)}
\end{bmatrix}
\tag{2.29}
$$

The elements in the shaded area above are reduced at step 2. We now show the start of step k and the operations at step k.

$$
\left[
\begin{array}{cccccccc}
a_{11} & a_{12} & a_{13} & \cdots\cdots\cdots\cdots\cdots & a_{1j} & \cdots & a_{1n} \\
0 & a_{22}^{(1)} & a_{23}^{(1)} & \cdots\cdots\cdots\cdots\cdots & a_{2j}^{(1)} & \cdots & a_{2n}^{(1)} \\
0 & 0 & a_{33}^{(2)} & \cdots\cdots\cdots\cdots\cdots & a_{3j}^{(2)} & \cdots & a_{3n}^{(2)} \\
\end{array}
\right.
$$

Start of step k

Row i

$$
\begin{array}{ccccccc}
0 & 0 & 0 & \cdots & a_{k+1,k+1}^{(k-1)} & \cdots & a_{k+1,j}^{(k-1)} & \cdots & a_{k+1,n}^{(k-1)} \\
0 & 0 & 0 & \cdots & a_{i,k+1}^{(k-1)} & \cdots & a_{ij}^{(k-1)} & \cdots & a_{in}^{(k-1)} \\
0 & 0 & 0 & \cdots & a_{n,k+1}^{(k-1)} & \cdots & a_{nj}^{(k-1)} & \cdots & a_{nn}^{(k-1)}
\end{array}
$$

$$
\begin{array}{c}
b_1 \\ b_2^{(1)} \\ b_3^{(2)} \\ \vdots \\ b_{k+1}^{(k-1)} \\ \vdots \\ b_i^{(k-1)} \\ \vdots \\ b_n^{(k-1)}
\end{array}
$$

Column j

$$(2.30)$$

At step k, elements in the shaded area are reduced. The general reduction scheme with limits on indices may be put as follows.

Step k

$$
a_{ij}^{(k)} = a_{ij}^{(k-1)} - \frac{a_{ik}^{(k-1)}}{a_{kk}^{(k-1)}} a_{kj}^{(k-1)} \qquad i, j = k + 1, \ldots, n
$$

$$
b_i^{(k)} = b_i^{(k-1)} - \frac{a_{ik}^{(k-1)}}{a_{kk}^{(k-1)}} b_k^{(k-1)} \qquad i = k + 1, \ldots, n
$$

$$(2.31)$$

After $(n - 1)$ steps, we get

$$
\left[
\begin{array}{cccccc}
a_{11} & a_{12} & a_{13} & a_{14} & \cdots & a_{1n} \\
 & a_{22}^{(1)} & a_{23}^{(1)} & a_{24}^{(1)} & \cdots & a_{2n}^{(1)} \\
 & & a_{33}^{(2)} & a_{34}^{(2)} & \cdots & a_{3n}^{(2)} \\
 & & & a_{44}^{(3)} & \cdots & a_{4n}^{(3)} \\
 & & \mathbf{0} & & & \vdots \\
 & & & & & a_{nn}^{(n-1)}
\end{array}
\right]
\left\{
\begin{array}{c}
x_1 \\ x_2 \\ x_3 \\ x_4 \\ \vdots \\ x_n
\end{array}
\right\}
=
\left\{
\begin{array}{c}
b_1 \\ b_2^{(1)} \\ b_3^{(2)} \\ b_4^{(3)} \\ \vdots \\ b_n^{(n-1)}
\end{array}
\right\}
$$

$$(2.32)$$

The superscripts are for the convenience of presentation. In the computer implementation these superscripts can be avoided. We now drop the superscripts for convenience, and the back-substitution process is given by

$$
x_n = \frac{b_n}{a_{nn}}
$$

$$(2.33)$$

and then

$$
x_i = \frac{b_i - \displaystyle\sum_{j=i+1}^{n} a_{ij} x_j}{a_{ii}} \qquad i = n - 1, n - 2, \ldots, 1
$$

$$(2.34)$$

This completes the Gauss elimination algorithm.

The algorithm discussed above is given below in the form of computer logic.

Algorithm 1: General Matrix

Forward elimination (reduction of A, b)

$$\text{DO} \quad k = 1, n - 1$$
$$\text{DO} \quad i = k + 1, n$$
$$c = \frac{a_{ik}}{a_{kk}}$$
$$\text{DO} \quad j = k + 1, n$$
$$a_{ij} = a_{ij} - c a_{kj}$$
$$b_i = b_i - c b_k$$

Back-substitution

$$b_n = \frac{b_n}{a_{nn}}$$
$$\text{DO} \quad ii = 1, n - 1$$
$$i = n - ii$$
$$\text{sum} = 0$$
$$\text{DO} \quad j = i + 1, n$$
$$\text{sum} = \text{sum} + a_{ij} b_j$$
$$b_i = \frac{b_i - \text{sum}}{a_{ii}}$$

Note. **b** contains the solution to **Ax = b**.

Symmetric Matrix

If **A** is symmetric, then the algorithm above needs two modifications. One is that the multiplier is defined as

$$c = \frac{a_{ki}}{a_{kk}} \tag{2.35}$$

The other modification is related to the DO LOOP index (the third DO LOOP in the previous algorithm):

$$\text{DO} \quad j = i, n \tag{2.36}$$

Symmetric Banded Matrices

In a **banded** matrix, all of the nonzero elements are contained within a band; outside of the band all elements are zero. The stiffness matrix that we will come across in subsequent chapters is a symmetric and banded matrix.

     Consider an $(n \times n)$ symmetric banded matrix:

$$
\begin{bmatrix}
\overset{\displaystyle \leftarrow \text{ nbw } \rightarrow}{X \quad X \quad X \quad X \quad X} & & & \\
\quad X \quad X \quad X \quad X \quad X & & 0 & \\
\quad\quad X \quad X \quad X \quad X \quad X & & & \\
\quad\quad\quad X \quad X \quad X \quad X \quad X & & & \\
\quad\quad\quad\quad X \quad X \quad X \quad X \quad X & & & \\
\text{Symmetric} \quad\quad X \quad X \quad X \quad X \quad X & & & \\
\quad\quad\quad\quad\quad\quad\quad X \quad X \quad X \quad X \quad X & & & \\
\quad\quad\quad\quad\quad\quad\quad\quad X \quad X \quad X \quad X & & & \\
\quad\quad\quad\quad\quad\quad\quad\quad\quad X \quad X \quad X & & & \\
\quad\quad\quad\quad\quad\quad\quad\quad\quad\quad X \quad X & & & \\
\quad\quad\quad\quad\quad\quad\quad\quad\quad\quad\quad X & & &
\end{bmatrix}
\tag{2.37}
$$

2nd diagonal
Main (1st) diagonal

Above, nbw is called the **half-bandwidth**. Since only the nonzero elements need to be stored, the elements of the above matrix are compactly stored in the $(n \times \text{nbw})$ matrix below

$$
\begin{array}{cccc}
\text{1st} \;\; \text{2nd} & & \text{nbw} \\
\end{array}
$$
$$
\begin{bmatrix}
X & X & X & X & X \\
X & X & X & X & X \\
X & X & X & X & X \\
X & X & X & X & X \\
X & X & X & X & X \\
X & X & X & X & X \\
X & X & X & X & X \\
X & X & X & X & \\
X & X & X & & \\
X & X & & 0 & \\
X & & & &
\end{bmatrix}
\tag{2.38}
$$

The principal diagonal or 1st diagonal of Eq. 2.37 is the first column of Eq. 2.38. In general, the pth diagonal of Eq. 2.37 is stored as the pth column of Eq. 2.38 as shown. The correspondence between the elements of Eqs. 2.37 and 2.38 is given by

$$\left. a_{ij} \right|_{(2.37)}^{(j > i)} = \left. a_{i(j-i+1)} \right|_{(2.38)} \qquad (2.39)$$

Also, we note that $a_{ij} = a_{ji}$ in (2.37), and that the number of elements in the kth row of Eq. 2.38 is $\min(n - k + 1, \text{nbw})$. We can now present the Gaussian elimination algorithm for symmetric banded matrix.

Algorithm 2: Symmetric Banded Matrix

Forward elimination

DO $k = 1, n - 1$

nbk $= \min(n - k + 1, \text{nbw})$

DO $i = k + 1, \text{nbk} + k - 1$

$i1 = i - k + 1$

$c = a_{k,i1}/a_{k,1}$

DO $j = i, \text{nbk} + k - 1$

$j1 = j - i + 1$

$j2 = j - k + 1$

$a_{i,j1} = a_{i,j1} - c a_{k,j2}$

$b_i = b_i - c b_k$

Back-substitution

$$b_n = \frac{b_n}{a_{n,1}}$$

DO $ii = 1, n - 1$

$i = n - ii$

nbi $= \min(n - i + 1, \text{nbw})$

sum $= 0$

DO $j = 2, \text{nbi}$

sum $= \text{sum} + a_{i,j} b_{i+j-1}$

$$b_i = \frac{b_i - \text{sum}}{a_{i,1}}$$

Comment. Above, the DO LOOP indices are based on the original matrix as in (2.37); the correspondence in (2.39) is then used while referring to elements of the banded matrix **A**. Alternatively, it is possible to express the DO LOOP indices directly as they refer to the banded **A** matrix. Both approaches are used in the computer programs.

Solution with Multiple Right Sides

Often, we need to solve $\mathbf{Ax} = \mathbf{b}$ with the same **A** but with different **b**'s. This happens in the finite element method when we wish to analyze the same structure for different loading conditions. In this situation, it is computationally more economical to separate the calculations associated with **A** from those associated with **b**. The reason for this is that the number of operations in reduction of an $(n \times n)$ matrix **A** to its triangular form is proportional to n^3, while the number of operations for reduction of **b** and back-substitution is proportional only to n^2. For large n, this difference is significant.

The previous algorithm for a symmetric banded matrix is modified accordingly as follows.

Algorithm 3: Symmetric Banded, Multiple Right Sides

Forward elimination for A

$$
\begin{array}{l}
\text{DO} \quad k = 1, n - 1 \\
\quad \text{nbk} = \min(n - k + 1, \text{nbw}) \\
\quad \text{DO} \quad i = k + 1, \text{nbk} + k - 1 \\
\quad\quad i1 = i - k + 1 \\
\quad\quad c = a_{k,i1}/a_{k,1} \\
\quad\quad \text{DO} \quad j = i, \text{nbk} + k - 1 \\
\quad\quad\quad j1 = j - i + 1 \\
\quad\quad\quad j2 = j - k + 1 \\
\quad\quad\quad a_{i,j1} = a_{i,j1} - ca_{k,j2}
\end{array}
$$

Forward elimination of each b

$$
\begin{array}{l}
\text{DO} \quad k = 1, n - 1 \\
\quad \text{nbk} = \min(n - k + 1, \text{nbw}) \\
\quad \text{DO} \quad i = k + 1, \text{nbk} + k - 1 \\
\quad\quad i1 = i - k + 1 \\
\quad\quad c = a_{k,i1}/a_{k,1} \\
\quad\quad b_i = b_i - cb_k
\end{array}
$$

Back-substitution   Same as in Algorithm 2.

Gauss Elimination with Column Reduction

A careful observation of the Gauss elimination process shows us a way to reduce the coefficients column after column. This process leads to the simplest procedure for **skyline solution**, which we present later. We consider the column reduction procedure for symmetric matrices. Let the coefficients in the upper triangular matrix and the vector **b** be stored.

We can understand the motivation behind the column approach by referring back to Eqs. 2.32 which are repeated below

$$
\begin{bmatrix}
a_{11} & a_{12} & a_{13} & a_{14} & \cdots & a_{1n} \\
 & a_{22}^{(1)} & a_{23}^{(1)} & a_{24}^{(1)} & \cdots & a_{2n}^{(1)} \\
 & & a_{33}^{(2)} & a_{34}^{(2)} & \cdots & a_{3n}^{(2)} \\
 & & & a_{44}^{(3)} & \cdots & a_{4n}^{(3)} \\
 & & & & \vdots & \\
 & & & & & a_{nn}^{(n-1)}
\end{bmatrix}
\begin{Bmatrix} x_1 \\ x_2 \\ x_3 \\ x_4 \\ \vdots \\ x_n \end{Bmatrix}
=
\begin{Bmatrix} b_1 \\ b_2^{(1)} \\ b_3^{(2)} \\ b_4^{(3)} \\ \vdots \\ b_n^{(n-1)} \end{Bmatrix}
$$

Let us focus our attention on, say, column 3 of the reduced matrix above. The first element in this column is unmodified, the second element is modified once, and the third element is modified twice. Further, from Eqs. 2.31, and using the fact that $a_{ij} = a_{ji}$ since **A** is assumed to be symmetric, we have

$$
a_{23}^{(1)} = a_{23} - \frac{a_{12}}{a_{11}} a_{13}
$$

$$
a_{33}^{(1)} = a_{33} - \frac{a_{13}}{a_{11}} a_{13} \tag{2.40}
$$

$$
a_{33}^{(2)} = a_{33}^{(1)} - \frac{a_{23}^{(1)}}{a_{22}^{(1)}} a_{23}^{(1)}
$$

From the above equations, we make the critical observation that *the reduction of column 3 can be done using only the elements in columns 1 and 2 and the already reduced elements in column 3*. This idea whereby column 3 is obtained using only elements in previous columns that have already been reduced is shown schematically:

$$
\begin{bmatrix}
a_{11} & a_{12} & a_{13} \\
 & a_{22}^{(1)} & a_{23} \\
 & & a_{33}
\end{bmatrix}
\rightarrow
\begin{bmatrix}
a_{11} & a_{12} & a_{13} \\
 & a_{22}^{(1)} & a_{23}^{(1)} \\
 & & a_{33}^{(1)}
\end{bmatrix}
\rightarrow
\begin{bmatrix}
a_{11} & a_{12} & a_{13} \\
 & a_{22}^{(1)} & a_{23}^{(1)} \\
 & & a_{33}^{(2)}
\end{bmatrix}
\tag{2.41}
$$

The reduction of other columns is similarly done. For instance, the reduction of column 4 can be done in three steps as shown schematically below:

$$\begin{Bmatrix} a_{14} \\ a_{24} \\ a_{34} \\ a_{44} \end{Bmatrix} \rightarrow \begin{Bmatrix} a_{14} \\ a_{24}^{(1)} \\ a_{34}^{(1)} \\ a_{44}^{(1)} \end{Bmatrix} \rightarrow \begin{Bmatrix} a_{14} \\ a_{24}^{(1)} \\ a_{34}^{(2)} \\ a_{44}^{(2)} \end{Bmatrix} \rightarrow \begin{Bmatrix} a_{14} \\ a_{24}^{(1)} \\ a_{34}^{(2)} \\ a_{44}^{(3)} \end{Bmatrix} \tag{2.42}$$

We now discuss the reduction of column j, $2 \le j \le n$, assuming that columns to the left of j have been fully reduced. The coefficients can be represented in the following form:

$$\tag{2.43}$$

The reduction of column j requires only elements from columns to the left of j and appropriately reduced elements from column j. We note that for column j, the number of steps needed are $j - 1$. Also, since a_{11} is not reduced, we need to reduce columns 2 to n only. The logic is now given below.

$$\begin{array}{l} \text{DO} \quad j = 2 \text{ to } n \\ \quad \text{DO} \quad k = 1 \text{ to } j - 1 \\ \qquad \text{DO} \quad i = k + 1 \text{ to } j \\ \qquad\quad a_{ij}^{(k)} = a_{ij}^{(k-1)} - \dfrac{a_{ki}^{(k-1)}}{a_{kk}^{(k-1)}} a_{kj}^{(k-1)} \end{array} \tag{2.44}$$

Interestingly, the reduction of the right side **b** can be considered as the reduction of one more column. Thus,

$$\begin{array}{l} \text{DO} \quad k = 1 \text{ to } n - 1 \\ \quad \text{DO} \quad i = k + 1 \text{ to } n \\ \qquad b_i^{(k)} = b_i^{(k-1)} - \dfrac{a_{ki}^{(k-1)}}{a_{kk}^{(k-1)}} b_k^{(k-1)} \end{array} \tag{2.45}$$

From Eqs. 2.44, we observe that if there are a set of zeroes at the top of a column, the operations need to be carried out only on the elements ranging from the first nonzero element to the diagonal. This leads naturally to the skyline solution.

Skyline Solution

If there are zeroes at the top of a column, only the elements starting from the first nonzero value need be stored. The line separating the top zeroes from the first nonzero element is called the skyline. Consider the example:

$$
\begin{array}{c}
\text{Column} \rightarrow \;\; |\;\; 1\;\; |\;\; 2\;\; |\;\; 2\;\; |\;\; 4\;\; |\;\; 4\;\; |\;\; 4\;\; |\;\; 3\;\; |\;\; 5\;\; | \\[2pt]
\text{height}
\end{array}
$$

$$
\begin{bmatrix}
a_{11} & a_{12} & 0 & a_{14} & 0 & 0 & 0 & 0 \\
 & a_{22} & a_{23} & a_{24} & a_{25} & 0 & 0 & 0 \\
 & & a_{33} & a_{34} & 0 & a_{36} & 0 & 0 \\
 & & & a_{44} & 0 & a_{46} & 0 & a_{48} \\
 & & & & a_{55} & a_{56} & a_{57} & 0 \\
 & & & & & a_{66} & a_{67} & a_{68} \\
 & & & & & & a_{77} & a_{78} \\
 & & & & & & & a_{88}
\end{bmatrix}
\qquad \text{Skyline}
\tag{2.46}
$$

For efficiency, only the active columns above need be stored. These can be stored in a column vector **A**, and a diagonal pointer vector **ID** as

$$
\mathbf{A} =
\begin{bmatrix}
a_{11} \\
a_{12} \\
a_{22} \\
a_{23} \\
a_{33} \\
a_{14} \\
a_{24} \\
a_{34} \\
a_{44} \\
\vdots \\
a_{88}
\end{bmatrix}
\begin{array}{l}
\leftarrow 1 \\ \\
\leftarrow 3 \\ \\
\leftarrow 5 \\ \\ \\ \\
\leftarrow 9 \\ \\ \\
\leftarrow 25
\end{array}
\qquad
\text{Diagonal pointer (ID)}
\qquad
\mathbf{ID} =
\begin{bmatrix}
1 \\
3 \\
5 \\
9 \\
13 \\
17 \\
20 \\
25
\end{bmatrix}
\tag{2.47}
$$

The height of column I is given by $\mathrm{ID}(I) - \mathrm{ID}(I - 1)$. The right side, **b**, is stored in a separate column. The column reduction scheme of the Gauss elimination method can be applied for solution of a set of equations. A skyline solver program is given.

```
1000 CLS
1010 REM ***** Gauss Elimination Method (General Matrix) *****
1020 REM *****            Row-wise Elimination            *****
1030 PRINT "======================================="
1040 PRINT "       Program written by          "
1050 PRINT " T.R.Chandrupatla and A.D.Belegundu   "
1060 PRINT "======================================="
1070 READ N
1080 DIM A(N, N), B(N)
1090 FOR I = 1 TO N
1100 FOR J = 1 TO N
1110 READ A(I, J): NEXT J: NEXT I
1120 FOR I = 1 TO N
1130 READ B(I): NEXT I
1140 FOR K = 1 TO N - 1
1150 FOR I = K + 1 TO N
1160 C = A(I, K) / A(K, K)
1170 FOR J = K + 1 TO N
1180 A(I, J) = A(I, J) - C * A(K, J): NEXT J
1190 B(I) = B(I) - C * B(K): NEXT I: NEXT K
1200 REM *** Back Substitution ***
1210 B(N) = B(N) / A(N, N)
1220 FOR II = 1 TO N - 1
1230 I = N - II
1240 C = 1 / A(I, I): B(I) = C * B(I)
1250 FOR K = I + 1 TO N
1260 B(I) = B(I) - C * A(I, K) * B(K): NEXT K: NEXT II
1270 PRINT "Solution Vector"
1280 FOR I = 1 TO N: PRINT USING "##.#  "; B(I), : NEXT I: PRINT
1290 END
3000 DATA 8
3010 DATA 6,0,1,2,0,0,2,1
3020 DATA 0,5,1,1,0,0,3,0
3030 DATA 1,1,6,1,2,0,1,2
3040 DATA 2,1,1,7,1,2,1,1
3050 DATA 0,0,2,1,6,0,2,1
3060 DATA 0,0,0,2,0,4,1,0
3070 DATA 2,3,1,1,2,1,5,1
3080 DATA 1,0,2,1,1,0,1,3
3090 DATA 1,1,1,1,1,1,1,1
```

```
1000 CLS
1010 REM **** Sky Line Method ****
1020 REM **** Gauss Elimination ****
1022 PRINT "======================================="
1024 PRINT "       Program written by          "
1026 PRINT " T.R.Chandrupatla and A.D.Belegundu   "
1028 PRINT "======================================="
1030 READ N
1040 NSUM = 0
1050 DIM ID(N)
1060 REM *** Read Column Heights then convert to Diagonal Pointers ***
1070 READ ID(1)
1080 FOR I = 2 TO N
1090 READ ID(I)
1100 ID(I) = ID(I) + ID(I - 1): NEXT I
1110 NSUM = ID(N)
```

```
1120 DIM A(NSUM), B(N)
1130 NT = 0: NI = 1
1140 FOR I = 1 TO N
1150 IF I = 1 GOTO 1170
1160 NT = ID(I - 1): NI = ID(I) - ID(I - 1)
1170 FOR J = 1 TO NI
1180 READ A(NT + J): NEXT J: NEXT I
1190 FOR I = 1 TO N
1200 READ B(I): NEXT I
1210 PRINT "Original Matrix and Right Hand Side"
1220 NI = 1
1230 FOR I = 1 TO N
1240 IF I = 1 GOTO 1260
1250 NI = ID(I) - ID(I - 1)
1260 FOR J = 1 TO NI: IJ = ID(I) - NI + J
1270 PRINT A(IJ), : NEXT J: PRINT : NEXT I
1280 PRINT : PRINT : FOR I = 1 TO N: PRINT B(I), : NEXT I: PRINT
1290 FOR J = 2 TO N
1300 NJ = ID(J) - ID(J - 1)
1310 IF NJ = 1 GOTO 1440
1320 K1 = 0: NJ = J - NJ + 1
1330 FOR K = NJ TO J - 1
1340 K1 = K1 + 1: KJ = ID(J - 1) + K1: KK = ID(K)
1350 C = A(KJ) / A(KK)
1360 FOR I = K + 1 TO J
1370 NI = ID(I) - ID(I - 1)
1380 IF (I - K + 1) > NI GOTO 1420
1390 IJ = ID(J) - J + I
1400 KI = ID(I) - I + K
1410 A(IJ) = A(IJ) - C * A(KI)
1420 NEXT I
1430 NEXT K
1440 NEXT J
1450 FOR K = 1 TO N - 1: KK = ID(K)
1460 C = B(K) / A(KK)
1470 FOR I = K + 1 TO N
1480 NI = ID(I) - ID(I - 1)
1490 IF (I - K + 1) > NI GOTO 1520
1500 KI = ID(I) - I + K
1510 B(I) = B(I) - C * A(KI)
1520 NEXT I: NEXT K
1530 PRINT "Reduced Matrix and Right Hand Side"
1540 NI = 1: NT = 0
1550 FOR I = 1 TO N
1560 IF I = 1 GOTO 1580
1570 NI = ID(I) - ID(I - 1): NT = ID(I - 1)
1580 FOR J = 1 TO NI
1590 PRINT USING "##.#  "; A(NT + J), : NEXT J: PRINT : NEXT I
1600 PRINT : PRINT : FOR I = 1 TO N: PRINT USING "##.#  "; B(I), : NEXT I: PRINT
1610 REM  **** Back -substitution ****
1620 NS = ID(N): B(N) = B(N) / A(NS)
1630 FOR I1 = 1 TO N - 1
1640 I = N - I1
1650 II = ID(I)
1660 C = 1 / A(II): B(I) = C * B(I)
1670 FOR J = I + 1 TO N
1680 J1 = J - I + 1: NJ = ID(J) - ID(J - 1)
1690 IF J1 > NJ GOTO 1720
1700 IJ = ID(J) - J + I
```

```
1710 B(I) = B(I) – C * A(IJ) * B(J)
1720 NEXT J
1730 NEXT I1
1740 PRINT : PRINT : FOR I = 1 TO N: PRINT USING "##.#  "; B(I), : NEXT I: PRINT
1750 END
3000 DATA 8
3010 DATA 1,1,3,4,3,3,7,8
3020 DATA 6
3030 DATA 5
3040 DATA 1,1,6
3050 DATA 2,1,1,7
3060 DATA 2,1,6
3070 DATA 2,0,4
3080 DATA 2,3,1,1,2,1,5
3090 DATA 1,0,2,1,1,0,1,3
3100 DATA 1,1,1,1,1,1,1,1
```

PROBLEMS

2.1. Given that

$$
\mathbf{A} = \begin{bmatrix} 8 & -2 & 0 \\ -2 & 4 & -3 \\ 0 & -3 & 3 \end{bmatrix} \qquad \mathbf{d} = \begin{Bmatrix} 2 \\ -1 \\ 3 \end{Bmatrix}
$$

determine
  (a) $I - \mathbf{dd}^T$
  (b) det $\mathbf{A}$
  (c) The eigenvalues and eigenvectors of $\mathbf{A}$. Is $\mathbf{A}$ positive definite?
  (d) The solution to $\mathbf{Ax} = \mathbf{d}$ using Algorithms 1 and 2, by hand calculation.

2.2. Given that

$$
\mathbf{N} = [\xi, 1 - \xi^2]
$$

find
  (a) $\displaystyle\int_{-1}^{1} \mathbf{N}d\xi$

  (b) $\displaystyle\int_{-1}^{1} \mathbf{N}^T \mathbf{N}d\xi$

2.3. Express $q = x_1 - 6x_2 + 3x_1^2 + 5x_1x_2$ in the matrix form $\frac{1}{2}\mathbf{x}^T\mathbf{Qx} + \mathbf{c}^T\mathbf{x}$.

2.4. Implement Algorithm 3 in BASIC. Hence, solve $\mathbf{Ax} = \mathbf{b}$ with $\mathbf{A}$ as in Problem 2.1, and each of the following $\mathbf{b}$'s:

$$
\mathbf{b} = [5, -10, 3]^T
$$

$$
\mathbf{b} = [2.2, -1, 3]^T
$$

2.5. Using the cofactor approach, determine the inverse of the matrix

$$
\begin{bmatrix} 2 & 1 & 1 \\ 1 & 2 & 1 \\ 1 & 1 & 2 \end{bmatrix}
$$

2.6. Given that the area of a triangle with corners at (x_1, y_1), (x_2, y_2), and (x_3, y_3) can be written in the form

$$\text{Area} = \frac{1}{2} \det \begin{bmatrix} 1 & x_1 & y_1 \\ 1 & x_2 & y_2 \\ 1 & x_3 & y_3 \end{bmatrix}$$

determine the area of the triangle with corners at $(1, 1)$, $(4, 2)$, and $(2, 4)$.

2.7. For the triangle in Fig. P2.7, the interior point P at $(2, 2)$ divides it into three areas, A_1, A_2, and A_3, as shown. Determine A_1/A, A_2/A, and A_3/A.

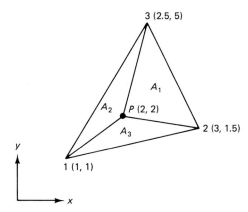

Figure P2.7

2.8. A symmetric matrix $[\mathbf{A}]_{n \times n}$ has a bandwidth nbw and is stored in the matrix $[\mathbf{B}]_{n \times \text{nbw}}$.
  (a) Find the location in $\mathbf{B}$ that corresponds to $A_{11,14}$
  (b) Find the location in $\mathbf{A}$ that corresponds to $B_{6,1}$.

2.9. For a symmetric (10×10) matrix with all nonzero elements, determine the number of locations needed for banded and skyline storage methods.

chapter three ———————————————————

One-Dimensional Problems

3.1 INTRODUCTION

The total potential energy and the stress–strain and strain–displacement relationships are now used in developing the finite element method for a one-dimensional problem. The basic procedure is the same for two- and three-dimensional problems discussed later in the book. For the one-dimensional problem, the stress, strain, displacement, and loading depend only on the variable x. That is, the vectors $\mathbf{u}$, $\boldsymbol{\sigma}$, $\boldsymbol{\epsilon}$, $\mathbf{T}$, and $\mathbf{f}$ in Chapter 1 now reduce to

$$u = u(x) \qquad \sigma = \sigma(x) \qquad \epsilon = \epsilon(x)$$
$$T = T(x) \qquad f = f(x) \tag{3.1}$$

Furthermore, the stress–strain and strain–displacement relations are

$$\sigma = E\epsilon \qquad \epsilon = \frac{du}{dx} \tag{3.2}$$

For one-dimensional problems, the differential volume dV can be written as

$$dV = A\, dx \tag{3.3}$$

The loading consists of three types: the **body force** f, the **traction force** T, and the **point load** P_i. These forces are shown acting on a body in Fig. 3.1. A body force is a distributed force acting on every elemental volume of the body, and has the units of force per unit volume. The self-weight due to gravity is an example of a body force. A traction force is a distributed load acting on the surface of the body. In Chapter 1, the traction force is defined as force per unit area. For the one-dimen-

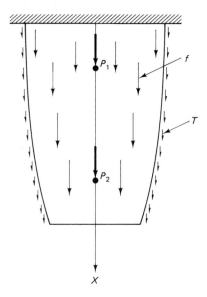

Figure 3.1 One-dimensional bar loaded by traction, body, and point loads.

sional problem considered here, however, the traction force is defined as force per unit length. This is done by taking the traction force to be the product of the force per unit area with the perimeter of the cross section. Frictional resistance, viscous drag, and surface shear are examples of traction forces in one-dimensional problems. Finally, P_i is a force acting at a point i and u_i is the x displacement at that point.

The finite element modeling of a one-dimensional body is considered below. The basic idea is to discretize the region and express the displacement field in terms of values at discrete points. Linear elements are introduced first. Stiffness and load concepts are developed using potential energy and Galerkin approaches. Boundary conditions are then considered. Temperature effects and quadratic elements are discussed later in this chapter.

3.2 FINITE ELEMENT MODELING

Element Division

Consider the bar in Fig. 3.1. The first step is to **model** the bar as a *stepped shaft,* consisting of a discrete number of elements, each having a uniform cross section. Specifically, let us model the bar using four finite elements. A simple scheme for doing this is to divide the bar into four regions, as shown in Fig. 3.2a. The average cross-sectional area within each region is evaluated and then used to define an element with uniform cross section. The resulting four-element, five-node finite element model is shown in Fig. 3.2b. In the finite element model, every element connects to two nodes. In Fig. 3.2b, the element numbers are circled to distinguish

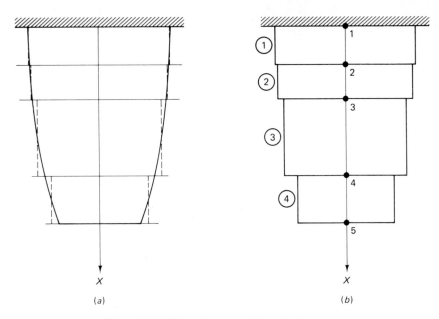

Figure 3.2 Finite element modeling of a bar.

them from node numbers. In addition to the cross section, traction and body forces are also (normally) treated as constant within each element. However, cross-sectional area, traction, and body forces can differ in magnitude from element to element. Better approximations are obtained by increasing the number of elements. It is convenient to define a node at each location where a point load is applied.

Numbering Scheme

We have shown how a rather complicated looking bar has been modeled using a discrete number of elements, each element having a simple geometry. The similarity of the various elements is one reason why the finite element method is easily amenable to computer implementation. For easy implementation, an orderly numbering scheme for the model has to be adopted, as discussed below.

In a one-dimensional problem, every node is permitted to displace only in the $\pm x$ direction. Thus, each node has only one *degree of freedom (dof)*. The five-node finite element model in Fig. 3.2b has five dof's. The displacements along each dof are denoted by $Q_1, Q_2, \ldots, Q_5$. In fact, the column vector $\mathbf{Q} = [Q_1, Q_2, \ldots, Q_5]^T$ is called the *global displacement vector*. The *global load vector* is denoted by $\mathbf{F} = [F_1, F_2, \ldots, F_5]^T$. The vectors $\mathbf{Q}$ and $\mathbf{F}$ are shown in Fig. 3.3. The sign convention used is that a displacement or load has a positive value if acting along the $+x$ direction. At this stage, conditions at the boundary are not imposed. For example, node 1 in Fig. 3.3 is fixed, which implies $Q_1 = 0$. These conditions are discussed later.

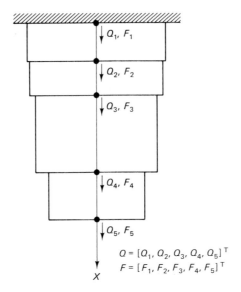

$$Q = [Q_1, Q_2, Q_3, Q_4, Q_5]^T$$
$$F = [F_1, F_2, F_3, F_4, F_5]^T$$

Figure 3.3   Q and F vectors.

Each element has two nodes; therefore, the **element connectivity** information can be conveniently represented as shown in Fig. 3.4. Further, the element connectivity table is also given. In the connectivity table, the headings 1 and 2 refer to *local node numbers* of an element and the corresponding node numbers on the body are called *global numbers*. Connectivity thus establishes the local–global correspondence. In this simple example, the connectivity can be easily generated since local node 1 is the same as the element number e and local node 2 is $e + 1$. Other ways of numbering nodes or more complex geometries suggest the need for a connectivity table. This will be more evident in two- and three-dimensional problems.

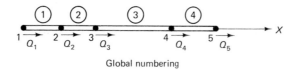

Global numbering

Figure 3.4   Element connectivity.

The concepts of dof, nodal displacements, nodal loads, and element connectivity are central to the finite element method and should be clearly understood.

3.3 COORDINATES AND SHAPE FUNCTIONS

Consider a typical finite element e in Fig. 3.5a. In the local number scheme, the first node will be numbered 1 and the second node 2. The notation $x_1 = x$ coordinate of node 1, $x_2 = x$ coordinate of node 2 is used. We define a **natural** or **intrinsic** coordinate system, denoted by ξ, as

$$\xi = \frac{2}{x_2 - x_1}(x - x_1) - 1 \tag{3.4}$$

From above, we see that $\xi = -1$ at node 1 and $\xi = 1$ at node 2 (Fig. 3.5b). The length of an element is covered when ξ changes from -1 to 1. We use this system of coordinates in defining shape functions, which are used in interpolating the displacement field.

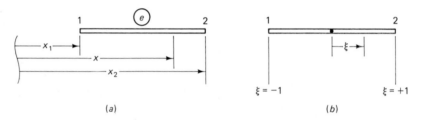

(a)   (b)

Figure 3.5   Typical element in x and ξ coordinates.

Now *the unknown displacement field within an element will be interpolated by a linear distribution* (Fig. 3.6). This approximation becomes increasingly accurate as more elements are considered in the model. To implement this linear interpolation,

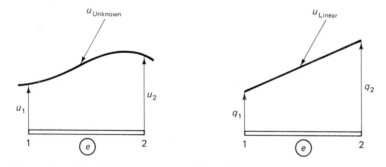

Figure 3.6   Linear interpolation of the displacement field within an element.

linear **shape functions** will be introduced as

$$N_1(\xi) = \frac{1 - \xi}{2} \tag{3.5}$$

$$N_2(\xi) = \frac{1 + \xi}{2} \tag{3.6}$$

The shape functions N_1 and N_2 are shown in Figs. 3.7a and b, respectively. The graph of the shape function N_1 in Fig. 3.7a is obtained from Eq. 3.5 by noting that $N_1 = 1$ at $\xi = -1$, $N_1 = 0$ at $\xi = 1$, and is a straight line between the two points. Similarly, the graph of N_2 in Fig. 3.7b is obtained from Eq. 3.6. Once the shape functions are defined, the linear displacement field within the element can be written in terms of the nodal displacements q_1 and q_2 as

$$u = N_1 q_1 + N_2 q_2 \tag{3.7a}$$

or, in matrix notation as

$$u = \mathbf{Nq} \tag{3.7b}$$

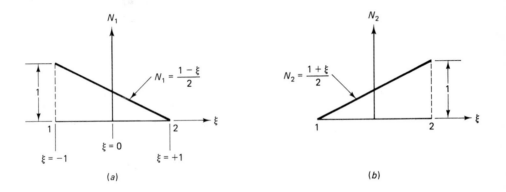

(a)   (b)

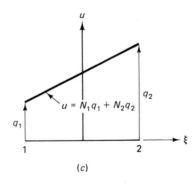

(c)

Figure 3.7 (a) Shape function N_1, (b) shape function N_2, and (c) linear interpolation using N_1 and N_2.

where

$$\mathbf{N} = [N_1, N_2] \quad \text{and} \quad \mathbf{q} = [q_1, q_2]^{\mathrm{T}} \tag{3.8}$$

In the equations above, $\mathbf{q}$ is referred to as the *element displacement vector*. It is readily verified from Eq. 3.7a that $u = q_1$ at node 1, $u = q_2$ at node 2, and that u varies linearly (Fig. 3.7c).

It may be noted that the transformation from x to ξ in Eq. 3.4 can be written in terms of N_1 and N_2 as

$$x = N_1 x_1 + N_2 x_2 \tag{3.9}$$

Comparing Eqs 3.7a and 3.9, we see that both the displacement u and the coordinate x are interpolated within the element using the *same* shape functions N_1 and N_2. This is referred to as the *isoparametric* formulation in the literature.

Though linear shape functions have been used above, other choices are possible. Quadratic shape functions are discussed in Section 3.9. In general, shape functions need to satisfy the following:

1. First derivatives must be finite within an element.
2. Displacements must be continuous across the element boundary.

Rigid body motion should not introduce any stresses in the element.

Example 3.1

Referring to Fig. E3.1, do the following:

(a) Evaluate ξ, N_1, and N_2 at point P.
(b) If $q_1 = 0.003$ in. and $q_2 = -0.005$ in., determine the value of the displacement q at point P.

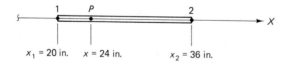

$x_1 = 20$ in. $x = 24$ in. $x_2 = 36$ in.   **Figure E3.1**

Solution

(a) Using Eq. 3.4, the ξ coordinate of point P is given by

$$\xi_p = \tfrac{2}{16}(24 - 20) - 1$$

$$= -0.5$$

Now Eqs. 3.5 and 3.6 yield

$$N_1 = 0.75 \quad \text{and} \quad N_2 = 0.25$$

(b) Using Eq. 3.7a, we get

$$u_p = 0.75(0.003) + 0.25(-0.005)$$

$$= 0.001 \text{ in.} \qquad \blacksquare$$

The strain–displacement relation in Eq. 3.2 is

$$\epsilon = \frac{du}{dx}$$

Upon using the chain rule of differentiation, we obtain

$$\epsilon = \frac{du}{d\xi} \frac{d\xi}{dx} \tag{3.10}$$

From the relation between x and ξ in Eq. 3.4, we have

$$\frac{d\xi}{dx} = \frac{2}{x_2 - x_1} \tag{3.11}$$

Also, since

$$u = N_1 q_1 + N_2 q_2 = \frac{1 - \xi}{2} q_1 + \frac{1 + \xi}{2} q_2$$

we have

$$\frac{du}{d\xi} = \frac{-q_1 + q_2}{2} \tag{3.12}$$

Thus, Eq. 3.10 yields

$$\epsilon = \frac{1}{x_2 - x_1} (-q_1 + q_2) \tag{3.13}$$

The above equation can be written as

$$\epsilon = \mathbf{Bq} \tag{3.14}$$

where the (1×2) matrix $\mathbf{B}$, called the *element strain–displacement matrix,* is given by

$$\mathbf{B} = \frac{1}{x_2 - x_1} [-1 \quad 1] \tag{3.15}$$

Note. Use of linear shape functions results in a constant $\mathbf{B}$ matrix and, hence, in a constant strain within the element. The stress, from Hooke's law, is

$$\sigma = E\mathbf{Bq} \tag{3.16}$$

The stress given by the above equation is also constant within the element. For in-

terpolation purposes, however, the stress obtained from Eq. 3.16 can be considered
to be the value at the centroid of the element.

The expressions $u = \mathbf{Nq}$, $\epsilon = \mathbf{Bq}$, $\sigma = E\mathbf{Bq}$ relate the displacement, strain
and stress, respectively, in terms of nodal values. These expressions will now be
substituted into the potential energy expression for the bar to obtain the element
stiffness and load matrices.

3.4 THE POTENTIAL ENERGY APPROACH

The general expression for the potential energy given in Chapter 1 is

$$\Pi = \frac{1}{2} \int_L \sigma^\mathrm{T} \epsilon A \, dx - \int_L u^\mathrm{T} f A \, dx - \int_L u^\mathrm{T} T \, dx - \sum_i u_i P_i \qquad (3.17)$$

The quantities σ, ϵ, u, f, and T above are discussed at the beginning of this
chapter. In the last term above, P_i, represents a force acting at point i, and u_i is the
x displacement at that point. The summation on i gives the potential energy due to
all point loads.

Since the continuum has been discretized into finite elements, the expression
for Π becomes

$$\Pi = \sum_e \frac{1}{2} \int_e \sigma^\mathrm{T} \epsilon A \, dx - \sum_e \int_e u^\mathrm{T} f A \, dx - \sum_e \int_e u^\mathrm{T} T \, dx - \sum_i Q_i P_i \qquad (3.18a)$$

The last term above assumes that point loads P_i are applied at the nodes. This as-
sumption makes the present derivation simpler with respect to notation, and is also a
common modeling practice. Equation 3.18a above can be written as

$$\Pi = \sum_e U_e - \sum_e \int_e u^\mathrm{T} f A \, dx - \sum_e \int_e u^\mathrm{T} T \, dx - \sum_i Q_i P_i \qquad (3.18b)$$

where

$$U_e = \frac{1}{2} \int_e \sigma^\mathrm{T} \epsilon A \, dx$$

is the element strain energy.

Element Stiffness Matrix

Consider the strain energy term

$$U_e = \frac{1}{2} \int_e \sigma^\mathrm{T} \epsilon A \, dx \qquad (3.19)$$

Substituting for $\sigma = E\mathbf{Bq}$ and $\epsilon = \mathbf{Bq}$ into the above yields

$$U_e = \frac{1}{2} \int_e \mathbf{q}^T \mathbf{B}^T E \mathbf{B} \mathbf{q} A \, dx \qquad (3.20a)$$

or

$$U_e = \frac{1}{2} \mathbf{q}^T \int_e [\mathbf{B}^T E \mathbf{B} A \, dx] \mathbf{q} \qquad (3.20b)$$

In the finite element model (Section 3.2), the cross-sectional area of element e, denoted by A_e, is constant. Also, $\mathbf{B}$ is a constant matrix. Further, the transformation from x to ξ in Eq. 3.4 yields

$$dx = \frac{x_2 - x_1}{2} \, d\xi \qquad (3.21a)$$

or

$$dx = \frac{\ell_e}{2} \, d\xi \qquad (3.21b)$$

where $-1 \le \xi \le 1$, and ℓ_e is the length of the element, $\ell_e = |x_2 - x_1|$.

The element strain energy U_e is now written as

$$U_e = \frac{1}{2} \mathbf{q}^T \left[A_e \frac{\ell_e}{2} E_e \mathbf{B}^T \mathbf{B} \int_{-1}^{1} d\xi \right] \mathbf{q} \qquad (3.22)$$

where E_e is Young's modulus of element e. Noting that $\int_{-1}^{1} d\xi = 2$, and substituting for $\mathbf{B}$ from Eq. 3.15, we get

$$U_e = \frac{1}{2} \mathbf{q}^T A_e \ell_e E_e \frac{1}{\ell_e^2} \begin{Bmatrix} -1 \\ 1 \end{Bmatrix} [-1 \quad 1] \mathbf{q} \qquad (3.23)$$

which results in

$$U_e = \frac{1}{2} \mathbf{q}^T \frac{A_e E_e}{\ell_e} \begin{bmatrix} 1 & -1 \\ -1 & 1 \end{bmatrix} \mathbf{q} \qquad (3.24)$$

The above equation is of the form

$$U_e = \frac{1}{2} \mathbf{q}^T \mathbf{k}^e \mathbf{q} \qquad (3.25)$$

where the **element stiffness matrix $\mathbf{k}^e$** is given by

$$\mathbf{k}^e = \frac{E_e A_e}{\ell_e} \begin{bmatrix} 1 & -1 \\ -1 & 1 \end{bmatrix} \qquad (3.26)$$

We note here the similarity of the strain energy expression in Eq. 3.26 with the strain energy in a simple spring, which is given as $U = \frac{1}{2} k Q^2$. Also observe that $\mathbf{k}^e$ is linearly proportional to the product $A_e E_e$ and inversely proportional to the length ℓ_e.

Force Terms

The element body force term $\int_e u^T f A\, dx$ appearing in the total potential energy is considered first. Substituting $u = N_1 q_1 + N_2 q_2$, we have

$$\int_e u^T f A\, dx = A_e f \int_e (N_1 q_1 + N_2 q_2)\, dx \tag{3.27}$$

Recall that the body force f has units of force per unit volume. In the above equation, A_e and f are constant within the element and were consequently brought outside the integral. The above equation can be written as

$$\int_e u^T f A\, dx = q^T \begin{Bmatrix} A_e f \int_e N_1 dx \\[2mm] A_e f \int_e N_2 dx \end{Bmatrix} \tag{3.28}$$

The integrals of the shape functions above can be readily evaluated by making the substitution $dx = (\ell_e/2)\, d\xi$. Thus,

$$\int_e N_1 dx = \frac{\ell_e}{2} \int_{-1}^{1} \frac{1-\xi}{2} d\xi = \frac{\ell_e}{2}$$

$$\int_e N_2 dx = \frac{\ell_e}{2} \int_{-1}^{1} \frac{1+\xi}{2} d\xi = \frac{\ell_e}{2} \tag{3.29}$$

Alternatively, $\int_e N_1 dx$ is simply the area under the N_1 curve as shown in Fig. 3.8, which equals $\frac{1}{2} \cdot \ell_e \cdot 1 = \ell_e/2$. Similarly, $\int N_2 dx = \frac{1}{2} \cdot \ell_e \cdot 1 = \ell_e/2$. The body force term in Eq. 3.28 reduces to

$$\int_e u^T f A\, dx = q^T \frac{A_e}{2} \ell_e f \begin{Bmatrix} 1 \\ 1 \end{Bmatrix} \tag{3.30a}$$

which is of the form

$$\int_e u^T f A\, dx = q^T f^e \tag{3.30b}$$

The right side of the above equation is of the form Displacement $\times$ Force.

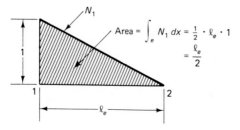

Figure 3.8 Integral of a shape function.

Thus, the **element body force vector**, $\mathbf{f}^e$, is identified as

$$\mathbf{f}^e = \frac{A_e \ell_e f}{2} \begin{Bmatrix} 1 \\ 1 \end{Bmatrix} \tag{3.31}$$

The element body force vector above has a simple physical explanation. Since $A_e \ell_e$ is the volume of the element and f is the body force per unit volume, we see that $A_e \ell_e f$ gives the total body force acting on the element. The factor $\frac{1}{2}$ in Eq. 3.31 tells us that this total body force is equally distributed to the two nodes of the element.

The element *traction force term* $\int_e u^{\mathrm{T}} T\, dx$ appearing in the total potential energy is now considered. We have,

$$\int_e u^{\mathrm{T}} T\, dx = \int_e (N_1 q_1 + N_2 q_2) T\, dx \tag{3.32}$$

Since the traction force T is constant within the element, we have

$$\int_e u^{\mathrm{T}} T\, dx = \mathbf{q}^{\mathrm{T}} \begin{Bmatrix} T \displaystyle\int_e N_1\, dx \\[2ex] T \displaystyle\int_e N_2\, dx \end{Bmatrix} \tag{3.33}$$

We have already shown that $\int_e N_1\, dx = \int_e N_2\, dx = \ell_e/2$. Thus, Eq. 3.33 is of the form

$$\int_e u^{\mathrm{T}} T\, dx = \mathbf{q}^{\mathrm{T}} \mathbf{T}^e \tag{3.34}$$

where the element traction force vector, $\mathbf{T}^e$, is given by

$$\mathbf{T}^e = \frac{T \ell_e}{2} \begin{Bmatrix} 1 \\ 1 \end{Bmatrix} \tag{3.35}$$

We can provide a physical explanation for the above equation as was given for the element body force vector.

At this stage, element matrices $\mathbf{k}^e$, $\mathbf{f}^e$, and $\mathbf{T}^e$ have been obtained. After we account for the element convectivity (in Fig. 3.3, for example, $\mathbf{q} = [Q_1, Q_2]^{\mathrm{T}}$ for element 1, $\mathbf{q} = [Q_2, Q_3]^{\mathrm{T}}$ for element 2, etc.), the total potential energy in Eq. 3.18b can be written as

$$\Pi = \tfrac{1}{2} \mathbf{Q}^{\mathrm{T}} \mathbf{K} \mathbf{Q} - \mathbf{Q}^{\mathrm{T}} \mathbf{F} \tag{3.36}$$

where $\mathbf{K}$ is the *global stiffness matrix,* $\mathbf{F}$ is the global load vector, and $\mathbf{Q}$ is the global displacement vector. For example, in the finite element model in Fig. 3.2b, $\mathbf{K}$ is a (5×5) matrix, and $\mathbf{Q}$ and $\mathbf{F}$ are each (5×1) vectors. $\mathbf{K}$ is obtained as follows: Using the element connectivity information, the elements of each $\mathbf{k}^e$ are placed in the appropriate locations in the larger $\mathbf{K}$ matrix and overlapping elements are then summed. The $\mathbf{F}$ vector is similarly assembled. This process of assembling $\mathbf{K}$ and $\mathbf{F}$ from element stiffness and force matrices is discussed in detail in Section 3.6.

3.5 THE GALERKIN APPROACH

Following the concepts introduced in Chapter 1, we introduce a virtual displacement field

$$\phi = \phi(x) \tag{3.37}$$

and associated virtual strain

$$\epsilon(\phi) = \frac{d\phi}{dx} \tag{3.38}$$

where ϕ is an arbitrary or virtual displacement consistent with the boundary conditions. Galerkin's variational form, given in Eq. 1.43, for the one-dimensional problem considered here, is

$$\int_L \sigma^T\epsilon(\phi)A \, dx - \int_L \phi^T fA \, dx - \int_L \phi^T T \, dx - \sum_i \phi_i P_i = 0 \tag{3.39a}$$

The above equation should hold for every ϕ consistent with the boundary conditions. The first term represents the internal virtual work, while the load terms represent the external virtual work.

On the discretized region, the equation above becomes

$$\sum_e \int_e \epsilon^T E\epsilon(\phi)A \, dx - \sum_e \int_e \phi^T fA \, dx - \sum_e \int_e \phi^T T \, dx - \sum_i \phi_i P_i = 0 \tag{3.39b}$$

Note that ϵ is the strain due to the actual loads in the problem, while $\epsilon(\phi)$ is a virtual strain. Similar to the interpolation steps in Eqs. 3.7b, 3.14, and 3.16, we express

$$\phi = \mathbf{N}\psi$$
$$\epsilon(\phi) = \mathbf{B}\psi \tag{3.40}$$

where $\psi = [\psi_1, \psi_2]^T$ represents the arbitrary nodal displacements of element e. Also, the global virtual displacements at the nodes are represented by

$$\Psi = [\Psi_1, \Psi_2, \ldots, \Psi_N]^T \tag{3.41}$$

Element Stiffness

Consider the first term, representing internal virtual work, in Eq 3.39b. Substituting Eq. 3.40 into Eq. 3.39b, and noting that $\epsilon = \mathbf{B}q$, we get

$$\int_e \epsilon^T E\epsilon(\phi)A \, dx = \int_e \mathbf{q}^T \mathbf{B}^T E\mathbf{B}\psi A \, dx \tag{3.42}$$

In the finite element model (Section 3.2), the cross-sectional area of element e, denoted by A_e, is constant. Also, $\mathbf{B}$ is a constant matrix. Further, $dx = (\ell_e/2) \, d\xi$. Thus,

$$\int_e \epsilon^{\mathsf{T}} E \epsilon (\phi) A \ dx = \mathbf{q}^{\mathsf{T}} \left[E_e A_e \frac{\ell_e}{2} \mathbf{B}^{\mathsf{T}} \mathbf{B} \int_{-1}^{1} d\xi \right] \psi \tag{3.43a}$$

$$= \mathbf{q}^{\mathsf{T}} \mathbf{k}^e \psi \tag{3.43b}$$

$$= \psi^{\mathsf{T}} \mathbf{k}^e \mathbf{q}$$

where $\mathbf{k}^e$ is the (symmetric) element stiffness matrix given by

$$\mathbf{k}^e = E_e A_e \ell_e \mathbf{B}^{\mathsf{T}} \mathbf{B} \tag{3.44}$$

Substituting $\mathbf{B}$ from Eq. 3.15, we have

$$\mathbf{k}^e = \frac{E_e A_e}{\ell_e} \begin{bmatrix} 1 & -1 \\ -1 & 1 \end{bmatrix} \tag{3.45}$$

Force Terms

Consider the second term in Eq. 3.39a, representing the virtual work done by the body force in an element. Using $\phi = \mathbf{N}\psi$, $dx = \ell_e/2 \ d\xi$, and noting that the body force in the element is assumed constant, we have

$$\int_e \phi^{\mathsf{T}} f A \ dx = \int_{-1}^{1} \psi^{\mathsf{T}} \mathbf{N}^{\mathsf{T}} f A_e \frac{\ell_e}{2} \ d\xi \tag{3.46a}$$

$$= \psi^{\mathsf{T}} \mathbf{f}^e \tag{3.46b}$$

where

$$\mathbf{f}^e = \frac{A_e \ell_e f}{2} \left\{ \begin{array}{c} \int_{-1}^{1} N_1 \ d\xi \\[2mm] \int_{-1}^{1} N_2 \ d\xi \end{array} \right\} \tag{3.47a}$$

is called the element body force vector. Substituting for $N_1 = (1 - \xi)/2$, $N_2 = (1 + \xi)/2$, we obtain $\int_{-1}^{1} N_1 \ d\xi = 1$. Alternatively, $\int_{-1}^{1} N_1 \ d\xi$ is the area under the N_1 curve $= \frac{1}{2} \times 2 \times 1 = 1$. Similarly, $\int_{-1}^{1} N_2 \ d\xi = 1$. Thus,

$$\mathbf{f}^e = \frac{A_e \ell_e f}{2} \left\{ \begin{array}{c} 1 \\ 1 \end{array} \right\} \tag{3.47b}$$

Similarly, the element traction term reduces to

$$\int_e \phi^{\mathsf{T}} T \ dx = \psi^{\mathsf{T}} \mathbf{T}^e \tag{3.48}$$

where the element traction force vector $\mathbf{T}^e$ is given by

$$\mathbf{T}^e = \frac{T \ell_e}{2} \left\{ \begin{array}{c} 1 \\ 1 \end{array} \right\} \tag{3.49}$$

At this stage, the element matrices $\mathbf{k}^e$, $\mathbf{f}^e$, and $\mathbf{T}^e$ have been obtained. After accounting for the element connectivity (in Fig. 3.3, for example, $\boldsymbol{\psi} = [\Psi_1, \Psi_2]^T$ for element 1, $\boldsymbol{\psi} = [\Psi_2, \Psi_3]^T$ for element 2, etc.) the variational form

$$\sum_e \boldsymbol{\psi}^T \mathbf{k}^e \mathbf{q} - \sum_e \boldsymbol{\psi}^T \mathbf{f}^e - \sum_e \boldsymbol{\psi}^T \mathbf{T}^e - \sum_i \Psi_i P_i = 0 \qquad (3.50)$$

can be written as

$$\boldsymbol{\Psi}^T(\mathbf{KQ} - \mathbf{F}) = 0 \qquad (3.51)$$

which should hold for every $\boldsymbol{\Psi}$ consistent with the boundary conditions. Methods for handling boundary conditions are discussed shortly. The global stiffness matrix $\mathbf{K}$ is assembled from element matrices $\mathbf{k}^e$ using element connectivity information. Likewise, $\mathbf{F}$ is assembled from element matrices $\mathbf{f}^e$ and $\mathbf{T}^e$. This assembly is discussed in detail in the next section.

3.6 ASSEMBLY OF THE GLOBAL STIFFNESS MATRIX AND LOAD VECTOR

We have noted above that the total potential energy written in the form

$$\Pi = \sum_e \frac{1}{2} \mathbf{q}^T \mathbf{k}^e \mathbf{q} - \sum_e \mathbf{q}^T \mathbf{f}^e - \sum_e \mathbf{q}^T \mathbf{T}^e - \sum_i P_i Q_i$$

can be written in the form

$$\Pi = \tfrac{1}{2} \mathbf{Q}^T \mathbf{K} \mathbf{Q} - \mathbf{Q}^T \mathbf{F}$$

by taking element connectivity into account. This step involves assembling $\mathbf{K}$ and $\mathbf{F}$ from element stiffness and force matrices. The assembly of the structural stiffness matrix $\mathbf{K}$ from element stiffness matrices $\mathbf{k}^e$ will first be shown below.

Referring to the finite element model in Fig. 3.2b, let us consider the strain energy in, say, element 3. We have

$$U_3 = \tfrac{1}{2} \mathbf{q}^T \mathbf{k}^3 \mathbf{q} \qquad (3.52a)$$

or, substituting for $\mathbf{k}^3$,

$$U_3 = \frac{1}{2} \mathbf{q}^T \frac{E_3 A_3}{\ell_3} \begin{bmatrix} 1 & -1 \\ -1 & 1 \end{bmatrix} \mathbf{q} \qquad (3.52b)$$

For element 3, we have $\mathbf{q} = [Q_3, Q_4]^T$. Thus, we can write U_3 as

$$U_3 = \frac{1}{2} [Q_1, Q_2, Q_3, Q_4, Q_5] \begin{bmatrix} 0 & 0 & 0 & 0 & 0 \\ 0 & 0 & 0 & 0 & 0 \\ 0 & 0 & \dfrac{E_3 A_3}{\ell_3} & \dfrac{-E_3 A_3}{\ell_3} & 0 \\ 0 & 0 & \dfrac{-E_3 A_3}{\ell_3} & \dfrac{E_3 A_3}{\ell_3} & 0 \\ 0 & 0 & 0 & 0 & 0 \end{bmatrix} \begin{Bmatrix} Q_1 \\ Q_2 \\ Q_3 \\ Q_4 \\ Q_5 \end{Bmatrix} \qquad (3.53)$$

From above, we see that elements of the matrix $\mathbf{k}^3$ occupy the third and fourth rows and columns of the $\mathbf{K}$ matrix. Consequently, when adding element strain energies, the elements of $\mathbf{k}^e$ are placed in the appropriate locations of the global $\mathbf{K}$ matrix, based on the element connectivity; overlapping elements are simply added. We can denote this assembly symbolically as

$$\mathbf{K} \leftarrow \sum_e \mathbf{k}^e \qquad (3.54a)$$

Similarly, the global load vector $\mathbf{F}$ is assembled from element force vectors and point loads, as

$$\mathbf{F} \leftarrow \sum_e (\mathbf{f}^e + \mathbf{T}^e) + \mathbf{P} \qquad (3.54b)$$

The Galerkin approach also gives us the same assembly procedure. An example is now given to illustrate this assembly procedure in detail. In actual computation, $\mathbf{K}$ is stored in banded or skyline form to take advantage of symmetry and sparsity. This aspect is discussed in Section 3.7 and in greater detail in Chapter 4.

Example 3.2

Consider the bar as shown in Fig. E3.2. For each element i, A_i, and ℓ_i are the cross-sectional area and length, respectively. Each element i is subjected to a traction force T_i per unit length and a body force f per unit volume. The units of T_i, f, A_i, and so on are assumed to be consistent. The Young's modulus of the material is E. A concentrated load P_2 is applied at node 2. The structural stiffness matrix and nodal load vector will not be assembled.

The element stiffness matrix for each element i is obtained from Eq. 3.26 as

$$[k^{(i)}] = \frac{EA_i}{\ell_i} \begin{bmatrix} 1 & -1 \\ -1 & 1 \end{bmatrix}$$

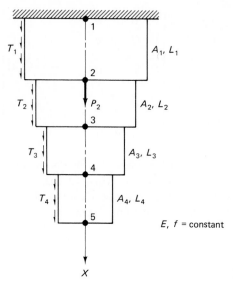

X

Figure E3.2

The element connectivity table is the following:

Element	1	2
1	1	2
2	2	3
3	3	4
4	4	5

The element stiffness matrices can be "expanded" using the connectivity table and then summed (or assembled) to obtain the structural stiffness matrix as follows:*

$$
\mathbf{K} = \frac{EA_1}{\ell_1}
\begin{bmatrix}
1 & -1 & 0 & 0 & 0 \\
-1 & 1 & 0 & 0 & 0 \\
0 & 0 & 0 & 0 & 0 \\
0 & 0 & 0 & 0 & 0 \\
0 & 0 & 0 & 0 & 0
\end{bmatrix}
+ \frac{EA_2}{\ell_2}
\begin{bmatrix}
0 & 0 & 0 & 0 & 0 \\
0 & 1 & -1 & 0 & 0 \\
0 & -1 & 1 & 0 & 0 \\
0 & 0 & 0 & 0 & 0 \\
0 & 0 & 0 & 0 & 0
\end{bmatrix}
$$

$$
+ \frac{EA_3}{\ell_3}
\begin{bmatrix}
0 & 0 & 0 & 0 & 0 \\
0 & 0 & 0 & 0 & 0 \\
0 & 0 & 1 & -1 & 0 \\
0 & 0 & -1 & 1 & 0 \\
0 & 0 & 0 & 0 & 0
\end{bmatrix}
+ \frac{EA_4}{\ell_4}
\begin{bmatrix}
0 & 0 & 0 & 0 & 0 \\
0 & 0 & 0 & 0 & 0 \\
0 & 0 & 0 & 0 & 0 \\
0 & 0 & 0 & 1 & -1 \\
0 & 0 & 0 & -1 & 1
\end{bmatrix}
$$

* This "expansion" of element stiffness matrices as shown in this example is merely for illustration purposes, and is never explicitly carried out in the computer, since storing zeroes is inefficient. Instead, $\mathbf{K}$ is assembled directly from $\mathbf{k}^e$ using the connectivity table.

which gives

$$
\mathbf{K} = E
\begin{bmatrix}
\dfrac{A_1}{\ell_1} & -\dfrac{A_1}{\ell_1} & 0 & 0 & 0 \\[2ex]
-\dfrac{A_1}{\ell_1} & \left(\dfrac{A_1}{\ell_1}+\dfrac{A_2}{\ell_2}\right) & -\dfrac{A_2}{\ell_2} & 0 & 0 \\[2ex]
0 & -\dfrac{A_2}{\ell_2} & \left(\dfrac{A_2}{\ell_2}+\dfrac{A_3}{\ell_3}\right) & -\dfrac{A_3}{\ell_3} & 0 \\[2ex]
0 & 0 & -\dfrac{A_3}{\ell_3} & \left(\dfrac{A_3}{\ell_3}+\dfrac{A_4}{\ell_4}\right) & -\dfrac{A_4}{\ell_4} \\[2ex]
0 & 0 & 0 & -\dfrac{A_4}{\ell_4} & \dfrac{A_4}{\ell_4}
\end{bmatrix}
$$

The global load vector is assembled as

$$
\mathbf{F} =
\left\{
\begin{array}{l}
\dfrac{A_1\ell_1 f}{2} + \dfrac{\ell_1 T_1}{2} \\[2ex]
\left(\dfrac{A_1\ell_1 f}{2} + \dfrac{\ell_1 T_1}{2}\right) + \left(\dfrac{A_2\ell_2 f}{2} + \dfrac{\ell_2 T_2}{2}\right) \\[2ex]
\left(\dfrac{A_2\ell_2 f}{2} + \dfrac{\ell_2 T_2}{2}\right) + \left(\dfrac{A_3\ell_3 f}{2} + \dfrac{\ell_3 T_3}{2}\right) \\[2ex]
\left(\dfrac{A_3\ell_3 f}{2} + \dfrac{\ell_3 T_3}{2}\right) + \left(\dfrac{A_4\ell_4 f}{2} + \dfrac{\ell_4 T_4}{2}\right) \\[2ex]
\dfrac{A_4\ell_4 f}{2} + \dfrac{\ell_4 T_4}{2}
\end{array}
\right\}
+
\left\{
\begin{array}{c}
0 \\[2ex]
P_2 \\[2ex]
0 \\[2ex]
0 \\[2ex]
0
\end{array}
\right\}
\qquad \blacksquare
$$

3.7 PROPERTIES OF K

Several important comments will now be made regarding the global stiffness matrix for the linear one-dimensional problem discussed above.

1. The dimension of the global stiffness $\mathbf{K}$ is $(N \times N)$, where N is the number of nodes. This follows from the fact that each node has only one degree of freedom.
2. $\mathbf{K}$ is symmetric.
3. $\mathbf{K}$ is a banded matrix. That is, all elements outside of the band are zero. This can be seen in Example 3.2, just considered. In this example, $\mathbf{K}$ can be compactly represented in banded form as

$$
\mathbf{K}_{\text{banded}} = E
\begin{bmatrix}
\dfrac{A_1}{\ell_1} & -\dfrac{A_1}{\ell_1} \\[2ex]
\dfrac{A_1}{\ell_1} + \dfrac{A_2}{\ell_2} & -\dfrac{A_2}{\ell_2} \\[2ex]
\dfrac{A_2}{\ell_2} + \dfrac{A_3}{\ell_3} & -\dfrac{A_3}{\ell_3} \\[2ex]
\dfrac{A_3}{\ell_3} + \dfrac{A_4}{\ell_4} & -\dfrac{A_4}{\ell_4} \\[2ex]
\dfrac{A_4}{\ell_4} & 0
\end{bmatrix}
$$

Note that $\mathbf{K}_{\text{banded}}$ is of dimension $[N \times \text{NBW}]$, where NBW is the half-bandwidth. In many one-dimensional problems such as the example considered above, the connectivity of element i is i, $i + 1$. In such cases, the banded matrix has only two columns (NBW = 2). In two and three dimensions, the direct formation of $\mathbf{K}$ in banded or skyline form from the element matrices involves some bookkeeping. This is discussed in detail at the end of Chapter 4. The reader should verify the following general formula for the half-bandwidth:

$$
\text{NBW} = \max \left(\begin{array}{c} \text{Difference between dof numbers} \\ \text{connecting an element} \end{array} \right) + 1 \tag{3.55}
$$

For example, consider a 4-element model of a bar that is numbered as shown in Fig. 3.9a. Using Eq. 3.55, we have

$$
\text{NBW} = \max \ (4 - 1, 5 - 4, 5 - 3, 3 - 2) + 1 = 4
$$

The numbering scheme in Fig. 3.9a is bad since $\mathbf{K}$ is almost "filled up" and consequently requires more computer storage and computation. Figure 3.9b shows the optimum numbering for minimum NBW.

Now, the potential energy or Galerkin's approach has to be applied, noting the boundary conditions of the problem, to yield the finite element (equilibrium) equations. Solution of these equations yields the global displacement vector $\mathbf{Q}$. The

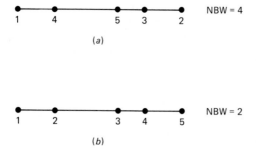

Figure 3.9   Node numbering and its effect on the half-bandwidth.

stresses and reaction forces can then be recovered. These steps will now be discussed in the next section.

3.8 THE FINITE ELEMENT EQUATIONS; TREATMENT OF BOUNDARY CONDITIONS

Types of Boundary Conditions

After using a discretization scheme to model the continuum, we have obtained an expression for the total potential energy in the body as

$$\Pi = \tfrac{1}{2}\mathbf{Q}^\mathrm{T}\mathbf{K}\mathbf{Q} - \mathbf{Q}^\mathrm{T}\mathbf{F}$$

where $\mathbf{K}$ is the structural stiffness matrix, $\mathbf{F}$ is the global load vector, and $\mathbf{Q}$ is the global displacement vector. As discussed previously, $\mathbf{K}$ and $\mathbf{F}$ are assembled from element stiffness and force matrices, respectively. We now have to arrive at the equations of equilibrium, from which we can determine nodal displacements, element stresses, and support reactions.

The minimum potential energy theorem (Chapter 1) is now invoked. This theorem is stated as follows: *Of all possible displacements that satisfy the boundary conditions of a structural system, those corresponding to equilibrium configurations make the total potential energy assume a minimum value.* Consequently, the equations of equilibrium can be obtained by minimizing, with respect to $\mathbf{Q}$, the potential energy $\Pi = \tfrac{1}{2}\mathbf{Q}^\mathrm{T}\mathbf{K}\mathbf{Q} - \mathbf{Q}^\mathrm{T}\mathbf{F}$ *subject to boundary conditions.* Boundary conditions are usually of the type

$$Q_{p_1} = a_1, \ Q_{p_2} = a_2, \ \ldots, \ Q_{p_r} = a_r \tag{3.56}$$

That is, the displacements along dof's $p_1, p_2, \ldots, p_r$ are specified to be equal to $a_1, a_2, \ldots, a_r$, respectively. In other words, there are r number of supports in the structure, with each support node given a specified displacement. For example, consider the bar in Fig. 3.2*b*. There is only one boundary condition in this problem, $Q_1 = 0$.

It is noted here that *the treatment of boundary conditions in this section is applicable to two- and three-dimensional problems as well.* For this reason, the term dof is used here instead of node, since a two-dimensional stress problem will have two degrees of freedom per node. The steps described in this section will be used in all subsequent chapters. Furthermore, a Galerkin-based argument leads to the same steps for handling boundary conditions as the energy approach used below.

There are *multipoint constraints* of the type

$$\beta_1 Q_{p_1} + \beta_2 Q_{p_2} = \beta_0 \tag{3.57}$$

where β_0, β_1, and β_2 are known constants. These types of boundary conditions are used in modeling inclined roller supports, rigid connections or shrink fits.

It should be emphasized that improper specification of boundary conditions can lead to erroneous results. Boundary conditions eliminate the possibility of the structure moving as a rigid body. Further, boundary conditions should accurately model the physical system. Two approaches will now be discussed for handling specified displacement boundary conditions of the type given in Eq. 3.56: the **elimination approach** and the **penalty approach**. For multipoint constraints in Eq. 3.57, only the penalty approach will be given, because it is simpler to implement.

Elimination Approach

To illustrate the basic idea, consider the single boundary condition $Q_1 = a_1$. The equilibrium equations are obtained by minimizing Π with respect to $\mathbf{Q}$, subject to the boundary condition $Q_1 = a_1$. For an $N -$ dof structure, we have

$$\mathbf{Q} = [Q_1, Q_2, \ldots, Q_N]^\mathrm{T}$$
$$\mathbf{F} = [F_1, F_2, \ldots, F_N]^\mathrm{T}$$

The global stiffness matrix is of the form

$$\mathbf{K} = \begin{bmatrix} K_{11} & K_{12} & \cdots & K_{1N} \\ K_{21} & K_{22} & \cdots & K_{2N} \\ \vdots & & & \\ K_{N1} & K_{N2} & \cdots & K_{NN} \end{bmatrix} \tag{3.58}$$

Note that K is a symmetric matrix. The potential energy $\Pi = \frac{1}{2}\mathbf{Q}^\mathrm{T}\mathbf{KQ} - \mathbf{Q}^\mathrm{T}\mathbf{F}$ can be written in expanded form as

$$\begin{aligned}\Pi = \tfrac{1}{2}(&Q_1K_{11}Q_1 + Q_1K_{12}Q_2 + \cdots + Q_1K_{1N}Q_N \\ &+ Q_2K_{21}Q_1 + Q_2K_{22}Q_2 + \cdots + Q_2K_{2N}Q_N \\ &\text{-------------------------} \\ &+ Q_NK_{N1}Q_1 + Q_NK_{N2}Q_2 + \cdots + Q_NK_{NN}Q_N) \\ -(&Q_1F_1 + Q_2F_2 + \cdots + Q_NF_N)\end{aligned} \tag{3.59}$$

If we now substitute the boundary condition $Q_1 = a_1$ into the expression for Π above, we obtain

$$\begin{aligned}\Pi = \tfrac{1}{2}(&a_1K_{11}a_1 + a_1K_{12}Q_2 + \cdots + a_1K_{1N}Q_N \\ &+ Q_2K_{21}a_1 + Q_2K_{22}Q_2 + \cdots + Q_2K_{2N}Q_N \\ &\text{-------------------------} \\ &+ Q_NK_{N1}a_1 + Q_NK_{N2}Q_2 + \cdots + Q_NK_{NN}Q_N) \\ -(&a_1F_1 + Q_2F_2 + \cdots + Q_NF_N)\end{aligned} \tag{3.60}$$

Note that the displacement Q_1 *has been eliminated* in the potential energy expression above. Consequently, the requirement that Π take on a minimum value implies that

$$\frac{d\Pi}{dQ_i} = 0 \qquad i = 2, 3, \ldots, N \tag{3.61}$$

We thus obtain, from Eqs. 3.60 and 3.61,

$$K_{22} Q_2 + K_{23} Q_3 + \cdots + K_{2N} Q_N = F_2 - K_{21} a_1$$

$$K_{32} Q_2 + K_{33} Q_3 + \cdots + K_{3N} Q_N = F_3 - K_{31} a_1 \tag{3.62}$$

$$- -$$

$$K_{N2} Q_2 + K_{N3} Q_3 + \cdots + K_{NN} Q_N = F_N - K_{N1} a_1$$

The above *finite element equations* can be expressed in matrix form as

$$\begin{bmatrix} K_{22} & K_{23} & \cdots & K_{2N} \\ K_{32} & K_{33} & \cdots & K_{3N} \\ \cdot & & & \\ \cdot & & & \\ \cdot & & & \\ K_{N2} & K_{N3} & \cdots & K_{NN} \end{bmatrix} \begin{Bmatrix} Q_2 \\ Q_3 \\ \cdot \\ \cdot \\ \cdot \\ Q_N \end{Bmatrix} = \begin{Bmatrix} F_2 - K_{21} a_1 \\ F_3 - K_{31} a_1 \\ \cdot \\ \cdot \\ \cdot \\ F_N - K_{N1} a_1 \end{Bmatrix} \tag{3.63}$$

We now observe that the $(N - 1 \times N - 1)$ stiffness matrix above is obtained simply be deleting or *eliminating* the first row and column (in view of $Q_1 = a_1$) from the original $(N \times N)$ stiffness matrix. Equation 3.63 may be denoted as

$$\mathbf{KQ} = \mathbf{F} \tag{3.64}$$

where $\mathbf{K}$ above is a reduced stiffness matrix obtained by eliminating the row and column corresponding to the specified or 'support' dof. Equations 3.64 can be solved for the displacement vector $\mathbf{Q}$ using Gaussian elimination. Note that the reduced $\mathbf{K}$ matrix is nonsingular, provided the boundary conditions have been specified properly; the original $\mathbf{K}$ matrix, on the other hand, is a singular matrix. Once $\mathbf{Q}$ has been determined, the element stress can be evaluated using Eq. 3.16: $\sigma = E\mathbf{Bq}$, where $\mathbf{q}$ for each element is extracted from $\mathbf{Q}$ using element connectivity information.

Assume that displacements and stresses have been determined. It is now necessary to calculate the **reaction force** R_1 at the support. This reaction force can be obtained from the finite element equation (or equilibrium equation) for node 1:

$$K_{11} Q_1 + K_{12} Q_2 + \cdots + K_{1N} Q_N = F_1 + R_1 \tag{3.65}$$

Here, $Q_1, Q_2, \ldots, Q_N$ are known. F_1, which equals the load applied at the support (if any), is also known. Consequently, the reaction force at the node that maintains equilibrium, is

$$R_1 = K_{11} Q_1 + K_{12} Q_2 + \cdots + K_{1N} Q_N - F_1 \tag{3.66}$$

Note that the elements $K_{11}, K_{12}, \ldots, K_{1N}$ used above, which form the first row of $\mathbf{K}$, need to be *stored separately*. This is because $\mathbf{K}$ in Eq. 3.64 is obtained by deleting this row and column from the original $\mathbf{K}$.

The modification to $\mathbf{K}$ and $\mathbf{F}$ discussed above are also derivable using Galerkin's variational formulation. We have Eq. 3.51:

$$\mathbf{\Psi}^{\mathrm{T}}(\mathbf{KQ} - \mathbf{F}) = 0 \qquad (3.67)$$

for every $\mathbf{\Psi}$ consistent with the boundary conditions of the problem. Specifically, consider the constraint

$$Q_1 = a_1 \qquad (3.68)$$

Then, we require

$$\Psi_1 = 0 \qquad (3.69)$$

Choosing virtual displacements $\mathbf{\Psi} = [0, 1, 0, \ldots, 0]$, $\mathbf{\Psi} = [0, 0, 1, 0, \ldots, 0]^{\mathrm{T}}$, $\ldots, \mathbf{\Psi} = [0, 0, \ldots, 0, 1]^{\mathrm{T}}$, and substituting each of these into Eq. 3.67, we obtain precisely the equilibrium equations given in Eqs. 3.63.

The preceding discussion addressed the boundary condition $Q_1 = a_1$. This procedure can readily be generalized to handle multiple boundary conditions. The general procedure is summarized below. Again, this procedure is also applicable to two- and three-dimensional problems.

Summary: Elimination Approach

Consider the boundary conditions

$$Q_{p_1} = a_1, \ Q_{p_2} = a_2, \ldots, \ Q_{p_r} = a_r$$

Step 1. Store the p_1th, p_2th, $\ldots$, and p_rth rows of the global stiffness matrix $\mathbf{K}$ and force vector $\mathbf{F}$. These rows will be used subsequently.

Step 2. Delete the p_1th row and column, the p_2th row and column, $\ldots$, and the p_rth row and column, from the $\mathbf{K}$ matrix. The resulting stiffness matrix $\mathbf{K}$ is of dimension $(N - r, N - r)$. Similarly, the corresponding load vector $\mathbf{F}$ is of dimension $(N - r, 1)$. Modify each load component as

$$F_i = F_i - (K_{i,p_1}a_1 + K_{i,p_2}a_2 + \cdots + K_{i,p_r}a_r) \qquad (3.70)$$

for each dof i that is not a support. Solve

$$\mathbf{KQ} = \mathbf{F}$$

for the displacement vector $\mathbf{Q}$.

Step 3. For each element, extract the element displacement vector $\mathbf{q}$ from the $\mathbf{Q}$ vector, using element connectivity, and determine element stresses.

Step 4. Using the information stored in step 1 above, evaluate the reaction forces at each support dof from

$$R_{p_1} = K_{p_11}Q_1 + K_{p_12}Q_2 + \cdots + K_{p_1N}Q_N - F_{p_1}$$

$$R_{p_2} = K_{p_21}Q_1 + K_{p_22}Q_2 + \cdots + K_{p_2N}Q_N - F_{p_2}$$

$$\text{-----------------------------} \qquad (3.71)$$

$$R_{p_r} = K_{p_r1}Q_1 + K_{p_r2}Q_2 + \cdots + K_{p_rN}Q_N - F_{p_r}$$

Example 3.3

Consider the thin (steel) plate in Fig. E3.3a. The plate has a uniform thickness $t = 1$ in., Young's modulus $E = 30 \times 10^6$ psi, and weight density $\rho = 0.2836$ lb/in.^{3}. In addition to its self-weight, the plate is subjected to a point load $P = 100$ lb at its midpoint.

(a) Model the plate with two finite elements.

(b) Write down expressions for the element stiffness matrices and element body force vectors.

(c) Assemble the structural stiffness matrix $\mathbf{K}$ and global load vector $\mathbf{F}$.

(d) Using the elimination approach, solve for the global displacement vector $\mathbf{Q}$.

(e) Evaluate the stresses in each element.

(f) Determine the reaction force at the support.

Solution

(a) Using two elements, each of length 12 in., we obtain the finite element model in Fig. E3.3b. Nodes and elements are numbered as shown. Note that the area at the midpoint of the plate in Fig. E3.3a is 4.5 in.^{2}. Consequently, the average area of element 1 is $A_1 = (6 + 4.5)/2 = 5.25$ in.^{2}, and the average area of element 2 is $A_2 = (4.5 + 3)/2 = 3.75$ in.^{2}. The boundary condition for this model is $Q_1 = 0$.

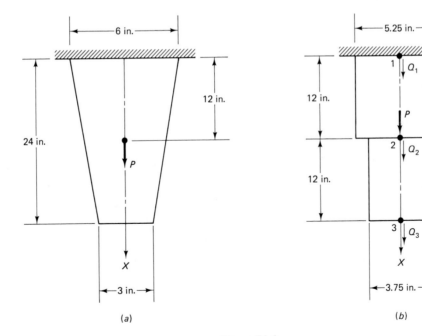

(a)   (b)

Figure E3.3

(b) From Eq. 3.26, we can write down expressions for the element stiffness matrices of the two elements as

$$\begin{array}{cc} & \quad 1 \qquad 2 \quad \leftarrow \text{Global dof} \\ & \qquad\qquad \downarrow \\ \mathbf{k}^1 = \dfrac{30 \times 10^6 \times 5.25}{12} \begin{bmatrix} \;\;1 & -1 \\ -1 & \;\;1 \end{bmatrix} \begin{matrix} 1 \\ 2 \end{matrix} \end{array}$$

and

$$\begin{array}{cc} & \quad 2 \qquad 3 \\ \mathbf{k}^2 = \dfrac{30 \times 10^6 \times 3.75}{12} \begin{bmatrix} \;\;1 & -1 \\ -1 & \;\;1 \end{bmatrix} \begin{matrix} 2 \\ 3 \end{matrix} \end{array}$$

Using Eq. 3.31, the element body force vectors are

$$\begin{array}{cc} & \qquad\qquad \text{Global dof} \\ & \qquad\qquad\qquad \downarrow \\ \mathbf{f}^1 = \dfrac{5.25 \times 12 \times 0.2836}{2} \begin{Bmatrix} 1 \\ 1 \end{Bmatrix} \begin{matrix} 1 \\ 2 \end{matrix} \end{array}$$

and

$$\mathbf{f}^2 = \dfrac{3.75 \times 12 \times 0.2836}{2} \begin{Bmatrix} 1 \\ 1 \end{Bmatrix} \begin{matrix} 2 \\ 3 \end{matrix}$$

(c) The global stiffness matrix $\mathbf{K}$ is assembled from $\mathbf{k}^1$ and $\mathbf{k}^2$ as

$$\begin{array}{cc} & \quad\;\; 1 \qquad\quad 2 \qquad\quad 3 \\ \mathbf{K} = \dfrac{30 \times 10^6}{12} \begin{bmatrix} \;\;5.25 & -5.25 & \;\;0 \\ -5.25 & \;\;9.00 & -3.75 \\ \;\;0 & -3.75 & \;\;3.75 \end{bmatrix} \begin{matrix} 1 \\ 2 \\ 3 \end{matrix} \end{array}$$

The externally applied global load vector $\mathbf{F}$ is assembled from $\mathbf{f}^1$, $\mathbf{f}^2$, and the point load $P = 100$ lb, as

$$\mathbf{F} = \begin{Bmatrix} 8.9334 \\ 15.3144 + 100 \\ 6.3810 \end{Bmatrix}$$

(d) In the elimination approach, the stiffness matrix $\mathbf{K}$ is obtained by deleting rows and columns corresponding to fixed dof's. In this problem, dof 1 is fixed. Thus, $\mathbf{K}$ is obtained by deleting the first row and column of the original $\mathbf{K}$. Also, $\mathbf{F}$ is obtained by deleting the first component of the original $\mathbf{F}$. The resulting operations are

$$\begin{array}{cc} & \quad\;\; 2 \qquad\quad 3 \\ \dfrac{30 \times 10^6}{12} \begin{bmatrix} \;\;9.00 & -3.75 \\ -3.75 & \;\;3.75 \end{bmatrix} \begin{Bmatrix} Q_2 \\ Q_3 \end{Bmatrix} = \begin{Bmatrix} 115.3144 \\ 6.3810 \end{Bmatrix} \end{array}$$

Solution of the above equations yields

$$Q_2 = 0.9272 \times 10^{-5} \text{ in.}$$

$$Q_3 = 0.9953 \times 10^{-5} \text{ in.}$$

Thus, $Q = [0, 0.9272 \times 10^{-5}, 0.9953 \times 10^{-5}]^T$ in.

(e) Using Eqs. 3.15 and 3.16, we obtain the stress in each element:

$$\sigma_1 = 30 \times 10^6 \times \tfrac{1}{12} [-1 \quad 1] \begin{Bmatrix} 0 \\ 0.9272 \times 10^{-5} \end{Bmatrix}$$

$$= 23.18 \text{ psi}$$

and

$$\sigma_2 = 30 \times 10^6 \times \tfrac{1}{12} [-1 \quad 1] \begin{Bmatrix} 0.9272 \times 10^{-5} \\ 0.9953 \times 10^{-5} \end{Bmatrix}$$

$$= 1.70 \text{ psi}$$

(f) The reaction force R_1 at node 1 is obtained from Eq. 3.71. This calculation requires the first row of **K** from part (c) above. Also, from part (c) above, note that the externally applied load (due to the self-weight) at note 1 is $F_1 = 8.9334$ lb. Thus,

$$R_1 = \frac{30 \times 10^6}{12} [5.25 \quad -5.25 \quad 0] \begin{Bmatrix} 0 \\ 0.9272 \times 10^{-5} \\ 0.9953 \times 10^{-5} \end{Bmatrix} - 8.9334$$

$$= -130.6 \text{ lb}$$

Evidently, the reaction is equal and opposite to the total downward load on the plate. ∎

Penalty Approach

A second approach for handling boundary conditions will now be discussed. This approach is easy to implement in a computer program, and retains its simplicity even when considering general boundary conditions as given in Eqs. 3.57. Specified displacement boundary conditions will first be discussed. The method will then be shown to apply to problems with multipoint constraints.

Specified displacement boundary conditions.   Consider the boundary condition

$$Q_1 = a_1$$

where a_1 is a known specified displacement along dof 1 of the support. The penalty approach for handling this boundary condition is now presented.

A spring with a large stiffness C is used to model the support. The magnitude of C is discussed subsequently. In this case, one end of the spring is displaced by an amount a_1, as shown in Fig. 3.10. The displacement Q_1 along dof 1 will be approxi-

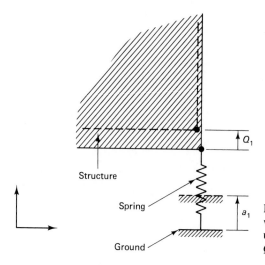

Figure 3.10 The penalty approach, where a spring with a large stiffness is used to model the boundary condition $Q_1 = a_1$.

mately equal to a_1, owing to the relatively small resistance offered by the structure. Consequently, the *net* extension of the spring is equal to $(Q_1 - a_1)$. The strain energy in the spring equals

$$U_s = \tfrac{1}{2} C (Q_1 - a_1)^2 \qquad\qquad (3.72)$$

This strain energy contributes to the total potential energy. As a result,

$$\Pi_M = \tfrac{1}{2} \mathbf{Q}^T \mathbf{K} \mathbf{Q} + \tfrac{1}{2} C (Q_1 - a_1)^2 - \mathbf{Q}^T \mathbf{F} \qquad\qquad (3.73)$$

The minimization of Π_M above can be carried out by setting $d\Pi_M/dQ_i = 0$, $i = 1,$ $2, \ldots, N$. The resulting finite element equations are

$$\begin{bmatrix} (K_{11} + C) & K_{12} & \cdots & K_{1N} \\ K_{21} & K_{22} & \cdots & K_{2N} \\ \vdots & \vdots & & \vdots \\ K_{N1} & K_{N2} & \cdots & K_{NN} \end{bmatrix} \begin{Bmatrix} Q_1 \\ Q_2 \\ \vdots \\ Q_N \end{Bmatrix} = \begin{Bmatrix} F_1 + Ca_1 \\ F_2 \\ \vdots \\ F_N \end{Bmatrix} \qquad (3.74)$$

Above, we see that the only modifications to handle $Q_1 = a_1$ are that a large number C gets added on to the first diagonal element of $\mathbf{K}$, and that Ca_1 gets added on to F_1. Solution of Eqs. 3.74 yields the displacement vector $\mathbf{Q}$.

The reaction force at node 1 equals the force exerted *by* the spring *on* the structure. Since the net extension of the spring is $(Q_1 - a_1)$, and the spring stiffness is C, the reaction force is given by

$$R_1 = -C (Q_1 - a_1) \qquad\qquad (3.75)$$

The modifications to $\mathbf{K}$ and $\mathbf{F}$ given in Eqs. 3.74 are also derivable using Galerkin's approach. Consider the boundary condition $Q_1 = a_1$. To handle this, we introduce a spring with a large stiffness C with the support given a displacement equal to a_1 (Fig. 3.10). The virtual work done by the spring as a result of an arbitrary displacement Ψ is

$$\delta W_s = \text{virtual displacement} \times \text{force in spring}$$

or

$$\delta W_s = \Psi_1 C(Q_1 - a_1) \qquad (3.76)$$

Thus, the variational form is

$$\Psi^T(KQ - F) + \Psi_1 C(Q_1 - a_1) = 0 \qquad (3.77)$$

which should be valid for any Ψ. Choosing $\Psi = [1, 0, \ldots, 0]^T$, $\Psi = [0, 1, 0, \ldots, 0]^T, \ldots, \Psi = [0, \ldots, 0, 1]^T$, and substituting each in turn into Eq. 3.77, we obtain precisely the modifications shown in Eqs. 3.74. The general procedure is now summarized below.

Summary: Penalty Approach

Consider the boundary conditions

$$Q_{p_1} = a_1, Q_{p_2} = a_2, \ldots, Q_{p_r} = a_r$$

Step 1. Modify the structural stiffness matrix **K** by adding a large number C to each of the p_1th, p_2th, . . . , and p_rth diagonal elements of **K**. Also, modify the global load vector **F** by adding Ca_1 to F_{p_1}, Ca_2 to F_{p_2}, . . . , and Ca_r to F_{p_r}. Solve **KQ = F** for the displacement **Q**, where **K** and **F** are the modified stiffness and load matrices.

Step 2. For each element, extract the element displacement vector **q** from the **Q** vector, using element connectivity, and determine the element stresses.

Step 3. Evaluate the reaction force at each support from

$$R_{p_i} = -C(Q_{p_i} - a_i) \qquad i = 1, 2, \ldots, r \qquad (3.78)$$

It should be noted that the penalty approach presented herein is an approximate approach. The accuracy of the solution, particularly the reaction forces, depends on the choice of C as discussed below.

Choice of C.   Let us expand the first equation in Eq. 3.74. We have

$$(K_{11} + C)Q_1 + K_{12}Q_2 + \cdots + K_{1N}Q_N = F_1 \qquad (3.79a)$$

Upon dividing by C, we get

$$\left(\frac{K_{11}}{C} + 1\right)Q_1 + \frac{K_{12}}{C}Q_2 + \cdots + \frac{K_{1N}}{C}Q_N = \frac{F_1}{C} + a_1 \qquad (3.79b)$$

From the equation above, we see that if C is chosen large enough, then $Q_1 \approx a_1$. Specifically, we see that if C is large compared to the stiffness coefficients K_{11},

$K_{12}, \ldots, K_{1N}$, then $Q_1 \approx a_1$. Note that F_1 is a load applied at the support (if any), and that F_1/C is generally of small magnitude.

A simple scheme suggests itself for choosing the magnitude of C:

$$C = \max |K_{ij}| \times 10^4$$

$$1 \le i \le N \qquad\qquad (3.80)$$

$$1 \le j \le N$$

The choice of 10^4 above has been found to be satisfactory on most computers. The reader may wish to choose a sample problem and experiment with this (using, say, 10^5 or 10^6), to check whether the reaction forces differ by much.

Example 3.4

Consider the bar shown in Fig. E3.4. An axial load $P = 200 \times 10^3$ N is applied as shown. Using the penalty approach for handling boundary conditions, do the following:

(a) Determine the nodal displacements.

(b) Determine the stress in each material.

(c) Determine the reaction forces.

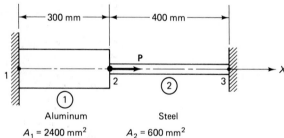

Aluminum   Steel

$A_1 = 2400$ mm^2 $A_2 = 600$ mm^2

$E_1 = 70 \times 10^9$ N/m^2 $E_2 = 200 \times 10^9$ N/m^2 **Figure E3.4**

Solution

(a) The element stiffness matrices are

$$\begin{matrix} & 1 & 2 & \leftarrow \text{Global dof} \end{matrix}$$
$$\mathbf{k}^1 = \frac{70 \times 10^3 \times 2400}{300} \begin{bmatrix} 1 & -1 \\ -1 & 1 \end{bmatrix}$$

and

$$\begin{matrix} & 2 & 3 \end{matrix}$$
$$\mathbf{k}^2 = \frac{200 \times 10^3 \times 600}{400} \begin{bmatrix} 1 & -1 \\ -1 & 1 \end{bmatrix}$$

The structural stiffness matrix that is assembled from $\mathbf{k}^1$ and $\mathbf{k}^2$ is

$$\mathbf{K} = 10^6 \begin{bmatrix} 0.56 & -0.56 & 0 \\ -0.56 & 0.86 & -0.30 \\ 0 & -0.30 & 0.30 \end{bmatrix} \begin{matrix} 1 \\ 2 \\ 3 \end{matrix}$$

The global load vector is

$$\mathbf{F} = [0,\ 200 \times 10^3,\ 0]^T$$

Now, dof's 1 and 3 are fixed. When using the penalty approach, therefore, a large number C is added to the first and third diagonal elements of $\mathbf{K}$. Choosing C based on Eq. 3.80, we get

$$C = [0.86 \times 10^6] \times 10^4$$

Thus, the modified stiffness matrix is

$$\mathbf{K} = 10^6 \begin{bmatrix} 8600.56 & -0.56 & 0 \\ -0.56 & 0.86 & -0.30 \\ 0 & -0.30 & 8600.30 \end{bmatrix}$$

The finite element equations are given by

$$10^6 \begin{bmatrix} 8600.56 & -0.56 & 0 \\ -0.56 & 0.86 & -0.30 \\ 0 & -0.30 & 8600.30 \end{bmatrix} \begin{Bmatrix} Q_1 \\ Q_2 \\ Q_3 \end{Bmatrix} = \begin{Bmatrix} 0 \\ 200 \times 10^3 \\ 0 \end{Bmatrix}$$

which yields the solution

$$\mathbf{Q} = [15.1432 \times 10^{-6},\ 0.23257,\ 8.1127 \times 10^{-6}]^T \text{ mm}$$

(b) The element stresses are (Eq. 3.16)

$$\sigma_1 = 70 \times 10^3 \times \frac{1}{300} [-1 \quad 1] \begin{Bmatrix} 15.1423 \times 10^{-6} \\ 0.23257 \end{Bmatrix}$$

$$= 54.27 \text{ MPa}$$

where 1 MPa $= 10^6$ N/m$^2 = 1$ N/mm^2. Also,

$$\sigma_2 = 200 \times 10^3 \times \frac{1}{400} [-1 \quad 1] \begin{Bmatrix} 0.23257 \\ 8.1127 \times 10^{-6} \end{Bmatrix}$$

$$= -116.29 \text{ MPa}$$

(c) The reaction forces are obtained from Eq. 3.78 as

$$R_1 = -CQ_1$$

$$= -[0.86 \times 10^{10}] \times 15.1432 \times 10^{-6} \text{ N}$$

$$= -130.23 \times 10^3 \text{ N}$$

Also,

$$R_3 = -CQ_3$$
$$= -[0.86 \times 10^{10}] \times 8.1127 \times 10^{-6}$$
$$= -69.77 \times 10^3 \text{ N} \qquad \blacksquare$$

Example 3.5

In Fig. E3.5a, a load $P = 60 \times 10^3$ N is applied as shown. Determine the displacement field, stress and support reactions in the body. Take $E = 20 \times 10^3$ N/mm^2.

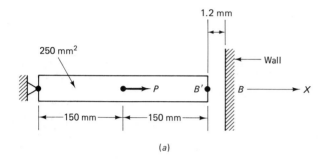

(a)

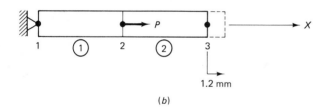

(b)

Figure E3.5

Solution   In this problem, we should first determine whether contact occurs between the bar and the wall, B. To do this, assume that the wall does not exist. Then, the solution to the problem can be verified to be

$$Q_{B'} = 1.8 \text{ mm}$$

where $Q_{B'}$ is the displacement of point B'. From this result, we see that contact does occur. The problem has to be re-solved, since the boundary conditions are now different: The displacement at B' is specified to be 1.2 mm. Consider the two-element finite element model in Fig. 3.5b. The boundary conditions are $Q_1 = 0$ and $Q_3 = 1.2$ mm. The structural stiffness matrix **K** is

$$\mathbf{K} = \frac{20 \times 10^3 \times 250}{150} \begin{bmatrix} 1 & -1 & 0 \\ -1 & 2 & -1 \\ 0 & -1 & 1 \end{bmatrix}$$

and the global load vector **F** is

$$\mathbf{F} = [0, \ 60 \times 10^3, \ 0]^{\mathrm{T}}$$

In the penalty approach, the boundary conditions $Q_1 = 0$ and $Q_3 = 1.2$ imply the following modifications: A large number C, chosen here as $C = (2/3) \times 10^{10}$, is added on to the 1st and 3rd diagonal elements of $\mathbf{K}$. Also, the number $(C \times 1.2)$ gets added on to the 3rd component of $\mathbf{F}$. Thus, the modified equations are

$$\frac{10^5}{3}\begin{bmatrix} 20001 & -1 & 0 \\ -1 & 2 & -1 \\ 0 & -1 & 20001 \end{bmatrix}\begin{Bmatrix} Q_1 \\ Q_2 \\ Q_3 \end{Bmatrix} = \begin{Bmatrix} 0 \\ 60.0 \times 10^3 \\ 80.0 \times 10^7 \end{Bmatrix}$$

The solution is

$$\mathbf{Q} = [7.49985 \times 10^{-5}, 1.500045, 1.200015]^T \text{ mm}$$

The element stresses are

$$\sigma_1 = 200 \times 10^3 \times \frac{1}{150}[-1 \quad 1]\begin{Bmatrix} 7.49985 \times 10^{-5} \\ 1.500045 \end{Bmatrix}$$

$$= 199.996 \text{ MPa}$$

$$\sigma_2 = 200 \times 10^3 \times \frac{1}{150}[-1 \quad 1]\begin{Bmatrix} 1.500045 \\ 1.200015 \end{Bmatrix}$$

$$= -40.004 \text{ MPa}$$

The reaction forces are

$$R_1 = -C \times 7.49985 \times 10^{-5}$$

$$= -49.999 \times 10^3 \text{ N}$$

and

$$R_3 = -C \times (1.200015 - 1.2)$$

$$= -10.001 \times 10^3 \text{ N}$$

The results obtained from penalty approach have a small approximation error due to the flexibility of the support introduced. In fact, the reader may verify that the elimination approach for handling boundary conditions yields the exact reactions, $R_1 = -500.0 \times 10^3$ N and $R_3 = -100.0 \times 10^3$ N. ■

Multipoint Constraints

In problems where, for example, inclined rollers or rigid connections are to be modeled, the boundary conditions take the form

$$\beta_1 Q_{p1} + \beta_2 Q_{p2} = \beta_0$$

where β_0, β_1, β_2 are known constants. Such boundary conditions are referred to as multipoint constraints in the literature. The penalty approach will now be shown to apply to this type of boundary condition.

Consider the modified total potential energy expression

$$\Pi_M = \tfrac{1}{2}\mathbf{Q}^T\mathbf{KQ} + \tfrac{1}{2}C(\beta_1 Q_{p1} + \beta_2 Q_{p2} - \beta_0)^2 - \mathbf{Q}^T\mathbf{F} \qquad (3.81)$$

where C is a large number. Since C is large, Π_M takes on a minimum value only when $(\beta_1 Q_{p1} + \beta_2 Q_{p2} - \beta_0)$ is very small, that is, when $\beta_1 Q_{p1} + \beta_2 Q_{p2} \approx \beta_0$, as desired. Setting $d\Pi_M/dQ_i = 0$, $i = 1, \ldots, N$ yields the the modified stiffness and force matrices. These modifications are given as

$$\begin{bmatrix} K_{p_1p_1} & K_{p_1p_2} \\ K_{p_2p_1} & K_{p_2p_2} \end{bmatrix} \longrightarrow \begin{bmatrix} K_{p_1p_1} + C\beta_1^2 & K_{p_1p_2} + C\beta_1\beta_2 \\ K_{p_2p_1} + C\beta_1\beta_2 & K_{p_2p_2} + C\beta_2^2 \end{bmatrix} \tag{3.82}$$

and

$$\begin{Bmatrix} F_{p_1} \\ F_{p_2} \end{Bmatrix} \longrightarrow \begin{Bmatrix} F_{p_1} + C\beta_0\beta_1 \\ F_{p_2} + C\beta_0\beta_2 \end{Bmatrix} \tag{3.83}$$

If we consider the equilibrium equations $\partial\Pi_M/\partial Q_{p1} = 0$ and $\partial\Pi_M/\partial Q_{p2} = 0$, and rearrange these in the form

$$\sum_j K_{p_1j} Q_j - F_{p_1} = R_{p_1} \quad \text{and} \quad \sum_j K_{p_2j} Q_j - F_{p_2} = R_{p_2}$$

we obtain the reaction forces R_{p_1} and R_{p_2} which are the reaction components along dof's p_1 and p_2, respectively, as

$$R_{p_1} = \frac{\partial}{\partial Q_{p_1}} [\tfrac{1}{2} C (\beta_1 Q_{p_1} + \beta_2 Q_{p_2} - \beta_0)^2] \tag{3.84a}$$

and

$$R_{p_2} = -\frac{\partial}{\partial Q_{p_2}} [\tfrac{1}{2} C (\beta_1 Q_{p_1} + \beta_2 Q_{p_2} - \beta_0)^2] \tag{3.84b}$$

Upon simplification, Eqs. 3.84 yield

$$R_{p_1} = -C\beta_1(\beta_1 Q_{p_1} + \beta_2 Q_{p_2} - \beta_0) \tag{3.85a}$$

and

$$R_{p_2} = -C\beta_2(\beta_1 Q_{p_1} + \beta_2 Q_{p_2} - \beta_0) \tag{3.85b}$$

We see that the penalty approach allows us to handle multipoint constraints, and is again easy to implement in a computer program. A nonphysical argument is used here to arrive at the modified potential energy in Eq. 3.81. Multipoint constraints are the most general types of boundary conditions, from which other types can be treated as special cases.

Example 3.6

Consider the structure shown in Fig. E3.6a. A *rigid* bar of negligible mass, pinned at one end, is supported by a steel rod and an aluminum rod. A load $P = 30 \times 10^3$ N is applied as shown.

(a) Model the structure using two finite elements. What are the boundary conditions for your model?

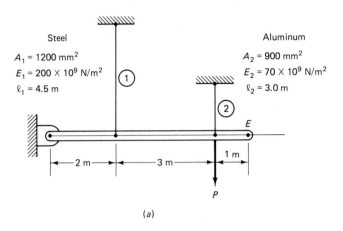

Steel

$A_1 = 1200$ mm^2
$E_1 = 200 \times 10^9$ N/m^2
$\ell_1 = 4.5$ m

Aluminum

$A_2 = 900$ mm^2
$E_2 = 70 \times 10^9$ N/m^2
$\ell_2 = 3.0$ m

(a)

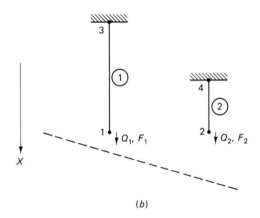

(b)

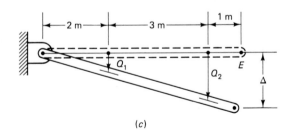

(c)

Figure E3.6

(b) Develop the modified stiffness matrix and modified load vector. Solve the equations for **Q**. Then determine element stresses.

Solution

(a) A two-element model of the structure is shown in Fig. E3.6b. The boundary conditions at nodes 3 and 4 are obvious: $Q_3 = Q_4 = 0$. Now, since the rigid bar has to remain straight, the displacements Q_1 anbd Q_2 are not independent. In fact, a look at the deformed configuration in Fig. E3.6c yields the boundary condition.

$$Q_1 - 0.4Q_2 = 0$$

which is a multipoint constraint.

(b) First, the element stiffness matrices are given by

$$\mathbf{k}^1 = \frac{200 \times 10^3 \times 1200}{4.5 \times 10^3} \begin{array}{cc} 3 & 1 \\ \begin{bmatrix} 1 & -1 \\ -1 & 1 \end{bmatrix} & \begin{array}{c} 3 \\ 1 \end{array} \end{array}$$

and

$$\mathbf{k}^2 = \frac{70 \times 10^3 \times 900}{3.0 \times 10^3} \begin{array}{cc} 4 & 2 \\ \begin{bmatrix} 1 & -1 \\ -1 & 1 \end{bmatrix} & \begin{array}{c} 4 \\ 2 \end{array} \end{array}$$

The global stiffness matrix $\mathbf{K}$ is

$$\mathbf{K} = 10^3 \begin{array}{cccc} 1 & 2 & 3 & 4 \\ \begin{bmatrix} 53.33 & 0 & -53.33 & 0 \\ 0 & 21.0 & 0 & -21.0 \\ -53.33 & 0 & 53.33 & 0 \\ 0 & -21.0 & 0 & 21.0 \end{bmatrix} & \begin{array}{c} 1 \\ 2 \\ 3 \\ 4 \end{array} \end{array}$$

The $\mathbf{K}$ matrix is modified as follows. First, since $Q_3 = Q_4 = 0$, a large number C is added on to the $(3, 3)$ and $(4, 4)$ locations of $\mathbf{K}$. Next, upon comparing the multipoint constraint given in part (a) above with the general form, we identify $\beta_0 = 0$, $\beta_1 = 1$, and $\beta_2 = -0.4$. From Eq. 3.82, we see that the terms C, $-0.4C$, and $0.16C$ have to be added to the $(1, 1)$, $(1, 2)$ and $(2, 2)$ locations of $\mathbf{K}$, respectively. Also, since $\beta_0 = 0$, we see from Eq. 3.83 that there are no modifications in $\mathbf{F}$. Now, C is chosen equal to $[53.33 \times 10^3] \times 10^4$. Thus, the equations are

$$10^3 \begin{bmatrix} 53.33533 \times 10^4 & -21.3320 \times 10^4 & -53.33 & 0 \\ -21.3320 \times 10^4 & 8.5349 \times 10^4 & 0 & -21.0 \\ -53.33 & 0 & 53.33533 \times 10^4 & 0 \\ 0 & -21.0 & 0 & 53.3321 \times 10^4 \end{bmatrix} \begin{Bmatrix} Q_1 \\ Q_2 \\ Q_3 \\ Q_4 \end{Bmatrix} = \begin{Bmatrix} 0 \\ 30 \times 10^3 \\ 0 \\ 0 \end{Bmatrix}$$

The solution, obtained from a computer program that solves matrix equations, is

$$\mathbf{Q} = [0.4206, 1.0517, 4.2059 \times 10^{-5}, 4.1411 \times 10^{-5}] \text{ mm}$$

The element stresses are now recovered from Eqs. 3.15 and 3.16 as

$$\sigma_1 = 200 \times 10^3 \frac{1}{4.5 \times 10^3} \begin{bmatrix} -1 & 1 \end{bmatrix} \begin{Bmatrix} 4.2059 \\ 0.4206 \end{Bmatrix}$$

$$= 18.693 \text{ MPa}$$

and

$$\sigma_2 = 24.540 \text{ MPa}$$

In this problem, consider what happens if the load P is applied at point E in Fig. E3.6a. In this case, equivalent nodal loads F_1 and F_2 have to be determined. The equivalent loads can be determined from the fact that the potential of P should equal that of F_1 and F_2. Thus, if Δ is the displacement along P (Fig. E3.6c), then F_1 and F_2 have to satisfy

$$P\Delta = F_1 Q_1 + F_2 Q_2$$

where $\Delta = 1.2Q_2 = 3Q_1$. Interestingly, any combination of F_1 and F_2 that satisfies the above equation will yield the same solution. ∎

3.9 QUADRATIC SHAPE FUNCTIONS

So far, the unknown displacement field was interpolated by linear shape functions within each element. In some problems, however, use of quadratic interpolation leads to far more accurate results. In this section, quadratic shape functions will be introduced, and the corresponding element stiffness matrix and load vectors will be derived. The reader should note that the basic procedure is the same as that used in the linear one dimensional element earlier.

Consider a typical three-node quadratic element, as shown in Fig. 3.11a. In the local numbering scheme, the left node will be numbered 1, the right node 2, and the midpoint 3. Node 3 has been introduced for the purposes of passing a quadratic fit, and is called an *internal node*. The notation $x_i = x$ coordinate of node i, $i = 1, 2, 3$, is used. Further, $\mathbf{q} = [q_1, q_2, q_3]^T$, where q_1, q_2, and q_3 are the displacements of nodes 1, 2, and 3, respectively. The x-coordinate system is mapped onto a ξ-coordinate system, which is given by the transformation

$$\xi = \frac{2(x - x_3)}{x_2 - x_1} \tag{3.86}$$

From Eq. 3.86 we see that $\xi = -1$, 0, and $+1$ at nodes 1, 3, and 2 (Fig. 3.11b).

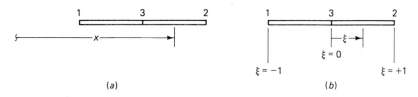

(a)   (b)

Figure 3.11   Quadratic element in x and ξ coordinates.

Now, in ξ coordinates, *quadratic shape functions*, N_1, N_2, and N_3 will be introduced as

$$N_1(\xi) = -\tfrac{1}{2}\xi(1 - \xi) \qquad (3.87a)$$

$$N_2(\xi) = \tfrac{1}{2}\xi(1 + \xi) \qquad (3.87b)$$

$$N_3(\xi) = (1 + \xi)(1 - \xi) \qquad (3.87c)$$

The shape function N_1 is equal to unity at node 1 and zero at nodes 2 and 3. Similarly, N_2 equals unity at node 2 and equals zero at the other two nodes; N_3 equals unity at node 3 and equals zero at nodes 1 and 2. The shape functions N_1, N_2, and N_3 are graphed in Fig. 3.12. The expressions for these shape functions can be written down by inspection. For example, since $N_1 = 0$ at $\xi = 0$ and $N_1 = 0$ at $\xi = 1$, we know that N_1 has to contain the product $\xi(1 - \xi)$. That is, N_1 is of the form

$$N_1 = c\xi(1 - \xi) \qquad (3.88)$$

The constant c is now obtained from the condition $N_1 = 1$ at $\xi = -1$, which yields $c = -\tfrac{1}{2}$, resulting in the formula given in Eq. 3.87a. These shape functions are called *Lagrange* shape functions.

Now the displacement field within the element is written in terms of the nodal displacements as

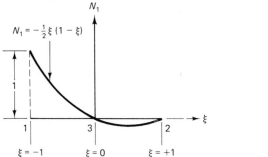

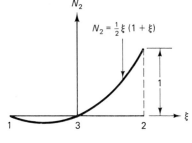

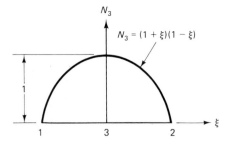

Figure 3.12   Shape functions N_1, N_2, and N_3.

$$u = N_1 q_1 + N_2 q_2 + N_3 q_3 \qquad (3.89a)$$

or

$$u = \mathbf{N}\mathbf{q} \qquad (3.89b)$$

where $\mathbf{N} = [N_1, N_2, N_3]$ is a (1×3) vector of shape functions and $\mathbf{q} = [q_1, q_2, q_3]^T$ is the (3×1) element displacement vector. At node 1 we see that $N_1 = 1$, $N_2 = N_3 = 0$, and hence $u = q_1$. Similarly, $u = q_2$ at node 2 and $u = q_3$ at node 3. Thus, u in Eq. 3.89a is a quadratic interpolation passing through q_1, q_2, and q_3 (Fig. 3.13).

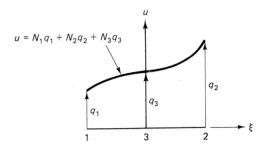

Figure 3.13   Interpolation using quadratic shape functions.

The strain ϵ is now given by

$$\epsilon = \frac{du}{dx} \qquad \text{(strain–displacement relation)}$$

$$= \frac{du}{d\xi}\frac{d\xi}{dx} \qquad \text{(chain rule)} \qquad (3.90)$$

$$= \frac{2}{x_2 - x_1}\frac{du}{d\xi} \qquad \text{(using Eq 3.86)}$$

$$= \frac{2}{x_2 - x_1}\left[\frac{dN_1}{d\xi}, \frac{dN_2}{d\xi}, \frac{dN_3}{d\xi}\right]\cdot\mathbf{q} \qquad \text{(using Eq. 3.89)}$$

Using Eqs 3.87, we have

$$\epsilon = \frac{2}{x_2 - x_1}\left[-\frac{1-2\xi}{2}, \frac{1+2\xi}{2}, -2\xi\right]\mathbf{q} \qquad (3.91)$$

which is of the form

$$\epsilon = \mathbf{B}\mathbf{q} \qquad (3.92)$$

where $\mathbf{B}$ is given by

$$\mathbf{B} = \frac{2}{x_2 - x_1}\left[-\frac{1-2\xi}{2}, \frac{1+2\xi}{2}, -2\xi\right] \qquad (3.93)$$

Using Hooke's law, we can write the stress as

$$\sigma = E\mathbf{B}\mathbf{q} \tag{3.94}$$

Note that since N_i are quadratic shape functions, $\mathbf{B}$ in Eq. 3.93 is linear in ξ. *This means that the strain and stress can vary linearly within the element.* Recall that when using linear shape functions, the strain and stress came out to be constant within the element.

We now have expressions for u, ϵ, and σ in Eqs. 3.89b, 3.92, and 3.94, respectively. Also, we have $dx = (\ell_e/2)\, d\xi$ from Eq. 3.86.

Again, in the finite element model considered here, it will be assumed that cross-sectional area A_e, body force F, and traction force T are constant within the element. Substituting for u, ϵ, σ, and dx from above into the potential energy expression yields

$$\Pi = \sum_e \frac{1}{2} \int_e \sigma^T \epsilon A\, dx - \sum_e \int_e u^T f A\, dx - \sum_e \int_e u^T T\, dx - \sum_i Q_i P_i$$

$$= \sum_e \frac{1}{2}\mathbf{q}^T\left(E_e A_e \frac{\ell_e}{2}\int_{-1}^{1}[\mathbf{B}^T\mathbf{B}]\,d\xi\right)\mathbf{q} - \sum_e \mathbf{q}^T\left(A_e \frac{\ell_e}{2}f\int_{-1}^{1}\mathbf{N}^T\,d\xi\right) \tag{3.95}$$

$$- \sum_e \mathbf{q}^T\left(\frac{\ell_e}{2}T\int_{-1}^{1}\mathbf{N}^T\,d\xi\right) - \sum_i Q_i P_i$$

Comparing the above equation with the general form

$$\Pi = \sum_e \frac{1}{2}\mathbf{q}^T\mathbf{k}^e\mathbf{q} - \sum_e \mathbf{q}^T\mathbf{f}^e - \sum_e \mathbf{q}^T\mathbf{T}^e - \sum_i Q_i P_i$$

yields

$$\mathbf{k}^e = \frac{E_e A_e \ell_e}{2}\int_{-1}^{1}[\mathbf{B}^T\mathbf{B}]\,d\xi \tag{3.96a}$$

which, upon substituting for $\mathbf{B}$ in Eq. 3.93, yields

$$\mathbf{k}^e = \frac{E_e A_e}{3\ell_e}\begin{bmatrix} 7 & 1 & -8 \\ 1 & 7 & -8 \\ -8 & -8 & 16 \end{bmatrix}\begin{matrix} 1 \\ 2 \\ 3 \end{matrix} \tag{3.96b}$$

$$\begin{matrix} 1 & 2 & 3 \leftarrow \text{Local dof} \end{matrix}$$

The element body force vector $\mathbf{f}^e$ is given by

$$\mathbf{f}^e = \frac{A_e \ell_e f}{2}\int_{-1}^{1}\mathbf{N}^T\,d\xi \tag{3.97a}$$

which, upon substituting for $\mathbf{N}$ in Eqs. 3.87, yields

↓ Local dof

$$\mathbf{f}^e = A_e \ell_e f \begin{Bmatrix} 1/6 \\ 1/6 \\ 2/3 \end{Bmatrix} \begin{matrix} 1 \\ 2 \\ 3 \end{matrix} \qquad (3.97b)$$

Similarly, the element traction force vector $\mathbf{T}^e$ is given by

$$\mathbf{T}^e = \frac{\ell_e T}{2} \int_{-1}^{1} \mathbf{N}^\mathsf{T} \, d\xi \qquad (3.98a)$$

which results in

↓ Local dof

$$\mathbf{T}^e = \ell_e T \begin{Bmatrix} 1/6 \\ 1/6 \\ 2/3 \end{Bmatrix} \begin{matrix} 1 \\ 2 \\ 3 \end{matrix} \qquad (3.98b)$$

The total potential energy is again of the form $\Pi = \frac{1}{2}\mathbf{Q}^\mathsf{T}\mathbf{K}\mathbf{Q} - \mathbf{Q}^\mathsf{T}\mathbf{F}$, where the structural stiffness matrix $\mathbf{K}$ and nodal load vector $\mathbf{F}$ are assembled from element stiffness matrices and load vectors, respectively.

Example 3.7

Consider the rod (a robot arm) in Fig. E3.7a, which is rotating at constant angular velocity $\omega = 30$ rad/s. Determine the axial stress distribution in the rod, using two quadratic elements. Consider only the centrifugal force. Ignore bending of the rod.

Solution   A finite element model of the rod, with two quadratic elements, is shown in Fig. E3.7b. The model has a total of five degrees of freedom. The element stiffness matrices are (from Eq. 3.96b)

$$\mathbf{k}^1 = \frac{10^7 \times 0.6}{3 \times 21} \begin{bmatrix} 7 & 1 & -8 \\ 1 & 7 & -8 \\ -8 & -8 & 16 \end{bmatrix} \begin{matrix} 1 \\ 3 \\ 2 \end{matrix}$$

with column headers $\begin{matrix} 1 & 3 & 2 \end{matrix}$ ← Global dof

and

$$\mathbf{k}^2 = \frac{10^7 \times 0.6}{3 \times 21} \begin{bmatrix} 7 & 1 & -8 \\ 1 & 7 & -8 \\ -8 & -8 & 16 \end{bmatrix} \begin{matrix} 3 \\ 5 \\ 4 \end{matrix}$$

with column headers $\begin{matrix} 3 & 5 & 4 \end{matrix}$

Thus,

$$\mathbf{K} = \frac{10^7 \times 0.6}{3 \times 21} \begin{bmatrix} 7 & -8 & 1 & 0 & 0 \\ -8 & 16 & -8 & 0 & 0 \\ 1 & -8 & 14 & -8 & 1 \\ 0 & 0 & -8 & 16 & -8 \\ 0 & 0 & 1 & -8 & 7 \end{bmatrix} \begin{matrix} 1 \\ 2 \\ 3 \\ 4 \\ 5 \end{matrix}$$

with column headers $\begin{matrix} 1 & 2 & 3 & 4 & 5 \end{matrix}$

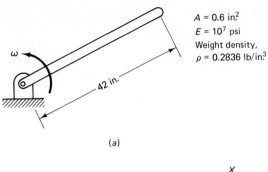

$A = 0.6$ in.^{2}
$E = 10^7$ psi
Weight density,
$\rho = 0.2836$ lb/in.^{3}

(a)

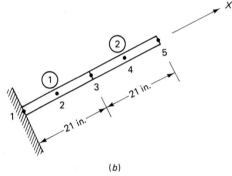

(b)

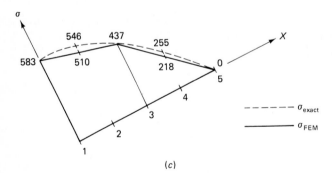

(c)   Figure E3.7

The body force f (lb/in.^{3}) is given by

$$f = \frac{\rho r \omega^2}{g} \quad \text{lb/in}^3$$

where ρ = weight density, and $g = 32.2$ ft/s^2. Note that f is a function of the distance r from the pin. Taking average values of f over each element, we have

$$f_1 = \frac{0.2836 \times 10.5 \times 30^2}{32.2 \times 12}$$

$$= 6.94$$

and

$$f_2 = \frac{0.2836 \times 31.5 \times 30^2}{32.2 \times 12}$$

$$= 20.81$$

Thus, the element body force vectors are (from Eq. 3.97b)

↓ Global dof

$$\mathbf{f}^1 = 0.6 \times 21 \times f_1 \begin{Bmatrix} \frac{1}{6} \\ \frac{1}{6} \\ \frac{2}{3} \end{Bmatrix} \begin{matrix} 1 \\ 3 \\ 2 \end{matrix}$$

and

↓ Global dof

$$\mathbf{f}^2 = 0.6 \times 21 \times f_2 \begin{Bmatrix} \frac{1}{6} \\ \frac{1}{6} \\ \frac{2}{3} \end{Bmatrix} \begin{matrix} 3 \\ 5 \\ 4 \end{matrix}$$

Assembling $\mathbf{f}^1$ and $\mathbf{f}^2$, we obtain

$$\mathbf{F} = [14.57, 58.26, 58.26, 174.79, 43.70]^T$$

Using the elimination method, the finite element equations are

$$\frac{10^7 \times 0.6}{63} \begin{bmatrix} 16 & -8 & 0 & 0 \\ -8 & 14 & -8 & 1 \\ 0 & -8 & 16 & -8 \\ 0 & 1 & -8 & 7 \end{bmatrix} \begin{Bmatrix} Q_2 \\ Q_3 \\ Q_4 \\ Q_5 \end{Bmatrix} = \begin{Bmatrix} 58.26 \\ 58.26 \\ 174.79 \\ 43.7 \end{Bmatrix}$$

which yields

$$\mathbf{Q} = 10^{-3} [0, .5735, 1.0706, 1.4147, 1.5294]^T \text{ mm}$$

The stresses can now be evaluated from Eqs. 3.93 and 3.94. The element connectivity table is

Element Number	1	2	3	← Local Node Nos.
1	1	3	2	↑ Global Node Nos.
2	3	5	4	↓

Thus,

$$\mathbf{q} = [Q_1, Q_3, Q_2]^T$$

for element 1 while

$$\mathbf{q} = [Q_3, Q_5, Q_4]^T$$

for element 2. Using Eqs. 3.93 and 3.94, we get

$$\sigma_1 = 10^7 \times \frac{2}{21} \left[-\frac{1 - 2\xi}{2}, \frac{1 + 2\xi}{2}, -2\xi \right] \begin{Bmatrix} Q_1 \\ Q_3 \\ Q_2 \end{Bmatrix}$$

where $-1 \le \xi \le 1$, and σ_1 denotes the stress in element 1. The stress at node 1 in element 1 is obtained by substituting $\xi = -1$ into the above, which results in

$$\sigma_1|_1 = 10^7 \times \tfrac{2}{21} \times 10^{-3}[-1.5, -0.5, +2.0] \begin{Bmatrix} 0 \\ 1.0706 \\ .5735 \end{Bmatrix}$$

$$= 583 \text{ psi}$$

The stress at node 2 (which corresponds to the midpoint of element 1 is obtained by substituting for $\xi = 0$:

$$\sigma_1|_2 = 10^7 \times \tfrac{2}{21} \times 10^{-3}[-0.5, 0.5, 0] \begin{Bmatrix} 0 \\ 1.0706 \\ .5735 \end{Bmatrix}$$

$$= 510 \text{ psi}$$

Similarly, we obtain

$$\sigma_1|_3 = \sigma_2|_1 = 437 \text{ psi} \qquad \sigma_2|_2 = 218 \text{ psi} \qquad \sigma_2|_3 = 0$$

The axial distribution is shown in Fig. E3.7c. The stresses obtained from the finite element model can be compared to the exact solution, given by

$$\sigma_{\text{exact}}(x) = \frac{\rho \omega^2}{2g}(L^2 - x^2)$$

The exact stress distribution based on the above equation is also shown in Fig. E3.7c.   ∎

3.10 TEMPERATURE EFFECTS

In this section, the stresses induced by temperature changes in an isotropic linearly elastic material will be considered. That is, the **thermal stress** problem will be considered. If the distribution of the change in temperature, $\Delta T(x)$, is known, then the strain due to this temperature change can be treated as an **initial strain,** ϵ_0, given as

$$\epsilon_0 = \alpha \, \Delta T \qquad\qquad (3.99)$$

where α is the coefficient of thermal expansion. Note that a positive ΔT implies a rise in temperature. The stress–strain law in the presence of ϵ_0 is shown in Fig. 3.14. From this figure, we see that the stress–strain relation is given by

$$\sigma = E(\epsilon - \epsilon_0) \qquad\qquad (3.100)$$

The strain energy per unit volume, u_0, is equal to the shaded area in Fig. 3.14 and is given by

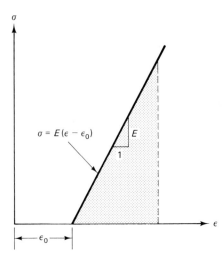

Figure 3.14   Stress–strain law in the presence of an initial strain.

$$u_0 = \tfrac{1}{2}\sigma(\epsilon - \epsilon_0) \tag{3.101}$$

By using Eq. 3.100, we find that Eq. 3.101 yields

$$u_0 = \tfrac{1}{2}(\epsilon - \epsilon_0)^T E(\epsilon - \epsilon_0) \tag{3.102a}$$

The total strain energy U in the structure is obtained by integrating u_0 over the volume of the structure:

$$U = \int_L \tfrac{1}{2}(\epsilon - \epsilon_0)^T E(\epsilon - \epsilon_0) A \, dx \tag{3.102b}$$

For a structure modeled using one-dimensional linear elements, the above equation becomes

$$U = \sum_e \frac{1}{2} A_e \frac{\ell_e}{2} \int_{-1}^{1} (\epsilon - \epsilon_0)^T E_e(\epsilon - \epsilon_0) \, d\xi \tag{3.102c}$$

Noting that $\epsilon = \mathbf{B}\mathbf{q}$, we get

$$U = \sum_e \frac{1}{2} \mathbf{q}^T \left(E_e A_e \frac{\ell_e}{2} \int_{-1}^{1} \mathbf{B}^T \mathbf{B} \, d\xi \right) \mathbf{q} - \sum_e \mathbf{q}^T E_e A_e \frac{\ell_e}{2} \epsilon_0 \int_{-1}^{1} \mathbf{B}^T \, d\xi$$

$$+ \sum_e \frac{1}{2} E_e A_e \frac{\ell_e}{2} \epsilon_0^2 \tag{3.102d}$$

Examining the strain energy expression above, we see that the first term on the right side yields the element stiffness matrix derived earlier in Section 3.4; the last term is a constant term and is of no consequence since it drops out of the equilibrium equations, which are obtained by setting $d\Pi/d\mathbf{Q} = 0$. The second term yields the desired element load vector $\boldsymbol{\theta}^e$, as a result of the temperature change:

$$\Theta^e = E_e A_e \frac{\ell_e}{2} \epsilon_0 \int_{-1}^{1} \mathbf{B}^T \, d\xi \tag{3.103a}$$

The above equation can be simplified by substituting for $\mathbf{B} = [-1 \quad 1]/(x_2 - x_1)$ and noting that $\epsilon_0 = \alpha \, \Delta T$. Thus,

$$\Theta^e = \frac{E_e A_e \ell_e \alpha \, \Delta T}{x_2 - x_1} \begin{Bmatrix} -1 \\ 1 \end{Bmatrix} \tag{3.103b}$$

In the expression above ΔT is the average change in temperature within the element. The temperature load vector in Eq. 3.103b can be assembled along with the body force, traction force, and point load vectors to yield the global load vector $\mathbf{F}$, for the structure. This assembly can be denoted as

$$\mathbf{F} = \sum_e (\mathbf{f}^e + \mathbf{T}^e + \Theta^e) + \mathbf{P} \tag{3.104}$$

After solving the finite element equations $\mathbf{KQ} = \mathbf{F}$ for the displacements $\mathbf{Q}$, the stress in each element can be obtained from Eq. (3.100) as

$$\sigma = E(\mathbf{Bq} - \alpha \, \Delta T) \tag{3.105a}$$

or

$$\sigma = \frac{E}{x_2 - x_1} [-1 \quad 1] \mathbf{q} - E\alpha \, \Delta T \tag{3.105b}$$

Example 3.8

An axial load $P = 300 \times 10^3$ N is applied at $20°$ C to the rod as shown in Fig. E3.8. The temperature is then raised to $60°$ C.

(a) Assemble the $\mathbf{K}$ and $\mathbf{F}$ matrices.

(b) Determine the nodal displacements and element stresses.

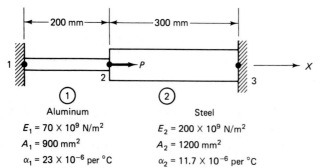

Aluminum	Steel
$E_1 = 70 \times 10^9$ N/m²	$E_2 = 200 \times 10^9$ N/m²
$A_1 = 900$ mm²	$A_2 = 1200$ mm²
$\alpha_1 = 23 \times 10^{-6}$ per °C	$\alpha_2 = 11.7 \times 10^{-6}$ per °C

Figure E3.8

Solution

(a) The element stiffness matrices are

$$\mathbf{k}^1 = \frac{70 \times 10^3 \times 900}{200} \begin{bmatrix} 1 & -1 \\ -1 & 1 \end{bmatrix} \text{N/mm}$$

$$\mathbf{k}^2 = \frac{200 \times 10^3 \times 1200}{300} \begin{bmatrix} 1 & -1 \\ -1 & 1 \end{bmatrix} \text{N/mm}$$

Thus,

$$\mathbf{K} = 10^3 \begin{bmatrix} 315 & -315 & 0 \\ -315 & 1115 & -800 \\ 0 & -800 & 800 \end{bmatrix} \text{N/mm}$$

Now, in assembing $\mathbf{F}$, both temperature and point load effects have to be considered. The element temperature forces due to $\Delta T = 40°\,\text{C}$ are obtained from Eq. 3.103b as

$$\qquad\qquad\qquad\qquad\qquad\qquad\qquad \downarrow \text{ Global dof}$$

$$\mathbf{\Theta}^1 = 70 \times 10^3 \times 900 \times 23 \times 10^{-6} \times 40 \begin{Bmatrix} -1 \\ 1 \end{Bmatrix} \begin{matrix} 1 \\ 2 \end{matrix} \quad \text{N}$$

and

$$\mathbf{\Theta}^2 = 200 \times 10^3 \times 1200 \times 11.7 \times 10^{-6} \times 40 \begin{Bmatrix} -1 \\ 1 \end{Bmatrix} \begin{matrix} 2 \\ 3 \end{matrix} \quad \text{N}$$

Upon assembling $\mathbf{\Theta}^1$, $\mathbf{\Theta}^2$, and the point load, we get

$$\mathbf{F} = 10^3 \begin{Bmatrix} -57.96 \\ 57.96 - 112.32 + 300 \\ 112.32 \end{Bmatrix}$$

or

$$\mathbf{F} = 10^3 \, [-57.96, \, 245.64, \, 112.32]^\mathsf{T} \text{ N}$$

(b) The elimination approach will now be used to solve for the displacements. Since dof's 1 and 3 are fixed, the first and third rows and columns of $\mathbf{K}$, together with the first and third components of $\mathbf{F}$, are deleted. This results in the scalar equation

$$10^3 [1115] Q_2 = 10^3 \times 245.64$$

yielding

$$Q_2 = 0.220 \text{ mm}$$

Thus,

$$\mathbf{Q} = [0, \, 0.220, \, 0]^\mathsf{T} \text{ mm}$$

In evaluating element stresses, we have to use Eq. 3.105b:

$$\sigma_1 = \frac{70 \times 10^3}{200}[-1 \quad 1]\begin{Bmatrix} 0 \\ 0.220 \end{Bmatrix} - 70 \times 10^3 \times 23 \times 10^{-6} \times 40$$

$$= 12.60 \text{ MPa}$$

and

$$\sigma_2 = \frac{200 \times 10^3}{300}[-1 \quad 1]\begin{Bmatrix} 0.220 \\ 0 \end{Bmatrix} - 200 \times 10^3 \times 11.7 \times 10^{-6} \times 40$$

$$= -240.27 \text{ MPa}$$ ■

Example 3.9

The solution of Example 3.3 using program FEM1D is presented below.

```
 Number of Elements =? 2
     Number of Loads =? 3
Num of Constr Nodes =? 1
Elem#, Area, E, Alpha, Temp Rise
? 1,5.25,30E6,0,0
? 2,3.75,30E6,0,0
Node#, Coordinate
? 1,0
? 2,12
? 3,24
Node#, Displacement
? 1,0
Node#, Applied Load
? 1,8.9334
? 2,115.3144
? 3,6.3810


NODE NO.        DISPLACEMENT
  1              9.952670476190476D-10
  2              9.273025743238095D-06
  3              9.953665743238095D-06
ELEM NO.        STRESS
  1              23.18007619047619
  2              1.7016
NODE NO.        REACTION
  1             -130.6288
Ok
```

■

```
1000 REM ******          FEM1D          ******
1010 REM  One Dimensional Problem with Temperature
1020 DEFINT I-N: CLS
1030 PRINT "======================================"
1040 PRINT "       Program written by        "
1050 PRINT "  T.R.Chandrupatla and A.D.Belegundu   "
1060 PRINT "======================================"
1070 INPUT " Number of Elements ="; NE
1080 INPUT "    Number of Loads ="; NL
1090 INPUT "Num of Constr Nodes ="; ND
1100 NN = NE + 1
1110 NBW = 2
1120 DIM X(NN), A(NE), NU(ND), U(ND), DT(NE),AL(NL),ILD(NL)
1130 DIM S(NN, NBW), F(NN), E(NE), ALP(NE),STRESS(NE),REACT(ND)
1140 PRINT "Elem#, Area, E, Alpha, Temp Rise"
1150 FOR I = 1 TO NE
1160 INPUT N, A(N), E(N), ALP(N), DT(N): NEXT I
1170 PRINT "Node#, Coordinate"
1180 FOR I = 1 TO NN
1190 INPUT N, X(N): NEXT I
1200 FOR I = 1 TO NN
1210 F(I) = 0
1220 FOR J = 1 TO NBW
1230 S(I, J) = 0: NEXT J: NEXT I
1240 PRINT "Node#, Displacement"
1250 FOR I = 1 TO ND
1260 INPUT NU(I), U(I): NEXT I
1270 IF NL = 0 THEN GOTO 1330
1280 PRINT "Node#, Applied Load"
1290 FOR I = 1 TO NL
1300 INPUT N, F(N)
1310 ILD(I)=N : AL(I)=F(N) : NEXT I
1320 REM  *** STIFFNESS MATRIX ***
1330 FOR I = 1 TO NE
1340 I1 = I: I2 = I + 1
1350 X21 = X(I2) - X(I1): EL = ABS(X21)
1360 EAL = E(I) * A(I) / EL
1370 TL = E(I) * ALP(I) * DT(I) * A(I) * EL / X21
1380 REM *** TEMPERATURE LOADS ***
1390 F(I1) = F(I1) - TL
1400 F(I2) = F(I2) + TL
1410 REM *** ELEMENT STIFFNESSES ***
1420 S(I1, 1) = S(I1, 1) + EAL
1430 S(I2, 1) = S(I2, 1) + EAL
1440 S(I1, 2) = S(I1, 2) - EAL: NEXT I
1450 REM *** MODIFY FOR BOUNDARY CONDITION ***
1460 CNST = S(1, 1) * 10000
1470 FOR I = 1 TO ND
1480 N = NU(I)
1490 S(N, 1) = S(N, 1) + CNST
1500 F(N) = F(N) + CNST * U(I): NEXT I
1510 REM *** EQUATION SOLVING ***
1520 GOSUB 5000
1530 PRINT "NODE NO.", "DISPLACEMENT"
```

```
1540 FOR I = 1 TO NN
1550 PRINT I, F(I): NEXT I
1560 REM *** STRESS CALCULATION ***
1570 PRINT "ELEM NO.", "STRESS"
1580 FOR I = 1 TO NE
1590 EPS = (F(I + 1) − F(I)) / (X(I + 1) − X(I))
1600 STRESS(I) = E(I) * (EPS − ALP(I) * DT(I))
1610 PRINT I, STRESS(I): NEXT I
1620 REM *** REACTION CALCULATION ***
1630 PRINT "NODE NO.", "REACTION"
1640 FOR I = 1 TO ND
1650 N = NU(I)
1660 R = CNST * (U(I) − F(N))
1670 REACT(I) = R
1680 PRINT N, R: NEXT I
1690 INPUT "DO YOU WISH TO SAVE THE OUTPUT (Y/N) ";ANS$
1700 IF ANS$ <> "y" AND ANS$ <> "Y" GOTO 1960
1710 INPUT "GIVE NAME OF OUTPUT FILE =";FILENAME$
1720 OPEN FILENAME$ FOR OUTPUT AS #2
1730 PRINT #2,"NE,NL,ND =",NE,NL,ND
1740 PRINT #2,"ELEM#, AREA, E, ALPHA, TEMP. RISE"
1750 FOR I = 1 TO NE
1760 PRINT #2,I,A(I),E(I),ALP(I),DT(I) :NEXT I
1770 PRINT #2,"NODE#, COORDINATE"
1780 FOR I = 1 TO NN
1790 PRINT #2,I,X(I) :NEXT I
1800 PRINT #2, "NODE#, SPECIFIED DISPLACEMENT"
1810 FOR I = 1 TO ND
1820 PRINT #2, NU(I),U(I) :NEXT I
1830 PRINT #2,"NODE#, APPLIED LOAD"
1840 FOR I=1 TO NL
1850 PRINT #2,ILD(I), AL(I) : NEXT I
1860 PRINT #2,"NODE#, DISPLACEMENT"
1870 FOR I = 1 TO NN
1880 PRINT #2, I, F(I) :NEXT I
1890 PRINT #2,"ELEMENT#, STRESS"
1900 FOR I=1 TO NE
1910 PRINT #2, I, STRESS(I) :NEXT I
1920 PRINT #2, "SUPPORT#, REACTION FORCE"
1930 FOR I=1 TO ND
1940 PRINT #2,NU(I),REACT(I) :NEXT I
1950 CLOSE #2
1960 END
5000 N=NN
5010 REM ** FORWARD ELIMINATION **
5020 FOR K = 1 TO N − 1
5030 NBK = N − K + 1
5040 IF N − K + 1 > NBW THEN NBK = NBW
5050 FOR I = K + 1 TO NBK + K − 1
5060 I1 = I − K + 1
5070 C = S(K, I1) / S(K, 1)
5080 FOR J = I TO NBK + K − 1
5090 J1 = J − I + 1
5100 J2 = J − K + 1
5110 S(I, J1) = S(I, J1) − C * S(K, J2): NEXT J
5120 F(I) = F(I) − C * F(K)
```

```
5130 NEXT I: NEXT K
5140 REM ** BACK SUBSTITUTION **
5150 F(N) = F(N) / S(N, 1)
5160 FOR II = 1 TO N - 1
5170 I = N - II
5180 NBI = N - I + 1
5190 IF N - I + 1 > NBW THEN NBI = NBW
5200 SUM = 0!
5210 FOR J = 2 TO NBI
5220 SUM = SUM + S(I, J) * F(I + J - 1): NEXT J
5230 F(I) = (F(I) - SUM) / S(I, 1): NEXT II
5240 RETURN
```

PROBLEMS

3.1. Consider the bar in Fig. P3.1. Cross-sectional area $A_e = 1.2$ in.^{2}, and Young's modulus $E = 30 \times 10^6$ psi. If $q_1 = 0.02$ in., and $q_2 = 0.025$ in., determine (by hand calculation):
   (a) The displacement at point P.
   (b) The strain ϵ and stress σ.
   (c) The element stiffness matrix.
   (d) The strain energy in the element.

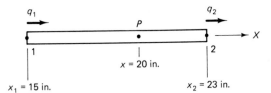

Figure P3.1

3.2. Consider the bar in Fig. P3.2 loaded as shown. Determine the nodal displacements, element stresses, and support reactions. Solve this problem by hand calculation, adopting the elimination method for handling boundary conditions. Verify your results using program FEM1D.

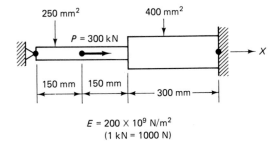

Figure P3.2

3.3. Repeat Example 3.5 in the text, but use the elimination approach for handling boundary conditions. Solve by hand calculation.

3.4 An axial load $P = 385$ kN is applied to the composite block shown in Fig. P3.4. Determine the stress in each material. (*Hint:* To use program FEM1D, use a 3-node model with two nodes identical).

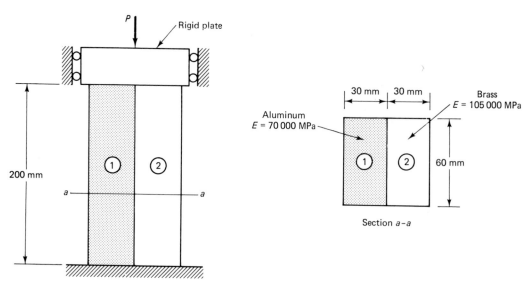

Figure P3.4

3.5. Consider the bar in Fig. P3.5. Determine the nodal displacements, element stresses and support reactions.

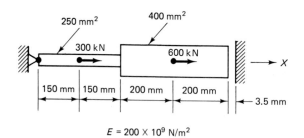

Figure P3.5

3.6. Do Example 3.7 in the text using:
   (a) Two linear finite elements.
   (b) Four linear finite elements.
   Plot the stress distributions on Fig. E3.7c.

3.7. The rigid beam in Fig. P3.7 was level before the load was applied. Find the stress in each vertical member. (*Hint:* The boundary condition is of the multipoint constraint type.)

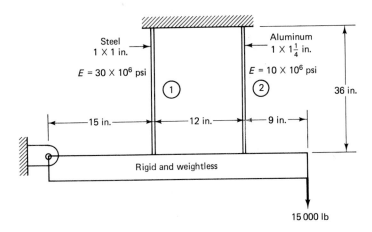

Figure P3.7

3.8. A brass bolt is fitted inside an aluminum tube, as shown in Fig. P3.8. After the nut has been fitted snugly, it is tightened one-quarter of a full turn. Given that the bolt is single threaded with a 2-mm pitch, determine the stress in bolt and tube. (*Hint:* The boundary condition is of the multipoint constraint type.)

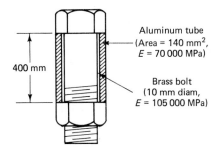

Figure P3.8

3.9. This problem reinforces the fact that once the shape functions are assumed, then all other element matrices can be derived. Certain arbitrary shape functions are given below, and the reader is asked to derive the **B** and **k** matrices.

Consider the one-dimensional element shown in Fig. P3.9. The transformation

$$\xi = \frac{2}{x_2 - x_1}(x - x_1) - 1$$

is used to relate x and ξ coordinates. Let the displacement field be interpolated as

$$u(\xi) = N_1 q_1 + N_2 q_2$$

where shape functions N_1, N_2 are assumed to be

$$N_1 = \cos \frac{\pi(1 + \xi)}{4} \qquad N_2 = \cos \frac{\pi(1 - \xi)}{4}$$

(a) Develop the relation $\epsilon = \mathbf{Bq}$. That is, develop the **B** matrix.

(b) Develop the stiffness matrix, $\mathbf{k}^e$. (You need not evaluate the integrals.)

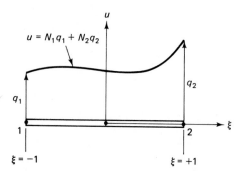

Figure P3.9

3.10. For plotting and extrapolation purposes (see Chapter 12), it is sometimes necessary to obtain nodal stress values from element stress values that are obtained from a computer run. Specifically, consider the element stresses, σ_1, σ_2, σ_3, which are constant within each element, as shown in Fig. P3.10. It is desired to obtain nodal stresses S_i, $i = 1$, 2, 3, 4, which best fit the elemental values. Obtain S_i from the least-squares criterion:

$$\text{Minimize } I = \sum_e \int_{l_e} \tfrac{1}{2}(\sigma - \sigma_e)^2 \, dx$$

where σ is expressed in terms of the nodal values s_i using linear shape functions as

$$\sigma = N_1 s_1 + N_2 s_2$$

where 1 and 2 are the local numbers.

Plot the distribution of stress from the nodal values.

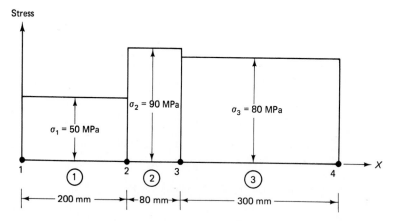

Figure P3.10

3.11. Determine the stresses in the 4 in. long bar in Fig. P3.11, using the following models:
  (a) One linear element.
  (b) Two linear elements. (Note: x in, T kips/in)

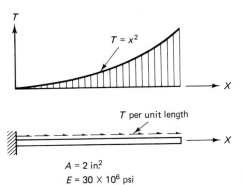

$A = 2$ in^2
$E = 30 \times 10^6$ psi

Figure P3.11

3.12 For the vertical rod shown in Fig. P3.12, find the deflection at A and the stress distribution. Use $E = 100$ MPa and weight per unit volume $= 0.06$ N/cm^3. (*Hint:* Introduce weight contribution to the nodal loads into the program and solve using two elements and four elements). Comment on the stress distribution.

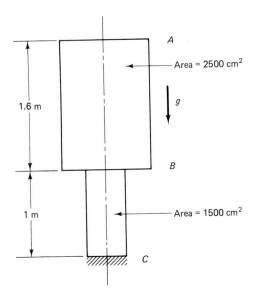

Figure P3.12

3.13 For Fig. P3.13, find the deflection at the free end under its own weight, using divisions of
  (a) 1 element   **(b)** 2 elements   **(c)** 4 elements   **(d)** 8 elements   **(e)** 16 elements
  Then plot number of elements vs. deflection.

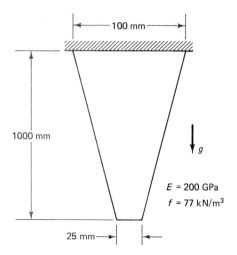

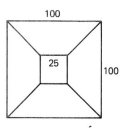

Figure P3.13

3.14. Consider the quadratic element shown in Fig. P3.14, subjected to a quadratically varying traction force (which is defined as force per unit length).
   (a) Express the traction force as a function of ξ, T_1, T_2, and T_3, using the shape functions N_1, N_2, and N_3.

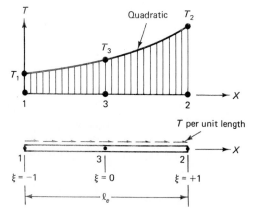

Figure P3.14

(b) Derive, from the potential term $\int_e u^T T \, dx$, an expression for the element traction force, $\mathbf{T}^e$. Leave your answer in terms of T_1, T_2, T_3, and ℓ_e.

(c) Re-solve Problem 3.11 using the exact traction load derived above, with one quadratic element, by hand calculations.

3.15. The structure in Fig. P3.15 is subjected to an increase in temperature, $\Delta T = 80°\,C$. Determine the displacements, stresses, and support reactions. Solve this problem by hand calculation, using the elimination method for handling boundary conditions.

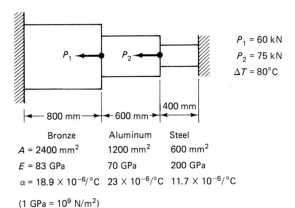

$P_1 = 60 \, kN$

$P_2 = 75 \, kN$

$\Delta T = 80°C$

	Bronze	Aluminum	Steel
$A =$	$2400 \, mm^2$	$1200 \, mm^2$	$600 \, mm^2$
$E =$	$83 \, GPa$	$70 \, GPa$	$200 \, GPa$
$\alpha =$	$18.9 \times 10^{-6}/°C$	$23 \times 10^{-6}/°C$	$11.7 \times 10^{-6}/°C$

$(1 \, GPa = 10^9 \, N/m^2)$

Figure P3.15

chapter four

Trusses

4.1 INTRODUCTION

The finite element analysis of truss structures is presented in this chapter. Two-dimensional trusses (or plane trusses) are treated in Section 4.2. In Section 4.3, this treatment is readily generalized to handle three-dimensional trusses. A typical plane truss is shown in Fig. 4.1. A truss structure consists only of two-force members. That is, every truss element is in direct tension or compression (Fig. 4.2). In a truss, it is required that all loads and reactions are applied only at the joints, and that all members are connected together at their ends by frictionless pin joints. Every engi-

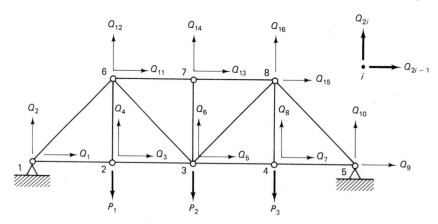

Figure 4.1   A two-dimensional truss.

Figure 4.2   A two-force member.

neering student has, in a course on statics, analyzed trusses using the method of joints and the method of sections. These methods, while illustrating the fundamentals of statics, become tedious when applied to large-scale statically indeterminate truss structures. Further, joint displacements are not readily obtainable. The finite element method on the other hand is applicable to statically determinate or indeterminate structures alike. The finite element method also provides joint deflections. Effects of temperature changes and support settlements can also be routinely handled.

4.2 PLANE TRUSSES

Local and Global Coordinate Systems

The main difference between the one-dimensional structures considered in Chapter 3 and trusses is that the elements of a truss have various orientations. To account for these different orientations, **local** and **global** coordinate systems are introduced as follows.

A typical plane truss element is shown in local and global coordinate systems in Fig. 4.3. In the local numbering scheme, the two nodes of the element are numbered 1 and 2. The local coordinate system consists of the x' axis, which runs along the element from node 1 toward node 2. All quantities in the local coordinate system will be denoted by a prime ('). The global x-, y-coordinate system is fixed and does not depend on the orientation of the element. Note that x, y, and z form a right-handed coordinate system with the z axis coming straight out of the paper. In the global coordinate system, every node has two degrees of freedom (dof's). A systematic numbering scheme is adopted here: A node whose global node number is j has associated with it dof's $2j - 1$ and $2j$. Further, the global displacements associated with node j are Q_{2j-1} and Q_{2j}, as shown in Fig. 4.1.

Let q_1' and q_2' be the displacements of nodes 1 and 2, respectively, in the local coordinate system. Thus, the element displacement vector in the local coordinate system is denoted by

$$\mathbf{q}' = [q_1', q_2']^{\mathrm{T}} \qquad (4.1)$$

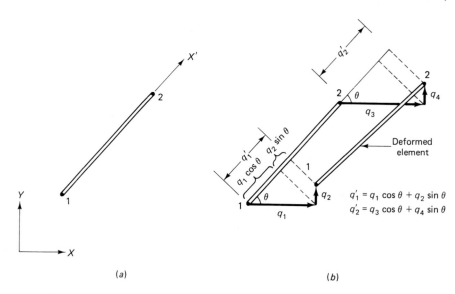

Figure 4.3 A two-dimensional truss element in (a) a local coordinate system, and (b) a global coordinate system.

The element displacement vector in the global coordinate system is a (4 × 1) vector denoted by

$$\mathbf{q} = [q_1, q_2, q_3, q_4]^T \qquad (4.2)$$

The relationship between $\mathbf{q}'$ and $\mathbf{q}$ is developed as follows. In Fig. 4.3b, we see that q_1' equals the sum of the projections of q_1 and q_2 onto the x' axis. Thus,

$$q_1' = q_1 \cos\theta + q_2 \sin\theta \qquad (4.3a)$$

Similarly,

$$q_2' = q_3 \cos\theta + q_4 \sin\theta \qquad (4.3b)$$

At this stage, the **direction cosines** ℓ and m are introduced as $\ell = \cos\theta$, $m = \cos\phi(= \sin\theta)$. These direction cosines are the cosines of the angles that the local x' axis makes with the global x, y axes, respectively. Equations 4.3a and 4.3b can now be written in matrix form as

$$\mathbf{q}' = \mathbf{Lq} \qquad (4.4)$$

where the **transformation matrix L** is given by

$$\mathbf{L} = \begin{bmatrix} \ell & m & 0 & 0 \\ 0 & 0 & \ell & m \end{bmatrix} \qquad (4.5)$$

Formulas for Calculating ℓ and m

Simple formulas are now given for calculating the direction cosines ℓ and m from nodal coordinate data. Referring to Fig. 4.4, let (x_1, y_1) and (x_2, y_2) be the coordinates of nodes 1 and 2, respectively. We then have

$$\ell = \frac{x_2 - x_1}{\ell_e} \qquad m = \frac{y_2 - y_1}{\ell_e} \tag{4.6}$$

where the length ℓ_e is obtained from

$$\ell_e = \sqrt{(x_2 - x_1)^2 + (y_2 - y_1)^2} \tag{4.7}$$

The expressions in Eqs. 4.6 and 4.7 are obtained from nodal coordinate data, and can readily be implemented in a computer program.

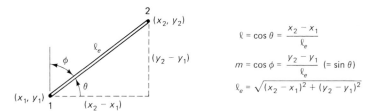

Figure 4.4   Direction cosines.

Element Stiffness Matrix

An important observation will now be made: *The truss element is a one-dimensional element when viewed in the local coordinate system.* This observation allows us to use previously developed results in Chapter 3 for one-dimensional elements. Consequently, from Eq. 3.26, the element stiffness matrix for a truss element in the local coordinate system is given by

$$\mathbf{k}' = \frac{E_e A_e}{\ell_e} \begin{bmatrix} 1 & -1 \\ -1 & 1 \end{bmatrix} \tag{4.8}$$

where A_e is the element cross-sectional area and E_e is Young's modulus. The problem at hand is to develop an expression for the element stiffness matrix in the global coordinate system. This is obtainable by considering the strain energy in the element. Specifically, the element strain energy in local coordinates is given by

$$U_e = \tfrac{1}{2}\mathbf{q}'^{\mathrm{T}}\mathbf{k}'\mathbf{q}' \tag{4.9}$$

Substituting for $\mathbf{q}' = \mathbf{L}\mathbf{q}$ into the above, we get

$$U_e = \tfrac{1}{2}\mathbf{q}^{\mathrm{T}}[\mathbf{L}^{\mathrm{T}}\mathbf{k}'\mathbf{L}]\mathbf{q} \tag{4.10}$$

The strain energy in global coordinates can be written as

$$U_e = \tfrac{1}{2}\mathbf{q}^{\mathrm{T}}\mathbf{k}\mathbf{q} \tag{4.11}$$

where **k** is the element stiffness matrix in global coordinates. From above, we obtain the element stiffness matrix in global coordinates as

$$\mathbf{k} = \mathbf{L}^T \mathbf{k}' \mathbf{L} \tag{4.12}$$

Substituting for **L** from Eq. 4.5 and for **k'** from Eq. 4.8, we get

$$\mathbf{k} = \frac{E_e A_e}{\ell_e} \begin{bmatrix} \ell^2 & \ell m & -\ell^2 & -\ell m \\ \ell m & m^2 & -\ell m & -m^2 \\ -\ell^2 & -\ell m & \ell^2 & \ell m \\ -\ell m & -m^2 & \ell m & m^2 \end{bmatrix} \tag{4.13}$$

The element stiffness matrices are assembled in the usual manner to obtain the structural stiffness matrix. This assembly is illustrated in Example 4.1. The computer logic for directly placing element stiffness matrices into global matrices for banded and skyline solutions is explained in Section 4.4.

The derivation of the result $\mathbf{k} = \mathbf{L}^T \mathbf{k}' \mathbf{L}$ above also follows from Galerkin's variational principle. The virtual work δW as a result of virtual displacement $\boldsymbol{\psi}'$ is

$$\delta W = \boldsymbol{\psi}'^T (\mathbf{k}' \mathbf{q}') \tag{4.14a}$$

Since $\boldsymbol{\psi}' = \mathbf{L}\boldsymbol{\psi}$ and $\mathbf{q}' = \mathbf{L}\mathbf{q}$, we have

$$\delta W = \boldsymbol{\psi}^T [\mathbf{L}^T \mathbf{k}' \mathbf{L}] \mathbf{q} \tag{4.14b}$$

$$= \boldsymbol{\psi}^T \mathbf{k} \mathbf{q}$$

Stress Calculations

Expressions for the element stresses can be obtained by noting that a truss element in local coordinates is a simple two-force member (Fig. 4.2). Thus, the stress σ in a truss element is given by

$$\sigma = E_e \epsilon \tag{4.15a}$$

Since the strain ϵ is the change in length per unit original length,

$$\sigma = E_e \frac{q_2' - q_1'}{\ell_e}$$

$$= \frac{E_e}{\ell_e} [-1 \quad 1] \begin{Bmatrix} q_1' \\ q_2' \end{Bmatrix} \tag{4.15b}$$

The above equation can be written in terms of the global displacements **q** using the transformation $\mathbf{q}' = \mathbf{L}\mathbf{q}$ as

$$\sigma = \frac{E_e}{\ell_e} [-1 \quad 1] \mathbf{L} \mathbf{q} \tag{4.15c}$$

Substituting for **L** from Eq. 4.5 yields

$$\sigma = \frac{E_e}{\ell_e}[-\ell \quad -m \quad \ell \quad m]\mathbf{q} \tag{4.16}$$

Once the displacements are determined by solving the finite element equations, the stresses can be recovered from Eq. 4.16 for each element. Note that a positive stress implies that the element is in tension and a negative stress implies compression.

Example 4.1

Consider the four-bar truss shown in Fig. E4.1a. It is given that $E = 29.5 \times 10^6$ psi and $A_e = 1$ in.^{2} for all elements.

(a) Determine the element stiffness matrix for each element.
(b) Assemble the structural stiffness matrix $\mathbf{K}$ for the entire truss.
(c) Using the elimination approach, solve for the nodal displacement.
(d) Recover the stresses in each element.
(e) Calculate the reaction forces.

Solution

(a) It is recommended that a *tabular* form be used for representing nodal coordinate data and element information as shown below. The nodal coordinate data is

Node	x	y
1	0	0
2	40	0
3	40	30
4	0	30

The element connectivity table is

Element	1	2
1	1	2
2	3	2
3	1	3
4	4	3

Note that the user has a choice in defining element connectivity. For example, the connectivity of element 2 can be defined as 2–3 instead of 3–2 as above. However, calculations of the direction cosines will be consistent with the adopted connectivity scheme. Using formulas in Eqs. 4.6 and 4.7, together with the nodal coordinate data and element connectivity information given above, we obtain the direction cosines table:

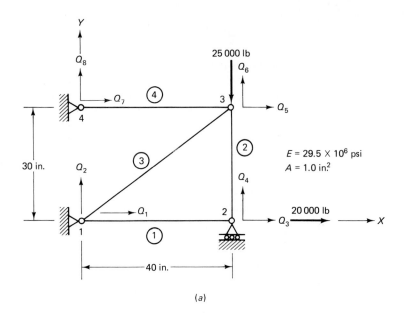

(a)

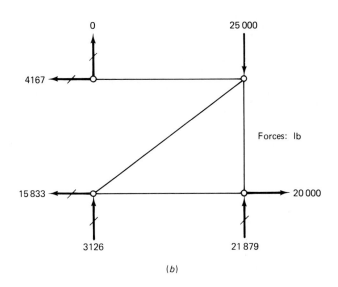

Forces: lb

(b)

Figure E4.1

Element	ℓ_e	ℓ	m
1	40	1	0
2	30	0	−1
3	50	0.8	0.6
4	40	1	0

For example, the direction cosines of elements 3 are obtained as $\ell = (x_3 - x_1)/\ell_e = (40 - 0)/50 = 0.8$, and $m = (y_3 - y_1)/\ell_e = (30 - 0)/50 = 0.6$. Now, using Eq. 4.13, the element stiffness matrices for element 1 can be written as

$$
\begin{array}{cccc}
1 & 2 & 3 & 4 \;\leftarrow_{\downarrow}\; \text{Global dof}
\end{array}
$$

$$
\mathbf{k}^1 = \frac{29.5 \times 10^6}{40}
\begin{bmatrix}
1 & 0 & -1 & 0 \\
0 & 0 & 0 & 0 \\
-1 & 0 & 1 & 0 \\
0 & 0 & 0 & 0
\end{bmatrix}
\begin{matrix} 1 \\ 2 \\ 3 \\ 4 \end{matrix}
$$

The global dof's associated with element 1, which is connected between nodes 1 and 2, are indicated in $\mathbf{k}^1$ above. These global dof's are shown in Fig. E4.1a, and assist in assembling the various element stiffness matrices.

The element stiffness matrices of elements 2, 3, and 4 are as follows:

$$
\begin{array}{cccc}
5 & 6 & 3 & 4
\end{array}
$$

$$
\mathbf{k}^2 = \frac{29.5 \times 10^6}{30}
\begin{bmatrix}
0 & 0 & 0 & 0 \\
0 & 1 & 0 & -1 \\
0 & 0 & 0 & 0 \\
0 & -1 & 0 & 1
\end{bmatrix}
\begin{matrix} 5 \\ 6 \\ 3 \\ 4 \end{matrix}
$$

$$
\begin{array}{cccc}
1 & 2 & 5 & 6
\end{array}
$$

$$
\mathbf{k}^3 = \frac{29.5 \times 10^6}{50}
\begin{bmatrix}
.64 & .48 & -.64 & -.48 \\
.48 & .36 & -.48 & -.36 \\
-.64 & -.48 & .64 & .48 \\
-.48 & -.36 & .48 & .36
\end{bmatrix}
\begin{matrix} 1 \\ 2 \\ 5 \\ 6 \end{matrix}
$$

$$
\begin{array}{cccc}
7 & 8 & 5 & 6
\end{array}
$$

$$
\mathbf{k}^4 = \frac{29.5 \times 10^6}{40}
\begin{bmatrix}
1 & 0 & -1 & 0 \\
0 & 0 & 0 & 0 \\
-1 & 0 & 1 & 0 \\
0 & 0 & 0 & 0
\end{bmatrix}
\begin{matrix} 7 \\ 8 \\ 5 \\ 6 \end{matrix}
$$

(b) The structural stiffness matrix $\mathbf{K}$ is now assembled from the element stiffness matrices. By adding the element stiffness contributions, noting the element connectivity, we get

$$
\mathbf{K} = \frac{29.5 \times 10^6}{600}
\begin{array}{c}
\begin{array}{cccccccc}
\;\;1 & \;\;2 & \;\;3 & \;\;4 & \;\;5 & \;\;6 & \;\;7 & \;8
\end{array} \\
\begin{bmatrix}
22.68 & 5.76 & -15.0 & 0 & -7.68 & -5.76 & 0 & 0 \\
5.76 & 4.32 & 0 & 0 & -5.76 & -4.32 & 0 & 0 \\
-15.0 & 0 & 15.0 & 0 & 0 & 0 & 0 & 0 \\
0 & 0 & 0 & 20.0 & 0 & -20.0 & 0 & 0 \\
-7.68 & -5.76 & 0 & 0 & 22.68 & 5.76 & -15.0 & 0 \\
-5.76 & -4.32 & 0 & -20.0 & 5.76 & 24.32 & 0 & 0 \\
0 & 0 & 0 & 0 & -15.0 & 0 & 15.0 & 0 \\
0 & 0 & 0 & 0 & 0 & 0 & 0 & 0
\end{bmatrix}
\end{array}
\begin{matrix} 1 \\ 2 \\ 3 \\ 4 \\ 5 \\ 6 \\ 7 \\ 8 \end{matrix}
$$

(c) The structural stiffness matrix **K** given above needs to be modified to account for the boundary conditions. The elimination approach discussed in Chapter 3 will be used here. The rows and columns corresponding to dof's 1, 2, 4, 7, and 8, which correspond to fixed supports, are deleted from the **K** matrix above. The reduced finite element equations are given as

$$\frac{29.5 \times 10^6}{600} \begin{bmatrix} 15 & 0 & 0 \\ 0 & 22.68 & 5.76 \\ 0 & 5.76 & 24.32 \end{bmatrix} \begin{Bmatrix} Q_3 \\ Q_5 \\ Q_6 \end{Bmatrix} = \begin{Bmatrix} 20\,000 \\ 0 \\ -25\,000 \end{Bmatrix}$$

Solution of the above equations yields the displacements

$$\begin{Bmatrix} Q_3 \\ Q_5 \\ Q_6 \end{Bmatrix} = \begin{Bmatrix} 26.12 \times 10^{-3} \\ 5.65 \times 10^{-3} \\ -22.25 \times 10^{-3} \end{Bmatrix} \text{ in.}$$

The nodal displacement vector for the entire structure can therefore be written as

$$\mathbf{Q} = [0, 0, 27.12 \times 10^{-3}, 0, 5.65 \times 10^{-3}, -22.25 \times 10^{-3}, 0, 0]^T \text{ in.}$$

(d) The stress in each element can now be determined from Eq. 4.16, as shown below.

The connectivity of element 1 is $1 - 2$. Consequently, the nodal displacement vector for element 1 is given by $\mathbf{q} = [0, 0, 27.12 \times 10^{-3}, 0]^T$, and Eq. 4.16 yields

$$\sigma_1 = \frac{29.5 \times 10^6}{40} [-1 \quad 0 \quad 1 \quad 0] \begin{Bmatrix} 0 \\ 0 \\ 27.12 \times 10^{-3} \\ 0 \end{Bmatrix}$$

$$= 20\,000.0 \text{ psi}$$

The stress in member 2 is given by

$$\sigma_2 = \frac{29.5 \times 10^6}{30} [0 \quad 1 \quad 0 \quad -1] \begin{Bmatrix} 5.65 \times 10^{-3} \\ -22.25 \times 10^{-3} \\ 27.12 \times 10^{-3} \\ 0 \end{Bmatrix}$$

$$= -21\,880.0 \text{ psi}$$

Following similar steps, we get

$$\sigma_3 = 5208.0 \text{ psi}$$

$$\sigma_4 = 4167.0 \text{ psi}$$

(e) The final step is to determine the support reactions. We need to determine the reaction forces along dof's 1, 2, 4, 7, and 8, which correspond to fixed supports. These are obtained by substituting for **Q** into the original finite element equation **R** = **KQ** − **F**. In this substitution only those rows of **K** corresponding to the support dof's are needed, and **F** = **O** for these dof's. Thus, we have

$$
\begin{Bmatrix} R_1 \\ R_2 \\ R_4 \\ R_7 \\ R_8 \end{Bmatrix} = \frac{29.5 \times 10^6}{600}
\begin{bmatrix}
22.68 & 5.76 & -15.0 & 0 & -7.68 & -5.76 & 0 & 0 \\
5.76 & 4.32 & 0 & 0 & -5.76 & -4.32 & 0 & 0 \\
0 & 0 & 0 & 20.0 & 0 & -20.0 & 0 & 0 \\
0 & 0 & 0 & 0 & -15.0 & 0 & 15.0 & 0 \\
0 & 0 & 0 & 0 & 0 & 0 & 0 & 0
\end{bmatrix}
\begin{Bmatrix}
0 \\ 0 \\ 27.12 \times 10^{-3} \\ 0 \\ 5.65 \times 10^{-3} \\ -22.25 \times 10^{-3} \\ 0 \\ 0
\end{Bmatrix}
$$

which results in

$$
\begin{Bmatrix} R_1 \\ R_2 \\ R_4 \\ R_7 \\ R_8 \end{Bmatrix} = \begin{Bmatrix} -15\,833.0 \\ 3\,126.0 \\ 21\,879.0 \\ -4\,167.0 \\ 0 \end{Bmatrix} \quad \text{lb}
$$

A free body diagram of the truss with reaction forces and applied loads is shown in Fig. E4.1b. ∎

Temperature Effects

The thermal stress problem is considered here. Since a truss element is simply a one-dimensional element when viewed in the local coordinate system, the element temperature load in the local coordinate system is given by (see Eq. 3.103b)

$$
\boldsymbol{\Theta}' = E_e A_e \epsilon_0 \begin{Bmatrix} -1 \\ 1 \end{Bmatrix} \tag{4.17}
$$

where the initial strain ϵ_0 associated with a temperature change is given by

$$
\epsilon_0 = \alpha\,\Delta T \tag{4.18}
$$

where α is the coefficient of thermal expansion, and ΔT is the average change in temperature in the element. It may be noted that the initial strain ϵ_0 can also be induced by forcing members into place that are either too long or too short, due to fabrication errors.

We will now express the load vector in Eq. 4.17 in the global coordinate system. Since the potential energy associated with this load is the same in magnitude whether measured in the local or global coordinate systems, we have

$$
\mathbf{q}'^{\mathrm{T}}\boldsymbol{\Theta}' = \mathbf{q}^{\mathrm{T}}\boldsymbol{\Theta} \tag{4.19}
$$

where $\boldsymbol{\Theta}$ is the load vector in the global coordinate system. Substituting for $\mathbf{q}' = \mathbf{Lq}$ into the above, we get

$$
\mathbf{q}^{\mathrm{T}}\mathbf{L}^{\mathrm{T}}\boldsymbol{\Theta}' = \mathbf{q}^{\mathrm{T}}\boldsymbol{\Theta} \tag{4.20}
$$

Comparing the left and right sides of the above equation, we obtain

$$\mathbf{\Theta} = \mathbf{L}^T \mathbf{\Theta}' \tag{4.21}$$

Substituting for $\mathbf{L}$ from Eq. 4.5, we can write down the expression for the element temperature load as

$$\mathbf{\Theta}^e = E_e A_e \epsilon_0 \begin{Bmatrix} -\ell \\ -m \\ \ell \\ m \end{Bmatrix} \tag{4.22}$$

The temperature loads, along with other externally applied loads, are assembled in the usual manner to obtain the nodal load vector $\mathbf{F}$. Once the displacements are obtained by solving the finite element equations, the stress in each truss element is obtained from (see Eq. 3.100)

$$\sigma = E(\epsilon - \epsilon_0) \tag{4.23}$$

This equation for the element stress can be simplified by using Eq. 4.16 and noting that $\epsilon_0 = \alpha \, \Delta T$, to obtain

$$\sigma = \frac{E_e}{\ell_e}[-\ell \quad -m \quad \ell \quad m]\mathbf{q} - E_e \alpha \, \Delta T \tag{4.24}$$

Example 4.2

The four-bar truss of Example 4.1 is considered here, but the loading is different. Take $E = 29.5 \times 10^6$ psi, and $\alpha = 1/150\,000$ per °F.

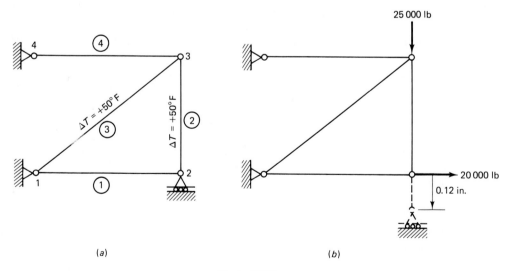

(a)   (b)

Figure E4.2

(a) There is an increase in temperature of 50 °F in bars 2 and 3 only (Fig. E4.2a). There are no other loads on the structure. Determine the nodal displacements and element stresses as a result of this temperature increase. Use the elimination approach.

(b) A support settlement effect is considered here. Node 2 settles by 0.12 in. vertically down, and in addition, two point loads are applied on the structure (Fig. E4.2b). Write down (without solving) the equilibrium equations $\mathbf{KQ} = \mathbf{F}$, where $\mathbf{K}$ and $\mathbf{F}$ are the modified structural stiffness matrix and load vector, respectively. Use the penalty approach.

(c) Use the program TRUSS to obtain the solution to part (b) above.

Solution

(a) The stiffness matrix for the truss structure has already been developed in Example 4.1. Only the load vector needs to be assembled due to the temperature increase. Using Eq. 4.22, the temperature loads as a result of temperature increases in elements 2 and 3 are, respectively,

$$\downarrow \text{Global dof}$$

$$\Theta^2 = \frac{29.5 \times 10^6 \times 50}{150\,000} \begin{Bmatrix} 0 \\ 1 \\ 0 \\ -1 \end{Bmatrix} \begin{matrix} 5 \\ 6 \\ 3 \\ 4 \end{matrix}$$

and

$$\Theta^3 = \frac{29.5 \times 10^6 \times 50}{150\,000} \begin{Bmatrix} -0.8 \\ -0.6 \\ 0.8 \\ 0.6 \end{Bmatrix} \begin{matrix} 1 \\ 2 \\ 5 \\ 6 \end{matrix}$$

The Θ^2 and Θ^3 vectors above contribute to the global load vector $\mathbf{F}$. Using the elimination approach, we can delete all rows and columns corresponding to support dof's in $\mathbf{K}$ and $\mathbf{F}$. The resulting finite element equations are

$$\frac{29.5 \times 10^6}{600} \begin{bmatrix} 15.0 & 0 & 0 \\ 0 & 22.68 & 5.76 \\ 0 & 5.76 & 24.32 \end{bmatrix} \begin{Bmatrix} Q_3 \\ Q_5 \\ Q_6 \end{Bmatrix} = \begin{Bmatrix} 0 \\ 7\,866.7 \\ 15\,733.3 \end{Bmatrix}$$

which yield

$$\begin{Bmatrix} Q_3 \\ Q_5 \\ Q_6 \end{Bmatrix} = \begin{Bmatrix} 0 \\ 0.003951 \\ 0.01222 \end{Bmatrix} \text{in.}$$

The element stresses can now be obtained from Eq. 4.24. For example, the stress in element 2 is given as

$$\sigma_2 = \frac{29.5 \times 10^6}{30}[0 \quad 1 \quad 0 \quad -1]\begin{Bmatrix} 0.003951 \\ 0.01222 \\ 0 \\ 0 \end{Bmatrix} - \frac{29.5 \times 10^6 \times 50}{150\,000}$$

$$= -8\,631.7 \text{ psi}$$

The complete stress solution is

$$\begin{Bmatrix} \sigma_1 \\ \sigma_2 \\ \sigma_3 \\ \sigma_4 \end{Bmatrix} = \begin{Bmatrix} 0 \\ 2183 \\ -3643 \\ 2914 \end{Bmatrix} \text{psi}$$

(b) Support 2 settles by 0.12 in. vertically down, and two concentrated forces are applied (Fig. E4.2b). In the penalty approach for handling boundary conditions (Chapter 3), recall that a large spring constant C is added to the diagonal elements in the structural stiffness matrix at those dof's where the displacements are specified. Typically C may be chosen 10^4 times to one of the diagonal elements of the unmodified stiffness matrix (see Eq. 3.80). Further, a force Ca is added to the force vector, where a is the specified displacement. In this example, for dof 4, $a = -0.12$ in., and consequently, a force equal to $-0.12C$ gets added to the fourth location in the force vector. Consequently, the modified finite element equations are given by

$$\frac{29.5 \times 10^6}{600}\begin{bmatrix} 22.68+C & 5.76 & -15.0 & 0 & -7.68 & -5.76 & 0 & 0 \\ & 4.32+C & 0 & 0 & -5.76 & -4.32 & 0 & 0 \\ & & 15.0 & 0 & 0 & 0 & 0 & 0 \\ & & & 20.0+C & 0 & -20.0 & 0 & 0 \\ & & & & 22.68 & 5.76 & -15.0 & 0 \\ & & & & & 24.32 & 0 & 0 \\ \text{Symmetric} & & & & & & 15.0+C & 0 \\ & & & & & & & C \end{bmatrix}\begin{Bmatrix} Q_1 \\ Q_2 \\ Q_3 \\ Q_4 \\ Q_5 \\ Q_6 \\ Q_7 \\ Q_8 \end{Bmatrix} = \begin{Bmatrix} 0 \\ 0 \\ 20\,000 \\ -0.12C \\ 0 \\ -25\,000.0 \\ 0 \\ 0 \end{Bmatrix}$$

(c) Obviously, the above equations are too large for hand calculations. In the program TRUSS, that is provided, the above equations are automatically generated and solved from the user's input data. The output from the program is given below:

$$\begin{Bmatrix} Q_3 \\ Q_4 \\ Q_5 \\ Q_6 \end{Bmatrix} = \begin{Bmatrix} 0.0271200 \\ -0.1200145 \\ 0.0323242 \\ -0.1272606 \end{Bmatrix} \text{in.}$$

and

$$\left\{\begin{array}{c} \sigma_1 \\ \sigma_2 \\ \sigma_3 \\ \sigma_4 \end{array}\right\} = \left\{\begin{array}{r} 20\,000.0 \\ -7\,125.3 \\ -29\,791.7 \\ 23\,833.3 \end{array}\right\} \text{psi}$$ ■

4.3 THREE-DIMENSIONAL TRUSSES

The 3-D truss element can be treated as a straightforward generalization of the 2-D truss element discussed above. The local and global coordinate systems for a 3-D truss element are shown in Fig. 4.5. Note that the local coordinate system is again the x' axis running along the element, since a truss element is simply a two-force member. Consequently, the nodal displacement vector in local coordinates is

$$\mathbf{q}' = [q_1', q_2']^T \tag{4.25}$$

The nodal displacement vector in global coordinates is now (Fig. 4.5b)

$$\mathbf{q} = [q_1, q_2, q_3, q_4, q_5, q_6]^T \tag{4.26}$$

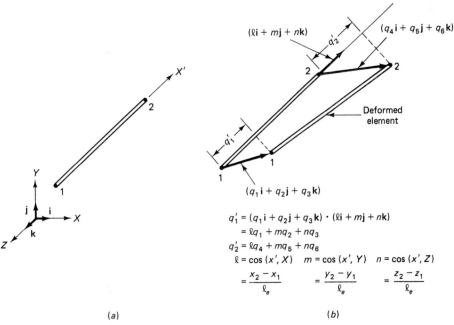

(a) (b)

Figure 4.5 A three-dimensional truss element in local and gobal coordinate systems.

Referring to Fig. 4.5, we find that the transformation between local and global coordinates is

$$\mathbf{q'} = \mathbf{Lq} \tag{4.27}$$

where the transformation matrix $\mathbf{L}$ is given by

$$\mathbf{L} = \begin{bmatrix} \ell & m & n & 0 & 0 & 0 \\ 0 & 0 & 0 & \ell & m & n \end{bmatrix} \tag{4.28}$$

where ℓ, m, and n are the direction cosines of the local x' axis with respect to the global x, y, and z axes, respectively. The element stiffness matrix in global coordinates is given by Eq. 4.12, which yields

$$\mathbf{k} = \frac{E_e A_e}{\ell_e} \begin{bmatrix} \ell^2 & \ell m & \ell n & -\ell^2 & -\ell m & -\ell n \\ \ell m & m^2 & mn & -\ell m & -m^2 & -mn \\ \ell n & mn & n^2 & -\ell n & -mn & -n^2 \\ -\ell^2 & -\ell m & -\ell n & \ell^2 & \ell m & \ell n \\ -\ell m & -m^2 & -mn & \ell m & m^2 & mn \\ -\ell n & -mn & -n^2 & \ell n & mn & n^2 \end{bmatrix} \tag{4.29}$$

The formulas for calculating ℓ, m, and n are

$$\ell = \frac{x_2 - x_1}{\ell_e} \qquad m = \frac{y_2 - y_1}{\ell_e} \qquad n = \frac{z_2 - z_1}{\ell_e} \tag{4.30}$$

where the length ℓ_e of the element is given by

$$\ell_e = \sqrt{(x_2 - x_1)^2 + (y_2 - y_1)^2 + (z_2 - z_1)^2} \tag{4.31}$$

Generalizations of the element stress and element temperature load expressions are left as an exercise.

4.4 ASSEMBLY OF GLOBAL STIFFNESS MATRIX FOR THE BANDED AND SKYLINE SOLUTIONS

The solution of the finite element equations should take advantage of symmetry and sparsity of the global stiffness matrix. Two methods, the banded approach and the skyline approach, are discussed in Chapter 2. In the banded approach, the elements of each element stiffness matrix $\mathbf{k}^e$ are directly placed in a banded matrix $\mathbf{S}$. In the skyline approach the elements of $\mathbf{k}^e$ are placed in a vector form with certain identification pointers. The bookkeeping aspects of this assembly procedure for banded and skyline solution are discussed below.

Assembly for Banded Solution

The assembly of elements of $\mathbf{k}^e$ into a banded global stiffness matrix $\mathbf{S}$ is now discussed for a two-dimensional truss element. Consider an element e whose connectivity is indicated below

Element	1	2	← Local Node Nos.
e	i	j	← Global Node Nos.

The element stiffness with its associated degrees of freedom are

$$\mathbf{k}^e = \begin{array}{cccc} 2i-1 & 2i & 2j-1 & 2j \end{array} \leftarrow \text{Global dof's} \\ \begin{bmatrix} k_{11} & k_{12} & k_{13} & k_{14} \\ & k_{22} & k_{23} & k_{24} \\ & & k_{33} & k_{34} \\ \text{Symmetric} & & & k_{44} \end{bmatrix} \begin{array}{c} 2i-1 \\ 2i \\ 2j-1 \\ 2j \end{array} \qquad (4.32)$$

The principal diagonal of $\mathbf{k}^e$ is placed in the first column of $\mathbf{S}$, the next-to-principal diagonal is placed in the second column, and so on. Thus, the correspondence between elements in $\mathbf{k}^e$ and $\mathbf{S}$ is given by (see Eq. 2.39)

$$k^e_{\alpha,\beta} \rightarrow S_{p,q-p+1} \qquad (4.33)$$

where α and β are the local dof's taking on values 1, 2, 3, and 4, while p and q are global dof's taking on values of $2i - 1$, $2i$, $2j - 1$, $2j$. For instance,

$$k^e_{1,3} \rightarrow S_{2i-1,2(j-i)+1}$$

and $\qquad (4.34)$

$$k^e_{4,4} \rightarrow S_{2j,1}$$

The assembly shown above is done only for elements in the upper triangle owing to symmetry. Thus, Eq. 4.33 is valid only for $q \geq p$. We can now follow the assembly steps given in program TRUSS.

A formula for the half-bandwidth, NBW, in 2-D truss structures can be readily derived. Consider a truss element e connected to, say, nodes 4 and 6. The degrees of freedom for the element are 7, 8, 11, and 12. Thus, the entries in the global stiffness matrix for this element will be

$$
\begin{array}{c}
\begin{array}{ccccccccc} 1 & 2 & \cdots & 7 & 8 & \cdots & 11 & 12 & \cdots & N \end{array} \\
\left[
\begin{array}{c}
\\
\hspace{2em} |\!\longleftarrow\!\!-\!\!-\!\!-\ m\ -\!\!-\!\!-\!\longrightarrow\! | \\
\\
\mathrm{X}\ \mathrm{X}\ .\ \ .\ \ \mathrm{X}\ \mathrm{X} \\
\mathrm{X}\ .\ \ .\ \ \mathrm{X}\ \mathrm{X} \\
\\
.\ \ .\ \ .\ \ . \\
\\
\mathrm{X}\ \mathrm{X} \\
\text{Symmetric} \hspace{6em} \mathrm{X} \\
\\
\\
\end{array}
\right]
\begin{array}{c}
1 \\ 2 \\ \cdot \\ \cdot \\ 7 \\ 8 \\ \cdot \\ \cdot \\ 11 \\ 12 \\ \cdot \\ \cdot \\ N
\end{array}
\end{array}
\qquad (4.35)
$$

We see that the span m of nonzero entries is equal to 6, which also follows from the connecting node numbers: $m = 2[6 - 4 + 1]$. In general the span associated with an element e connecting nodes i and j is

$$ m_e = 2[|i - j| + 1] \qquad (4.36) $$

Thus, the maximum span or half-bandwidth is

$$ \mathrm{NBW} = \max_{1 \le e < \mathrm{NE}} m_e \qquad (4.37) $$

In the banded approach, we see that differences in node numbers connecting an element should be kept to a minimum for computational efficiency.

Skyline assembly.   As discussed in Chapter 2, the first step involves the evaluation of the skyline height or the column height for each diagonal location. Consider the element e with the end nodes i and j shown in Fig. 4.6. Without loss of generality let i be the smaller node number; that is $i < j$. Then, starting with a vector of identifiers, **ID**, we look at the four degrees of freedom $2i - 1$, $2i$, $2j - 1$, and $2j$. At the location corresponding to one of these four dof's represented by I, the previous value is replaced by the larger of the two numbers $\mathrm{ID}(I)$ and $I - (2i - 1) + 1$. This is precisely represented in the table given in Fig. 4.6. The process is repeated over all the elements. At this stage all the skyline heights have been determined, and placed in the vector **ID**. Then, starting from location $I = 2$, replacing the location at I by the sum $\mathrm{ID}(I) + \mathrm{ID}(I - 1)$ gives the pointer numbers as discussed in Chapter 2.

The next step involves assembling the element stiffness values into the column vector **A**. The correspondence of the global locations of the square stiffness matrix coming from an element shown in Fig. 4.6 are clearly presented in Fig.4.7, using the diagonal pointers discussed above. Details presented above have been imple-

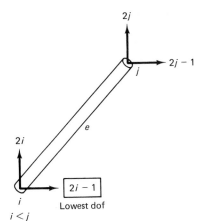

Element e	
Location No. I	Skyline Height ID(I)
$2i - 1$	max (1, OLD)
$2i$	max (2, OLD)
$2j - 1$	max $(2j - 2i + 1$, OLD)
$2j$	max $(2j - 2i + 2$, OLD)

max $(X,$ OLD$) \equiv$ REPLACE by X if $X >$ OLD
(start value of OLD = 0)

Figure 4.6 Skyline heights.

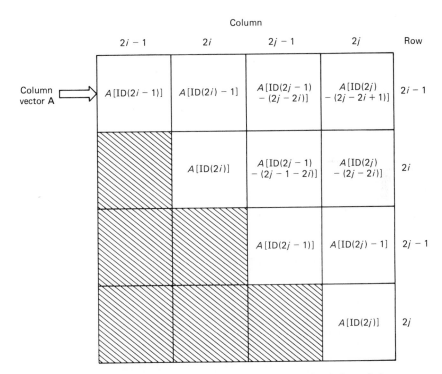

Figure 4.7 Stiffness locations in column vector form for skyline solution.

mented in program TRUSSKY. Other programs provided may be similarly modified for skyline solution instead of banded solution.

Example 4.3

The solution of Example 4.1 using program TRUSS is presented below.

```
Number of Elements =? 4
Number of Nodes =? 4
Number of Constrained DOF =? 5
Number of Component Loads =? 2
Number of Materials =? 1
Number of Areas of Cross Section =? 1
Number of Elements with Temperature Change =? 0
ELEM#, NODE1, NODE2, MATERIAL#, AREA#
? 1,1,2,1,1
? 2,3,2,1,1
? 3,3,1,1,1
? 4,4,3,1,1
Area#, Area
? 1,1.0
THE BANDWIDTH IS 6
Node#, X-Coord, Y-Coord
? 1,0,0
? 2,40,0
? 3,40,30
? 4,0,30
Material#, E, Alpha
? 1,29.5E6,0
DOF#, Displacement
? 1,0
? 2,0
? 4,0
? 7,0
? 8,0
DOF#, Applied Load
? 3,20000
? 6,-25000
```

```
NODE NUMBER, DISPLACEMENTS
 1
 1.419857E-06
-2.802779E-07
 2
 2.712006E-02
-1.961674E-06
 3
 5.650775E-03
-2.224734E-02
```

```
      4
      3.737038E-07
      0
      ELEMENT NUMBER, STRESS
      1              20000
      2            -21874.63
      3            -5208.964
      4             4167.171
```

```
1000 REM ******         TRUSS        ******
1010 REM **** Two Dimensional Truss Analysis ****
1020 CLS
1030 PRINT "======================================="
1040 PRINT "       Program written by            "
1050 PRINT "  T.R.Chandrupatla and A.D.Belegundu   "
1060 PRINT "======================================="
1070 INPUT "Number of Elements ="; NE
1080 INPUT "Number of Nodes ="; NN
1090 INPUT "Number of Constrained DOF ="; ND
1100 INPUT "Number of Component Loads ="; NL
1110 INPUT "Number of Materials ="; NM
1120 INPUT "Number of Areas of Cross Section ="; NA
1130 INPUT "Number of Elements with Temperature Change ="; NTEL
1140 REM TOTAL DOF IS "NQ"
1150 NQ = 2 * NN
1160 DIM X(NN, 2), NOC(NE, 4), PM(NM, 2), AREA(NA), NU(ND), U(ND)
1170 DIM TEMP(NTEL), NSET(NTEL),STRESSP(NE),REACT(ND),ILD(NL),AL(NL)
1180 REM BANDWIDTH NBW FROM CONNECTIVITY "NOC"
1190 REM STRESSP,REACT,ILD,AL HAVE BEEN DIMENSIONED FOR OUTPUT PURPOSES
1200 NBW = 0
1210 PRINT "ELEM#, NODE1, NODE2, MATERIAL#, AREA#"
1220 FOR I = 1 TO NE
1230 INPUT N, NOC(N, 1), NOC(N, 2), NOC(N, 3), NOC(N, 4)
1240 C = 2 * (ABS(NOC(N, 2) - NOC(N, 1))) + 2
1250 IF NBW < C THEN NBW = C
1260 NEXT I
1270 IF NTEL = 0 GOTO 1310
1280 PRINT "Element#, Temperature Change"
1290 FOR I = 1 TO NTEL
1300 INPUT NSET(I), TEMP(I): NEXT I
1310 PRINT "Area#, Area"
1320 FOR I = 1 TO NA
1330 INPUT N, AREA(N): NEXT I
1340 PRINT "THE BANDWIDTH IS"; NBW
1350 REM *** INITIALIZATION ***
1360 DIM S(NQ, NBW), F(NQ), SE(4, 4)
1370 FOR I = 1 TO NQ
1380 F(NQ) = 0
1390 FOR J = 1 TO NBW
1400 S(I, J) = 0: NEXT J: NEXT I
1410 PRINT "Node#, X-Coord, Y-Coord"
1420 FOR I = 1 TO NN
1430 INPUT N, X(N, 1), X(N, 2): NEXT I
1440 PRINT "Material#, E, Alpha"
1450 FOR I = 1 TO NM
```

```
1460 INPUT N, PM(N, 1), PM(N, 2): NEXT I
1470 PRINT "DOF#, Displacement"
1480 FOR I = 1 TO ND
1490 INPUT NU(I), U(I): NEXT I
1500 PRINT "DOF#, Applied Load"
1510 FOR I = 1 TO NL
1520 INPUT N, F(N)
1530 ILD(I)=N : AL(I)=F(N) : NEXT I
1540 REM *** GLOBAL STIFFNESS MATRIX ***
1550 FOR N = 1 TO NE
1560 '====== ELEMENT STIFFNESS ======
1570 I1 = NOC(N, 1): I2 = NOC(N, 2)
1580 I3 = NOC(N, 3): I4 = NOC(N, 4)
1590 X21 = X(I2, 1) - X(I1, 1)
1600 Y21 = X(I2, 2) - X(I1, 2)
1610 EL = SQR(X21 * X21 + Y21 * Y21)
1620 EAL = PM(I3, 1) * AREA(I4) / EL
1630 CS = X21 / EL: SN = Y21 / EL
1640 SE(1, 1) = CS * CS * EAL
1650 SE(1, 2) = CS * SN * EAL: SE(2, 1) = SE(1, 2)
1660 SE(1, 3) = -CS * CS * EAL: SE(3, 1) = SE(1, 3)
1670 SE(1, 4) = -CS * SN * EAL: SE(4, 1) = SE(1, 4)
1680 SE(2, 2) = SN * SN * EAL
1690 SE(2, 3) = -CS * SN * EAL: SE(3, 2) = SE(2, 3)
1700 SE(2, 4) = -SN * SN * EAL: SE(4, 2) = SE(2, 4)
1710 SE(3, 3) = CS * CS * EAL
1720 SE(3, 4) = CS * SN * EAL: SE(4, 3) = SE(3, 4)
1730 SE(4, 4) = SN * SN * EAL
1740 PRINT "++++Placing Stiffness in Global Locations++++"
1750 FOR II = 1 TO 2
1760 NRT = 2 * (NOC(N, II) - 1)
1770 FOR IT = 1 TO 2
1780 NR = NRT + IT
1790 I = 2 * (II - 1) + IT
1800 FOR JJ = 1 TO 2
1810 NCT = 2 * (NOC(N, JJ) - 1)
1820 FOR JT = 1 TO 2
1830 J = 2 * (JJ - 1) + JT
1840 NC = NCT + JT - NR + 1
1850 IF NC <= 0 GOTO 1870
1860 S(NR, NC) = S(NR, NC) + SE(I, J)
1870 NEXT JT: NEXT JJ: NEXT IT: NEXT II
1880 FOR IJT = 1 TO NTEL
1890 IF I = NSET(IJT) GOTO 1910
1900 GOTO 1940
1910 EE0 = PM(I3, 2) * TEMP(IJT) * PM(I3, 1) * AREA(I4)
1920 F(2 * I1 - 1) = F(2 * I1 - 1) - EE0 * CS: F(2 * I1) = F(2 * I1) - EE0 * SN
1930 F(2 * I2 - 1) = F(2 * I2 - 1) + EE0 * CS: F(2 * I2) = F(2 * I2) + EE0 * SN
1940 NEXT IJT
1950 NEXT N
1960 REM *** MODIFY FOR BOUNDARY CONDITIONS ***
1970 CNST = S(1, 1) * 10000
1980 FOR I = 1 TO ND
1990 N = NU(I)
2000 S(N, 1) = S(N, 1) + CNST
2010 F(N) = F(N) + CNST * U(I): NEXT I
2020 REM *** EQUATION SOLVING ***
2030 GOSUB 5010
```

```
2040 PRINT "Node#    X-Displ      Y-Displ"
2050 FOR I = 1 TO NN
2060 PRINT USING " ###"; I;
2070 PRINT USING "    ##.###^^^^"; F(2 * I - 1); F(2 * I)
2080 NEXT I
2090 REM *** STRESS CALCULATION ***
2100 PRINT "Elem#    Stress"
2110 FOR I = 1 TO NE
2120 I1 = NOC(I, 1): I2 = NOC(I, 2)
2130 I3 = NOC(I, 3)
2140 X21 = X(I2, 1) - X(I1, 1): Y21 = X(I2, 2) - X(I1, 2)
2150 EL = SQR(X21 * X21 + Y21 * Y21)
2160 CS = X21 / EL: SN = Y21 / EL
2170 J2 = 2 * I1: J1 = J2 - 1: K2 = 2 * I2: K1 = K2 - 1
2180 DT = (F(K1) - F(J1)) * CS + (F(K2) - F(J2)) * SN
2190 STRESS = DT * PM(I3, 1) / EL
2200 FOR IJT = 1 TO NTEL
2210 IF I = NSET(IJT) GOTO 2230
2220 GOTO 2240
2230 STRESS = STRESS - PM(I3, 1) * PM(I3, 2) * TEMP(IJT)
2240 NEXT IJT
2250 PRINT USING " ###"; I; : PRINT USING "    ##.###^^^^"; STRESS
2260 STRESSP(I)=STRESS
2270 NEXT I
2280 REM *** REACTION CALCULATION ***
2290 PRINT "DOF#", "REACTION"
2300 FOR I=1 TO ND
2310 N=NU(I)
2320 REACT(I)=CNST * (U(I) - F(N))
2330 PRINT N, REACT(I): NEXT I
2340 INPUT "DO YOU WISH TO SAVE THE OUTPUT (Y/N) ";ANS$
2350 IF ANS$ <> "y" AND ANS$ <> "Y" GOTO 2690
2360 INPUT "GIVE NAME OF OUTPUT FILE =";FILENAME$
2370 OPEN FILENAME$ FOR OUTPUT AS #2
2380 PRINT #2,"NE,NN,ND,NL,NM,NA,NTEL,NBW ="
2390 PRINT #2,NE; NN; ND; NL; NM; NA; NTEL; NBW
2400 PRINT #2,"ELEM#, NODE1, NODE2, MATERIAL#, AREA#"
2410 FOR I = 1 TO NE
2420 PRINT #2,I,NOC(I,1),NOC(I,2),NOC(I,3),NOC(I,4) : NEXT I
2430 IF NTEL = 0 GOTO 2470
2440 PRINT #2,"ELEM#, TEMPERATURE CHANGE"
2450 FOR I = 1 TO NTEL
2460 PRINT #2,NSET(I),TEMP(I) : NEXT I
2470 PRINT #2, "AREA#, AREA"
2480 FOR I=1 TO NA
2490 PRINT #2,I,AREA(I) : NEXT I
2500 PRINT #2, "NODE#, X-COORD., Y-COORD."
2510 FOR I=1 TO NN
2520 PRINT #2, I,X(I,1),X(I,2) : NEXT I
2530 PRINT #2, "DOF#, SPECIFIED DISPLACEMENT"
2540 FOR I=1 TO ND
2550 PRINT #2, NU(I),U(I) : NEXT I
2560 PRINT #2,"DOF#,APPLIED LOAD"
2570 FOR I=1 TO NL
2580 PRINT #2,ILD(I),AL(I) : NEXT I
2590 PRINT #2,"NODE#, X-DISPLACEMENT, Y-DISPLACEMENT"
2600 FOR I=1 TO NN
2610 PRINT #2, I,F(2*I-1),F(2*I) : NEXT I
```

```
2620 PRINT #2, "ELEMENT#, STRESS"
2630 FOR I=1 TO NE
2640 PRINT #2,I,STRESSP(I) : NEXT I
2650 PRINT #2," DOF#, REACTION FORCE"
2660 FOR I = 1 TO ND
2670 PRINT #2,NU(I),REACT(I) : NEXT I
2680 CLOSE #2
2690 END
5000 '======= EQUATION SOLVING  ========
5010 N1 = NQ - 1
5020 REM  *** FORWARD ELIMINATION ***
5030 FOR K = 1 TO N1
5040 NK = NQ - K + 1
5050 IF NK > NBW THEN NK = NBW
5060 FOR I = 2 TO NK
5070 C1 = S(K, I) / S(K, 1)
5080 I1 = K + I - 1
5090 FOR J = I TO NK
5100 J1 = J - I + 1
5110 S(I1, J1) = S(I1, J1) - C1 * S(K, J): NEXT J
5120 F(I1) = F(I1) - C1 * F(K): NEXT I: NEXT K
5130 REM ** BACK SUBSTITUTION **
5140 F(NQ) = F(NQ) / S(NQ, 1)
5150 FOR KK = 1 TO N1
5160 K = NQ - KK
5170 C1 = 1 / S(K, 1)
5180 F(K) = C1 * F(K)
5190 NK = NQ - K + 1
5200 IF NK > NBW THEN NK = NBW
5210 FOR J = 2 TO NK
5220 F(K) = F(K) - C1 * S(K, J) * F(K + J - 1)
5230 NEXT J
5240 NEXT KK
5250 RETURN


1000 REM ********          TRUSSKY        *********
1010 REM *** Truss with Temperature using Skyline Solver ***
1020 CLS
1030 PRINT "======================================"
1040 PRINT "      Program written by           "
1050 PRINT "  T.R.Chandrupatla and A.D.Belegundu   "
1060 PRINT "======================================"
1070 INPUT "Number of Elements ="; NE
1080 INPUT "Number of Nodes ="; NN
1090 INPUT "Number of Constrained DOF ="; ND
1100 INPUT "Number of Component Loads ="; NL
1110 INPUT "Number of Materials ="; NM
1120 INPUT "Number of Areas of Cross Section ="; NA
1130 INPUT "Number of Elements with Temperature Change ="; NTEL
1140 REM TOTAL DOF IS "NQ"
1150 NQ = 2 * NN
1160 DIM X(NN, 2), NOC(NE, 4), PM(NM, 2), AREA(NA), NU(ND), U(ND)
1170 DIM ID(NQ), TEMP(NTEL), NSET(NTEL)
1180 REM ======= DIAGONAL POINTERS FOR SKYLINE STORAGE  =======
1190 '++++ Initialization of Diagonal Pointers ++++
1200 FOR I = 1 TO NQ: ID(I) = 0: NEXT I
1210 PRINT "Element#, Node1, Node2, Material#, Area#"
1220 FOR I = 1 TO NE
1230 INPUT N, NOC(N, 1), NOC(N, 2), NOC(N, 3), NOC(N, 4)
```

```
1240 II = NOC(N, 1): IF II > NOC(N, 2) THEN II = NOC(N, 2)
1250 II1 = 2 * II - 1
1260 FOR J = 1 TO 2
1270 NIJ = 2 * (NOC(N, J) - 1)
1280 FOR JJ = 1 TO 2
1290 NDG = NIJ + JJ: NHT = NDG - II1 + 1
1300 IF NHT > ID(NDG) THEN ID(NDG) = NHT
1310 NEXT JJ: NEXT J: NEXT I
1320 FOR I = 2 TO NQ
1330 ID(I) = ID(I) + ID(I - 1): NEXT I
1340 NSUM = ID(NQ)
1350 DIM A(NSUM), F(NQ), SE(4, 4)
1360 IF NTEL = 0 GOTO 1400
1370 PRINT "Element#, Temperature Change"
1380 FOR I = 1 TO NTEL
1390 INPUT NSET(I), TEMP(I): NEXT I
1400 PRINT "Area#,  Area"
1410 FOR I = 1 TO NA
1420 INPUT N, AREA(N): NEXT I
1430 PRINT "Node#, X-Coord, Y-Coord"
1440 FOR I = 1 TO NN
1450 INPUT N, X(N, 1), X(N, 2): NEXT I
1460 PRINT "Material#,  E,  Alpha"
1470 FOR I = 1 TO NM
1480 INPUT N, PM(N, 1), PM(N, 2): NEXT I
1490 PRINT "DOF#,  Known Displacement"
1500 FOR I = 1 TO ND
1510 INPUT NU(I), U(I): NEXT I
1520 PRINT "DOF#,  Applied Load"
1530 FOR I = 1 TO NL
1540 INPUT N, F(N): NEXT I
1550 REM *** INITIALIZATION ***
1560 FOR I = 1 TO NQ: F(NQ) = 0: NEXT I
1570 FOR I = 1 TO NSUM: A(I) = 0: NEXT I
1580 REM *** GLOBAL STIFFNESS MATRIX ***
1590 FOR N = 1 TO NE
1600 '====== ELEMENT STIFFNESS ======
1610 I1 = NOC(N, 1): I2 = NOC(N, 2)
1620 I3 = NOC(N, 3): I4 = NOC(N, 4)
1630 X21 = X(I2, 1) - X(I1, 1)
1640 Y21 = X(I2, 2) - X(I1, 2)
1650 EL = SQR(X21 * X21 + Y21 * Y21)
1660 EAL = PM(I3, 1) * AREA(I4) / EL
1670 CS = X21 / EL: SN = Y21 / EL
1680 SE(1, 1) = CS * CS * EAL
1690 SE(1, 2) = CS * SN * EAL: SE(2, 1) = SE(1, 2)
1700 SE(1, 3) = -CS * CS * EAL: SE(3, 1) = SE(1, 3)
1710 SE(1, 4) = -CS * SN * EAL: SE(4, 1) = SE(1, 4)
1720 SE(2, 2) = SN * SN * EAL
1730 SE(2, 3) = -CS * SN * EAL: SE(3, 2) = SE(2, 3)
1740 SE(2, 4) = -SN * SN * EAL: SE(4, 2) = SE(2, 4)
1750 SE(3, 3) = CS * CS * EAL
1760 SE(3, 4) = CS * SN * EAL: SE(4, 3) = SE(3, 4)
1770 SE(4, 4) = SN * SN * EAL
1780 PRINT "++++PLACING STIFFNESS IN GLOBAL LOCATIONS++++"
1790 FOR II = 1 TO 2
1800 NCT = 2 * (NOC(N, II) - 1)
1810 FOR IT = 1 TO 2
1820 NC = NCT + IT: IID = ID(NC)
1830 I = 2 * (II - 1) + IT
1840 FOR JJ = 1 TO 2
1850 NRT = 2 * (NOC(N, JJ) - 1)
1860 FOR JT = 1 TO 2
1870 J = 2 * (JJ - 1) + JT
```

```
1880 NR = NRT + JT
1890 IF NR > NC GOTO 1920
1900 NLOC = IID - (NC - NR)
1910 A(NLOC) = A(NLOC) + SE(I, J)
1920 NEXT JT: NEXT JJ: NEXT IT: NEXT II
1930 FOR IJT = 1 TO NTEL
1940 IF I = NSET(IJT) GOTO 1960
1950 GOTO 2010
1960 EE0 = PM(I3, 2) * TEMP(IJT) * PM(I3, 1) * AREA(I4)
1970 F(2 * I1 - 1) = F(2 * I1 - 1) - EE0 * CS
1980 F(2 * I1) = F(2 * I1) - EE0 * SN
1990 F(2 * I2 - 1) = F(2 * I2 - 1) + EE0 * CS
2000 F(2 * I2) = F(2 * I2) + EE0 * SN
2010 NEXT IJT
2020 NEXT N
2030 REM  *** MODIFY FOR BOUNDARY CONDITIONS ***
2040 CNST = A(1) * 10000
2050 FOR I = 1 TO ND
2060 N = NU(I): II = ID(N)
2070 A(II) = A(II) + CNST
2080 F(N) = F(N) + CNST * U(I): NEXT I
2090 REM *** EQUATION SOLVING ***
2100 GOSUB 5020
2110 PRINT "Node#    X-Displ      Y-Displ"
2120 FOR I = 1 TO NN
2130 PRINT USING " ###"; I;
2140 PRINT USING "    ##.###^^^^"; F(2 * I - 1); F(2 * I)
2150 NEXT I
2160 REM *** STRESS CALCULATION ***
2170 PRINT "Elem#     Stress"
2180 FOR I = 1 TO NE
2190 I1 = NOC(I, 1): I2 = NOC(I, 2)
2200 I3 = NOC(I, 3)
2210 X21 = X(I2, 1) - X(I1, 1): Y21 = X(I2, 2) - X(I1, 2)
2220 EL = SQR(X21 * X21 + Y21 * Y21)
2230 CS = X21 / EL: SN = Y21 / EL
2240 J2 = 2 * I1: J1 = J2 - 1: K2 = 2 * I2: K1 = K2 - 1
2250 DT = (F(K1) - F(J1)) * CS + (F(K2) - F(J2)) * SN
2260 STRESS = DT * PM(I3, 1) / EL
2270 FOR IJT = 1 TO NTEL
2280 IF I = NSET(IJT) GOTO 2300
2290 GOTO 2310
2300 STRESS = STRESS - PM(I3, 1) * PM(I3, 2) * TEMP(IJT)
2310 NEXT IJT
2320 PRINT USING " ###"; I; : PRINT USING "    ##.###^^^^"; STRESS
2330 NEXT I
2340 END
5000 REM ****  Sky Line Method  ****
5010 REM **** Gauss Elimination ****
5020 FOR J = 2 TO NQ
5030 NJ = ID(J) - ID(J - 1)
5040 IF NJ = 1 GOTO 5170
5050 K1 = 0: NJ = J - NJ + 1
5060 FOR K = NJ TO J - 1
5070 K1 = K1 + 1: KJ = ID(J - 1) + K1: KK = ID(K)
5080 C = A(KJ) / A(KK)
5090 FOR I = K + 1 TO J
5100 NI = ID(I) - ID(I - 1)
5110 IF (I - K + 1) > NI GOTO 5150
5120 IJ = ID(J) - J + I
5130 KI = ID(I) - I + K
5140 A(IJ) = A(IJ) - C * A(KI)
5150 NEXT I
5160 NEXT K
5170 NEXT J
```

```
5180 FOR K = 1 TO NQ - 1: KK = ID(K)
5190 C = F(K) / A(KK)
5200 FOR I = K + 1 TO NQ
5210 NI = ID(I) - ID(I - 1)
5220 IF (I - K + 1) > NI GOTO 5250
5230 KI = ID(I) - I + K
5240 F(I) = F(I) - C * A(KI)
5250 NEXT I: NEXT K
5260 REM **** Back -substitution ****
5270 NS = ID(NQ): F(NQ) = F(NQ) / A(NS)
5280 FOR I1 = 1 TO NQ - 1
5290 I = NQ - I1
5300 II = ID(I)
5310 C = 1 / A(II): F(I) = C * F(I)
5320 FOR J = I + 1 TO N
5330 J1 = J - I + 1: NJ = ID(J) - ID(J - 1)
5340 IF J1 > NJ GOTO 5370
5350 IJ = ID(J) - J + I
5360 F(I) = F(I) - C * A(IJ) * F(J)
5370 NEXT J
5380 NEXT I1
5390 RETURN
```

PROBLEMS

4.1. Consider the truss element shown in Fig. P4.1. The x, y coordinates of the two nodes are indicated in the figure. If $\mathbf{q} = [1.5, 1.0, 2.1, 4.3]^T \times 10^2$ in.:

   (a) Determine the vector $\mathbf{q}'$.

   (b) Determine the stress in the element.

   (c) Determine the $\mathbf{k}$ matrix.

   (d) Determine the strain energy in the element.

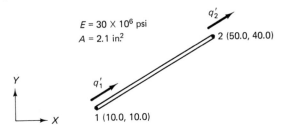

$E = 30 \times 10^6$ psi

$A = 2.1$ in.^{2}

Figure P4.1

4.2. For the pin-jointed configuration shown in Fig. P4.2, determine the stiffness values K_{11}, K_{12}, K_{22} of the global stiffness matrix.

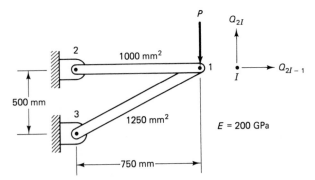

$E = 200$ GPa

Figure P4.2

4.3. For the truss in Fig. P4.3, a horizontal load of $P = 4000$ lb is applied in the x direction at node 2.

(a) Write down the element stiffness matrix **k** for each element.

(b) Assemble the **K** matrix.

(c) Using the elimination approach, solve for **Q**.

(d) Evaluate the stress in elements 2 and 3.

(e) Determine the reaction force at node 2 in the y direction.

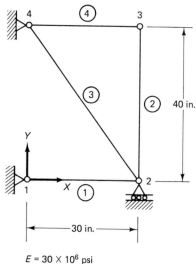

$E = 30 \times 10^6$ psi

$A = 1.5$ in.² for each member

Figure P4.3

4.4. In Fig. P4.3, determine the stresses in each element due to the following support movement: support at node 2, 0.24 in. down.

4.5. A small railroad bridge is constructed of steel members, all of which have a cross-sectional area of 3250 mm². A train stops on the bridge, and the loads applied to the truss on one side of the bridge are as shown in Fig. P4.5. Estimate how much the point R moves horizontally because of this loading. Also determine the nodal displacements and element stresses.

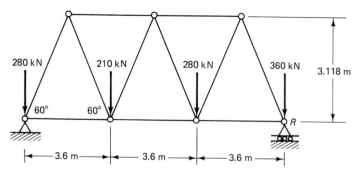

Figure P4.5

4.6. Consider the truss in Fig. P4.6 loaded as shown. Cross-sectional areas in square inches are shown in parentheses. Consider symmetry and model only one-half of the truss shown. Determine displacements and element stresses. $E = 30 \times 10^6$ psi.

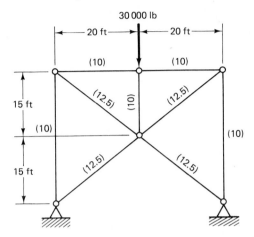

Figure P4.6

4.7. Determine the nodal displacements and element stresses in the truss in Fig. P4.7, due to each of the following conditions:

(a) Increase of temperature of 50°F in elements 1, 3, 7, and 8.

(b) Elements 9 and 10 are $\frac{1}{4}$ in. too short and element 6 is $\frac{1}{8}$ in. too long, owing to errors in fabrication, and it was necessary to force them into place.

(c) Support at node 6 moves 0.12 in. down. *Data*: Take $E = 30 \times 10^6$ psi, $\alpha = 1/150\,000$ per °F. Cross-sectional areas for each element are as follows:

Element	Area (in.²)
1, 3	25
2, 4	12
5	1
6	4
7, 8, 9	17
10	5

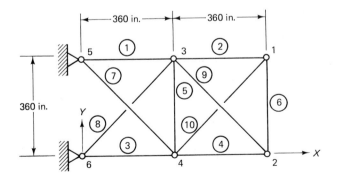

Figure P4.7

4.8. Expressions for the element stress (Eq. 4.16), and element temperature load (Eq. 4.22) were derived for a two-dimensional truss element. Generalize these expressions for a three-dimensional truss element.

4.9. Find deflections at nodes, stresses in members, and reactions at supports for the truss shown in Fig. P4.9 when the 150-kip load is applied.

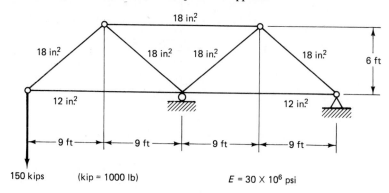

150 kips   (kip = 1000 lb) $E = 30 \times 10^6$ psi

Figure P4.9

4.10. Find the deflections at the nodes for the truss configuration shown in Fig. P4.10. Area = 8 in.^{2} for each member.

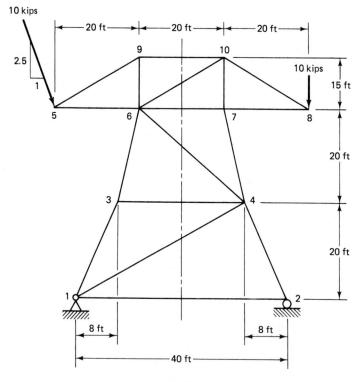

Figure P4.10

*4.11. Modify program TRUSS to handle 3-D trusses and solve the problem in Fig. P4.11.

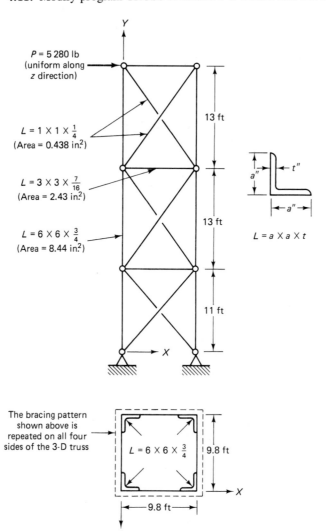

Figure P4.11   3-D truss model of a steel tower, supporting a water tank, and subjected to wind loads.

chapter five

Two-Dimensional Problems Using Constant Strain Triangles

5.1 INTRODUCTION

The two-dimensional finite element formulation in this chapter follows the steps used in the one-dimensional problem. The displacements, traction components, and distributed body force values are functions of the position indicated by (x, y). The displacement vector $\mathbf{u}$ is given as

$$\mathbf{u} = [u, v]^T \tag{5.1}$$

where u and v are the x and y components of $\mathbf{u}$, respectively. The stresses and strains are given by

$$\boldsymbol{\sigma} = [\sigma_x, \sigma_y, \tau_{xy}]^T \tag{5.2}$$

$$\boldsymbol{\epsilon} = [\epsilon_x, \epsilon_y, \gamma_{xy}]^T \tag{5.3}$$

From Fig. 5.1, representing the two-dimensional problem in a general setting, the body force, traction vector, and elemental volume are given by

$$\mathbf{f} = [f_x, f_y]^T \qquad \mathbf{T} = [T_x, T_y]^T \quad \text{and} \quad dV = t\, dA \tag{5.4}$$

where t is the thickness along the z direction. The body force $\mathbf{f}$ has the units force/unit volume, while the traction force $\mathbf{T}$ has the units force/unit area. The strain–displacement relations are given by

$$\boldsymbol{\epsilon} = \left[\frac{\partial u}{\partial x}, \frac{\partial v}{\partial y}, \left(\frac{\partial u}{\partial y} + \frac{\partial v}{\partial x} \right) \right]^T \tag{5.5}$$

Stresses and strains are related by (see Eqs. 1.18 and 1.19)

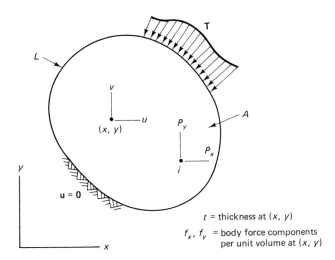

t = thickness at (x, y)

f_x, f_y = body force components per unit volume at (x, y)

Figure 5.1 Two-dimensional problem.

$$\boldsymbol{\sigma} = \mathbf{D}\boldsymbol{\epsilon} \tag{5.6}$$

The region is discretized with the idea of expressing the displacements in terms of values at discrete points. Triangular elements are introduced first. Stiffness and load concepts are then developed using energy and Galerkin approaches.

5.2 FINITE ELEMENT MODELING

The two-dimensional region is divided into straight-sided triangles. Figure 5.2 shows a typical triangulation. The points where the corners of the triangles meet are called *nodes,* and each triangle formed by three nodes and three sides is called an *element*. The elements fill the entire region except a small region at the boundary. This unfilled region exists for curved boundaries and it can be reduced by choosing smaller elements or elements with curved boundaries. The idea of the finite element method is to solve the continuous problem approximately, and this unfilled region contributes to some part of this approximation. For the triangulation shown in Fig. 5.2, the node numbers are indicated at the corners and element numbers are circled.

In the two-dimensional problem discussed here, each node is permitted to displace in the two directions x and y. Thus, each node has two degrees of freedom (dof's). As seen from the numbering scheme used in trusses, the displacement components of node j are taken as Q_{2j-1} in the x direction and Q_{2j} in the y direction. We denote the global displacement vector as

$$\mathbf{Q} = [Q_1, Q_2, \ldots, Q_N]^{\mathrm{T}} \tag{5.7}$$

where N is the number of degrees of freedom.

Computationally, the information on the triangulation is to be represented in the form of *nodal coordinates* and *connectivity*. The nodal coordinates are stored in a two-dimensional array represented by the total number of nodes and the two coor-

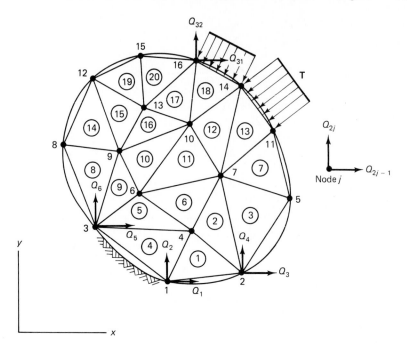

Figure 5.2   Finite element discretization.

dinates per node. The connectivity may be clearly seen by isolating a typical element, as shown in Fig. 5.3. For the three nodes designated locally as 1, 2, and 3, the corresponding global node numbers are defined in Fig. 5.2. This element connectivity information becomes an array of the size of number of elements and three

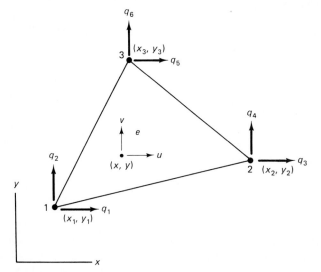

Figure 5.3   Triangular element.

nodes per element. A typical connectivity representation is shown in Table 5.1. Most standard finite element codes use the convention of going around the element in a counterclockwise direction to avoid calculating a negative area. However, in the program that accompanies this chapter, ordering is not necessary.

Table 5.1 establishes the correspondence of local and global node numbers and the corresponding degrees of freedom. The displacement components of a local node j in Fig. 5.3 are represented as q_{2j-1} and q_{2j} in the x and y directions, respectively. We denote the element displacement vector as

$$\mathbf{q} = [q_1, q_2, \ldots, q_6] \tag{5.8}$$

Note that from the connectivity matrix in Table 5.1, we can extract the $\mathbf{q}$ vector from the global $\mathbf{Q}$ vector, an operation performed frequently in a finite element program. Also, the nodal coordinates designated by (x_1, y_1), (x_2, y_2), and (x_3, y_3) have the global correspondence established through Table 5.1. The local representation of nodal coordinates and degrees of freedom provides a setting for a simple and clear representation of element characteristics.

TABLE 5.1 ELEMENT CONNECTIVITY

Element number e	Three nodes		
	1	2	3
1	1	2	4
2	4	2	7
⋮			
11	6	7	10
⋮			
20	13	16	15

5.3 CONSTANT STRAIN TRIANGLE (CST)

The displacements at points inside an element need to be represented in terms of the nodal displacements of the element. As discussed earlier, the finite element method uses the concept of shape functions in systematically developing these interpolations. For the constant strain triangle, the shape functions are linear over the element. The three shape functions N_1, N_2, and N_3 corresponding to nodes 1, 2, and 3, respectively, are shown in Fig. 5.4. Shape function N_1 is 1 at node 1 and linearly reduces to 0 at nodes 2 and 3. The values of shape function N_1 thus define a plane surface shown shaded in Fig. 5.4a. N_2 and N_3 are represented by similar surfaces having values of 1 at nodes 2 and 3, respectively and dropping to 0 at the opposite edges. Any linear combination of these shape functions also represents a plane surface. In particular, $N_1 + N_2 + N_3$ represents a plane at a height of 1 at nodes 1, 2, and 3, and,

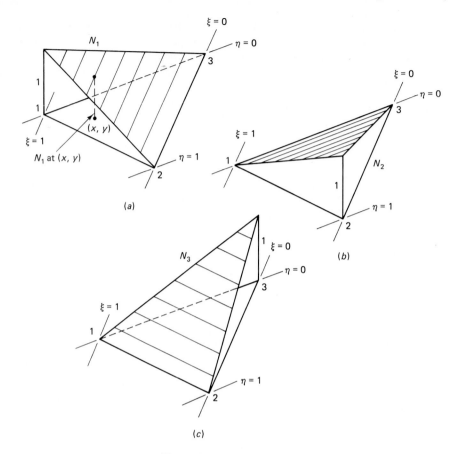

Figure 5.4 Shape functions.

thus, it is parallel to the triangle 123. Consequently, for every N_1, N_2, and N_3,

$$N_1 + N_2 + N_3 = 1 \tag{5.9}$$

N_1, N_2, and N_3 are therefore not linearly independent; only two of these are independent. The independent shape functions are conveniently represented by the pair ξ, η, as follows

$$N_1 = \xi \quad N_2 = \eta \quad N_3 = 1 - \xi - \eta \tag{5.10}$$

where ξ, η are natural coordinates (Fig. 5.4). At this stage, the similarity with the one-dimensional element (Chapter 3) should be noted: In the one-dimensional problem the x coordinates were mapped onto the ξ coordinates and shape functions were defined as functions of ξ. Here, in the two-dimensional problem, the x, y coordinates are mapped onto the ξ, η coordinates and shape functions are defined as functions of ξ and η.

The shape functions can be physically represented by **area coordinates**. A point (x, y) in a triangle divides it into three areas, A_1, A_2, and A_3, as shown in Fig. 5.5. The shape functions N_1, N_2, and N_3 are precisely represented by

$$N_1 = \frac{A_1}{A} \qquad N_2 = \frac{A_2}{A} \qquad N_3 = \frac{A_3}{A} \qquad (5.11)$$

where A is the area of the element. Clearly, $N_1 + N_2 + N_3 = 1$ at every point inside the triangle.

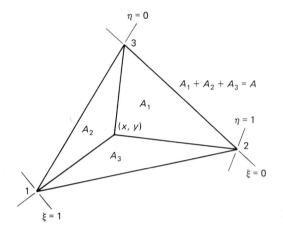

Figure 5.5   Area coordinates.

Isoparametric Representation

The displacements inside the element are now written using the shape functions and the nodal values of the unknown displacement field.

$$u = N_1 q_1 + N_2 q_3 + N_3 q_5$$
$$v = N_1 q_2 + N_2 q_4 + N_3 q_6 \qquad (5.12a)$$

or using Eq. 5.10,

$$u = (q_1 - q_5)\xi + (q_3 - q_5)\eta + q_5$$
$$v = (q_2 - q_6)\xi + (q_4 - q_6)\eta + q_6 \qquad (5.12b)$$

The relations 5.12a can be expressed in a matrix form by defining a shape function matrix **N**,

$$\mathbf{N} = \begin{bmatrix} N_1 & 0 & N_2 & 0 & N_3 & 0 \\ 0 & N_1 & 0 & N_2 & 0 & N_3 \end{bmatrix} \qquad (5.13)$$

and

$$\mathbf{u} = \mathbf{Nq} \qquad (5.14)$$

For the triangular element, the coordinates x, y can also be represented in terms of nodal coordinates using the same shape functions. This is **isoparametric representation**. This approach lends to simplicity of development and retains the uniformity with other complex elements. We have

$$x = N_1 x_1 + N_2 x_2 + N_3 x_3$$
$$y = N_1 y_1 + N_2 y_2 + N_3 y_3 \qquad (5.15a)$$

or

$$x = (x_1 - x_3)\xi + (x_2 - x_3)\eta + x_3$$
$$y = (y_1 - y_3)\xi + (y_2 - y_3)\eta + y_3 \qquad (5.15b)$$

Using the notation, $x_{ij} = x_i - x_j$ and $y_{ij} = y_i - y_j$, we can write Eq. 5.15b as

$$x = x_{13}\xi + x_{23}\eta + x_3$$
$$y = y_{13}\xi + y_{23}\eta + y_3 \qquad (5.15c)$$

This equation relates x and y coordinates to the ξ and η coordinates. Equation 5.12 expresses u and v as functions of ξ, η.

Example 5.1

Evaluate the shape functions N_1, N_2, and N_3 at the interior point P for the triangular element shown in Fig. E5.1.

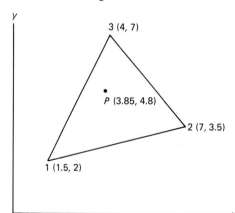

Figure E5.1 Examples 5.1 and 5.2.

Solution Using the isoparametric representation (Eqs. 5.15), we have

$$3.85 = 1.5N_1 + 7N_2 + 4N_3 = -2.5\xi + 3\eta + 4$$
$$4.8 = 2N_1 + 3.5N_2 + 7N_3 = -5\xi - 3.5\eta + 7$$

The two equations above are rearranged in the form

$$2.5\xi - 3\eta = 0.15$$
$$5\xi + 3.5\eta = 2.2$$

Solving the equations, we obtain $\xi = 0.3$ and $\eta = 0.2$, which implies

$$N_1 = 0.3 \qquad N_2 = 0.2 \qquad N_3 = 0.5 \qquad \blacksquare$$

In evaluating the strains, partial derivatives of u and v are to be taken with respect to x and y. From Eqs. 5.12 and 5.15, we see that u, v and x, y are functions of ξ and η. That is, $u = u(x(\xi, \eta), y(\xi, \eta))$ and similarly $v = v(x(\xi, \eta), y(\xi, \eta))$. Using the chain rule for partial derivatives of u, we have

$$\frac{\partial u}{\partial \xi} = \frac{\partial u}{\partial x}\frac{\partial x}{\partial \xi} + \frac{\partial u}{\partial y}\frac{\partial y}{\partial \xi}$$

$$\frac{\partial u}{\partial \eta} = \frac{\partial u}{\partial x}\frac{\partial x}{\partial \eta} + \frac{\partial u}{\partial y}\frac{\partial y}{\partial \eta}$$

which can be written in matrix notation as

$$\left\{ \begin{matrix} \dfrac{\partial u}{\partial \xi} \\ \dfrac{\partial u}{\partial \eta} \end{matrix} \right\} = \begin{bmatrix} \dfrac{\partial x}{\partial \xi} & \dfrac{\partial y}{\partial \xi} \\ \dfrac{\partial x}{\partial \eta} & \dfrac{\partial y}{\partial \eta} \end{bmatrix} \left\{ \begin{matrix} \dfrac{\partial u}{\partial x} \\ \dfrac{\partial u}{\partial y} \end{matrix} \right\} \tag{5.16}$$

where the (2×2) square matrix is denoted as the Jacobian of the transformation, $\mathbf{J}$:

$$\mathbf{J} = \begin{bmatrix} \dfrac{\partial x}{\partial \xi} & \dfrac{\partial y}{\partial \xi} \\ \dfrac{\partial x}{\partial \eta} & \dfrac{\partial y}{\partial \eta} \end{bmatrix} \tag{5.17}$$

Some additional properties of the Jacobian are given in the Appendix. On taking the derivative of x and y,

$$\mathbf{J} = \begin{bmatrix} x_{13} & y_{13} \\ x_{23} & y_{23} \end{bmatrix} \tag{5.18}$$

Also, from Eq. 5.16,

$$\left\{ \begin{matrix} \dfrac{\partial u}{\partial x} \\ \dfrac{\partial u}{\partial y} \end{matrix} \right\} = \mathbf{J}^{-1} \left\{ \begin{matrix} \dfrac{\partial u}{\partial \xi} \\ \dfrac{\partial u}{\partial \eta} \end{matrix} \right\} \tag{5.19}$$

where $\mathbf{J}^{-1}$ is the inverse of the Jacobian $\mathbf{J}$, given by

$$\mathbf{J}^{-1} = \frac{1}{\det \mathbf{J}} \begin{bmatrix} y_{23} & -y_{13} \\ -x_{23} & x_{13} \end{bmatrix} \tag{5.20}$$

$$\det \mathbf{J} = x_{13}y_{23} - x_{23}y_{13} \tag{5.21}$$

From the knowledge of the area of the triangle, it can be seen that the magnitude of det **J** is twice the area of the triangle. If the points 1, 2, and 3 are ordered in a counterclockwise manner, det **J** is positive in sign. We have

$$A = \tfrac{1}{2}|\det \mathbf{J}| \tag{5.22}$$

where $|\ |$ represents the magnitude. Most computer codes use a counterclockwise order for the nodes and use det **J** for evaluating the area.

Example 5.2

Determine the Jacobian of the transformation **J** for the triangular element shown in Fig. E5.1.

Solution We have

$$\mathbf{J} = \begin{bmatrix} x_{13} & y_{13} \\ x_{23} & y_{23} \end{bmatrix} = \begin{bmatrix} -2.5 & -5.0 \\ 3.0 & -3.5 \end{bmatrix}$$

Thus, det $\mathbf{J} = 23.75$ units. This is twice the area of the triangle. If 1, 2, 3 are in a clockwise order, then det **J** will be negative. ∎

From Eqs. 5.19 and 5.20, it follows that

$$\begin{Bmatrix} \dfrac{\partial u}{\partial x} \\[2mm] \dfrac{\partial u}{\partial y} \end{Bmatrix} = \frac{1}{\det \mathbf{J}} \begin{Bmatrix} y_{23}\dfrac{\partial u}{\partial \xi} - y_{13}\dfrac{\partial u}{\partial \eta} \\[3mm] -x_{23}\dfrac{\partial u}{\partial \xi} + x_{13}\dfrac{\partial u}{\partial \eta} \end{Bmatrix} \tag{5.23a}$$

Replacing u by the displacement v, we get a similar expression

$$\begin{Bmatrix} \dfrac{\partial v}{\partial x} \\[2mm] \dfrac{\partial v}{\partial y} \end{Bmatrix} = \frac{1}{\det \mathbf{J}} \begin{Bmatrix} y_{23}\dfrac{\partial v}{\partial \xi} - y_{13}\dfrac{\partial v}{\partial \eta} \\[3mm] -x_{23}\dfrac{\partial v}{\partial \xi} + x_{13}\dfrac{\partial v}{\partial \eta} \end{Bmatrix} \tag{5.23b}$$

Using the strain–displacement relations (5.5) and Eqs. 5.12b and 5.23, we get

$$\boldsymbol{\epsilon} = \begin{Bmatrix} \dfrac{\partial u}{\partial x} \\[2mm] \dfrac{\partial v}{\partial y} \\[2mm] \dfrac{\partial u}{\partial y} + \dfrac{\partial v}{\partial x} \end{Bmatrix}$$

$$= \frac{1}{\det \mathbf{J}} \begin{Bmatrix} y_{23}(q_1 - q_5) - y_{13}(q_3 - q_5) \\ -x_{23}(q_2 - q_6) + x_{13}(q_4 - q_6) \\ -x_{23}(q_1 - q_5) + x_{13}(q_3 - q_5) + y_{23}(q_2 - q_6) - y_{13}(q_4 - q_6) \end{Bmatrix} \tag{5.24a}$$

From the definition of x_{ij} and y_{ij}, we can write $y_{31} = -y_{13}$ and $y_{12} = y_{13} - y_{23}$, and so on. The above equation can be written in the form

$$\boldsymbol{\epsilon} = \frac{1}{\det \mathbf{J}} \left\{ \begin{array}{l} y_{23}q_1 + y_{31}q_3 + y_{12}q_5 \\ x_{32}q_2 + x_{13}q_4 + x_{21}q_6 \\ x_{32}q_1 + y_{23}q_2 + x_{13}q_3 + y_{31}q_4 + x_{21}q_5 + y_{12}q_6 \end{array} \right\} \tag{5.24b}$$

The above equation can be written in matrix form as

$$\boldsymbol{\epsilon} = \mathbf{Bq} \tag{5.25}$$

where $\mathbf{B}$ is a (3×6) element strain–displacement matrix relating the three strains to the six nodal displacements and is given by

$$\mathbf{B} = \frac{1}{\det \mathbf{J}} \begin{bmatrix} y_{23} & 0 & y_{31} & 0 & y_{12} & 0 \\ 0 & x_{32} & 0 & x_{13} & 0 & x_{21} \\ x_{32} & y_{23} & x_{13} & y_{31} & x_{21} & y_{12} \end{bmatrix} \tag{5.26}$$

It may be noted that all the elements of the $\mathbf{B}$ matrix are constants expressed in terms of the nodal coordinates.

Example 5.3

Find the strain–nodal displacement matrices $\mathbf{B}^e$ for the elements shown in Fig. E5.3. Use local numbers given at the corners.

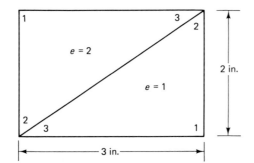

Figure E5.3

Solution   We have

$$\mathbf{B}^1 = \frac{1}{\det \mathbf{J}} \begin{bmatrix} y_{23} & 0 & y_{31} & 0 & y_{12} & 0 \\ 0 & x_{32} & 0 & x_{13} & 0 & x_{21} \\ x_{32} & y_{23} & x_{13} & y_{31} & x_{21} & y_{12} \end{bmatrix}$$

$$= \frac{1}{6} \begin{bmatrix} 2 & 0 & 0 & 0 & -2 & 0 \\ 0 & -3 & 0 & 3 & 0 & 0 \\ -3 & 2 & 3 & 0 & 0 & -2 \end{bmatrix}$$

where $\det \mathbf{J}$ is obtained from $x_{13}y_{23} - x_{23}y_{13} = (3)(2) - (3)(0) = 6$. Using the local numbers at the corners, $\mathbf{B}^2$ can be written using the relationship as

$$\mathbf{B}^2 = \frac{1}{6}\begin{bmatrix} -2 & 0 & 0 & 0 & 2 & 0 \\ 0 & 3 & 0 & -3 & 0 & 0 \\ 3 & -2 & -3 & 0 & 0 & 2 \end{bmatrix}$$

■

Potential Energy Approach

The potential energy of the system, Π, is given by

$$\Pi = \frac{1}{2}\int_A \boldsymbol{\epsilon}^\mathsf{T}\mathbf{D}\boldsymbol{\epsilon}t \, dA - \int_A \mathbf{u}^\mathsf{T}\mathbf{f}t \, dA - \int_L \mathbf{u}^\mathsf{T}\mathbf{T}t \, d\ell - \sum_i \mathbf{u}_i^\mathsf{T}\mathbf{P}_i \qquad (5.27)$$

In the last term above, i indicates the point of application of a point load $\mathbf{P}_i$ and $\mathbf{P}_i = [P_x, P_y]_i^\mathsf{T}$. The summation on i gives the potential energy due to all point loads.

Using the triangulation shown in Fig. 5.2, the total potential energy can be written in the form

$$\Pi = \sum_e \frac{1}{2}\int_e \boldsymbol{\epsilon}^\mathsf{T}\mathbf{D}\boldsymbol{\epsilon}t \, dA - \sum_e \int_e \mathbf{u}^\mathsf{T}\mathbf{f}t \, dA - \int_L \mathbf{u}^\mathsf{T}\mathbf{T}t \, d\ell - \sum_i \mathbf{u}_i^\mathsf{T}\mathbf{P}_i \qquad (5.28a)$$

or

$$\Pi = \sum_e U_e - \sum_e \int_e \mathbf{u}^\mathsf{T}\mathbf{f}t \, dA - \int_L \mathbf{u}^\mathsf{T}\mathbf{T}t \, d\ell - \sum_i \mathbf{u}_i^\mathsf{T}\mathbf{P}_i \qquad (5.28b)$$

where $U_e = \frac{1}{2}\int_e \boldsymbol{\epsilon}^\mathsf{T}\mathbf{D}\boldsymbol{\epsilon}t \, dA$ is the element strain energy.

Element Stiffness

We now substitute for the strain from the element strain–displacement relationship in Eq. 5.25 into the element strain energy U_e in Eq. 5.28b, to obtain

$$U_e = \frac{1}{2}\int_e \boldsymbol{\epsilon}^\mathsf{T}\mathbf{D}\boldsymbol{\epsilon}t \, dA$$

$$= \frac{1}{2}\int_e \mathbf{q}^\mathsf{T}\mathbf{B}^\mathsf{T}\mathbf{D}\mathbf{B}\mathbf{q}t \, dA \qquad (5.29a)$$

Taking the element thickness t_e as constant over the element, and since all terms in the $\mathbf{D}$ and $\mathbf{B}$ matrices are constants, we have

$$U_e = \tfrac{1}{2}\mathbf{q}^\mathsf{T}\mathbf{B}^\mathsf{T}\mathbf{D}\mathbf{B}t_e\left(\int_e dA\right)\mathbf{q} \qquad (5.29b)$$

Now, $\int_e dA = A_e$, where A_e is the area of the element. Thus,

$$U_e = \tfrac{1}{2}\mathbf{q}^\mathsf{T}t_e A_e\mathbf{B}^\mathsf{T}\mathbf{D}\mathbf{B}\mathbf{q} \qquad (5.29c)$$

or

$$U_e = \tfrac{1}{2}\mathbf{q}^\mathsf{T}\mathbf{k}^e\mathbf{q} \qquad (5.29d)$$

where $\mathbf{k}^e$ is the element stiffness matrix given by

$$\mathbf{k}^e = t_e A_e \mathbf{B}^\mathrm{T} \mathbf{D} \mathbf{B} \tag{5.30}$$

For plane stress or plane strain, the element stiffness matrix can be obtained by taking the appropriate material property matrix $\mathbf{D}$ defined in Chapter 1, and carrying out the above multiplication on the computer. We note that $\mathbf{k}^e$ is symmetric since $\mathbf{D}$ is symmetric. The element connectivity as established in Table 5.1 is now used to add the element stiffness values in $\mathbf{k}^e$ into the corresponding global locations in the global stiffness matrix $\mathbf{K}$, so that

$$\begin{aligned} U &= \sum_e \frac{1}{2} \mathbf{q}^\mathrm{T} \mathbf{k}^e \mathbf{q} \\ &= \frac{1}{2} \mathbf{Q}^\mathrm{T} \mathbf{K} \mathbf{Q} \end{aligned} \tag{5.31}$$

The global stiffness matrix $\mathbf{K}$ is symmetric and banded or sparse. The stiffness value K_{ij} is zero when the degrees of freedom i and j are not connected through an element. If i and j are connected through one or more elements, stiffness values accumulate from these elements. For the global dof numbering shown in Fig. 5.2, the bandwidth is related to the maximum difference in node numbers of an element, over all the elements. If i_1, i_2, and i_3 are node numbers of an element e, the maximum element node number difference is given by

$$m_e = \max \left(|i_1 - i_2|, |i_2 - i_3|, |i_3 - i_1| \right) \tag{5.32a}$$

The half-bandwidth is then given by

$$\mathrm{NBW} = 2 \left[\max_{1 \le e \le \mathrm{NE}} (m_e) + 1 \right] \tag{5.32b}$$

where NE is the number of elements and 2 is the number of degrees of freedom per node.

The global stiffness $\mathbf{K}$ is in a form where all the degrees of freedom $\mathbf{Q}$ are free. It needs to be modified to account for the boundary conditions.

Force Terms

The **body force term** $\int_e \mathbf{u}^\mathrm{T} \mathbf{f} t \, dA$ appearing in the total potential energy in Eq. 5.28b is considered first. We have,

$$\int_e \mathbf{u}^\mathrm{T} \mathbf{f} t \, dA = t_e \int_e (u f_x + v f_y) \, dA$$

Using the interpolation relations given in Eq. 5.12a, we find

$$\int_e \mathbf{u}^\mathrm{T} \mathbf{f} t \, dA = q_1 \left(t_e f_x \int_e N_1 \, dA \right) + q_2 \left(t_e f_y \int_e N_1 \, dA \right)$$

$$+ q_3\left(t_e f_x \int_e N_2\, dA\right) + q_4\left(t_e f_y \int_e N_2\, dA\right) \qquad (5.33)$$

$$+ q_5\left(t_e f_x \int_e N_3\, dA\right) + q_6\left(t_e f_y \int_e N_3\, dA\right)$$

From the definition of shape functions on a triangle, shown in Fig 5.4, $\int_e N_1\, dA$ represents the volume of a tetrahedron with base area A_e and height of corner equal to 1 (nondimensional). The volume of this tetrahedron is given by $\frac{1}{3} \times$ Base area $\times$ Height (Fig. 5.6).

$$\int_e N_i\, dA = \tfrac{1}{3} A_e \qquad (5.34)$$

Similarly, $\int_e N_2\, dA = \int_e N_3\, dA = \tfrac{1}{3} A_e$. Equation 5.33 can now be written in the form

$$\int_e \mathbf{u}^T \mathbf{f} t\, dA = \mathbf{q}^T \mathbf{f}^e \qquad (5.35)$$

where $\mathbf{f}^e$ is the element body force vector, given as

$$\mathbf{f}^e = \frac{t_e A_e}{3}[f_x, f_y, f_x, f_y, f_x, f_y]^T \qquad (5.36)$$

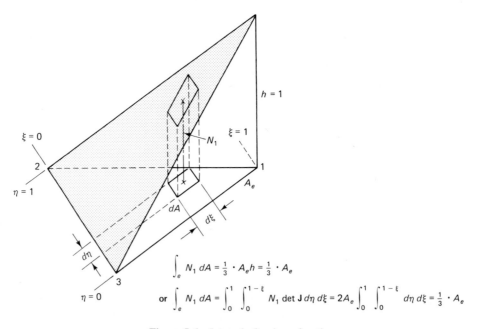

Figure 5.6 Integral of a shape function.

These element nodal forces contribute to the global load vector $\mathbf{F}$. The connectivity in Table 5.1 needs to be used again to add $\mathbf{f}^e$ to the global force vector $\mathbf{F}$. The vector $\mathbf{f}^e$ is of dimension (6×1), whereas $\mathbf{F}$ is $(N \times 1)$. This assembly procedure is discussed in Chapters 3 and 4. Stating this symbolically,

$$\mathbf{F} \longleftarrow \sum_e \mathbf{f}^e \tag{5.37}$$

A **traction force** is a distributed load acting on the surface of the body. Such a force acts on edges connecting boundary nodes. A traction force acting on the edge of an element contributes to the global load vector $\mathbf{F}$. This contribution can be determined by considering the traction force term $\int \mathbf{u}^T \mathbf{T} t \, d\ell$. Consider an edge ℓ_{1-2}, acted on by a traction T_x, T_y in units of force per unit surface area, shown in Fig. 5.7. We have

$$\int_L \mathbf{u}^T \mathbf{T} t \, d\ell = \int_{\ell_{1-2}} (uT_x + vT_y)t \, d\ell \tag{5.38a}$$

Using $\mathbf{u} - \mathbf{Nq}$, we get

$$\int_{\ell_{1-2}} \mathbf{u}^T \mathbf{T} t \, d\ell = q_1 \left(t_e T_x \int N_1 \, d\ell \right) + q_2 \left(t_e T_y \int N_1 \, d\ell \right)$$
$$+ q_3 \left(t_e T_x \int N_2 \, d\ell \right) + q_4 \left(t_e T_y \int N_2 \, d\ell \right) \tag{5.38b}$$

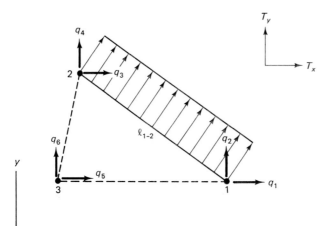

Figure 5.7   Traction load.

We note here that N_3 is zero along the edge 1–2, and N_1 and N_2 are similar to the shape functions in one dimension satisfying $N_1 + N_2 = 1$. Each of the integrals in Eq. 5.38b above equals one-half the base length ℓ_e times the height $(=1)$:

$$\int_{\ell_{1-2}} N_i \, d\ell = \tfrac{1}{2}\ell_{1-2} \tag{5.39}$$

where

$$\ell_{1-2} = \sqrt{(x_2 - x_1)^2 + (y_2 - y_1)^2}$$

The traction term is now given by

$$\int_{\ell_{1-2}} \mathbf{u}^T \mathbf{T} t \, d\ell = \bar{\mathbf{q}}^T \mathbf{T}^e \tag{5.40}$$

where $\bar{\mathbf{q}}$ is the set of nodal degrees of freedom corresponding to the edge 1–2. That is,

$$\bar{\mathbf{q}} = [q_1, q_2, q_3, q_4]^T \tag{5.41}$$

and

$$\mathbf{T}^e = \frac{t_e \ell_{1-2}}{2} [T_x, T_y, T_x, T_y]^T \tag{5.42}$$

The traction load contributions need to be added to the global force vector $\mathbf{F}$.

The **point load** term is easily considered by having a node at the point of application of the point load. If i is the node at which $\mathbf{P}_i = [P_x, P_y]^T$ is applied, then

$$\mathbf{u}_i^T \mathbf{P}_i = Q_{2i-1} P_x + Q_{2i} P_y \tag{5.43}$$

Thus, P_x and P_y, the x and y components of $\mathbf{P}_i$, get added to the $(2i - 1)$th and $(2i)$th components of the global force $\mathbf{F}$.

The contribution of body forces, traction forces, and point loads to the global force $\mathbf{F}$ can be represented as $\mathbf{F} \leftarrow \Sigma_e(\mathbf{f}^e + \mathbf{T}^e) + \mathbf{P}$.

Consideration of the strain energy and the force terms gives us the total potential energy in the form

$$\Pi = \tfrac{1}{2}\mathbf{Q}^T\mathbf{K}\mathbf{Q} - \mathbf{Q}^T\mathbf{F} \tag{5.44}$$

The stiffness and force modifications are made to account for the boundary conditions. Using the methods presented in Chapters 3 and 4, we have

$$\mathbf{KQ} = \mathbf{F} \tag{5.45}$$

where $\mathbf{K}$ and $\mathbf{F}$ are modified stiffness matrix and force vector, respectively. These equations are solved by Gaussian elimination or other techniques, to yield the displacement vector $\mathbf{Q}$.

Galerkin Approach

Following the steps presented in Chapter 1, we introduce

$$\boldsymbol{\phi} = [\phi_x, \phi_y]^T \tag{5.46}$$

and

$$\boldsymbol{\epsilon}(\boldsymbol{\phi}) = \left[\frac{\partial \phi_x}{\partial x}, \frac{\partial \phi_y}{\partial y}, \frac{\partial \phi_x}{\partial y} + \frac{\partial \phi_y}{\partial x}\right]^T \tag{5.47}$$

where $\boldsymbol{\phi}$ is an arbitrary (virtual) displacement vector, consistent with the boundary conditions. The variational form is given by

$$\int_A \boldsymbol{\sigma}^T \boldsymbol{\epsilon}(\boldsymbol{\phi}) t \, dA - \left(\int_A \boldsymbol{\phi}^T \mathbf{f} t \, dA + \int_L \boldsymbol{\phi}^T \mathbf{T} t \, d\ell + \sum_i \boldsymbol{\phi}_i^T \mathbf{P}_i \right) = 0 \qquad (5.48)$$

where the first term represents the internal virtual work. The expression in parentheses represents the external virtual work. On the discretized region, the above equation becomes

$$\sum_e \int_e \boldsymbol{\epsilon}^T \mathbf{D} \boldsymbol{\epsilon}(\boldsymbol{\phi}) t \, dA - \left(\sum_e \int_e \boldsymbol{\phi}^T \mathbf{f} t \, dA + \int_L \boldsymbol{\phi}^T \mathbf{T} t \, d\ell + \sum_i \boldsymbol{\phi}_i^T \mathbf{P}_i \right) = 0 \qquad (5.49)$$

Using the interpolation steps of Eqs. 5.12–5.14, we express

$$\boldsymbol{\phi} = \mathbf{N} \boldsymbol{\psi} \qquad (5.50)$$

$$\boldsymbol{\epsilon}(\boldsymbol{\phi}) = \mathbf{B} \boldsymbol{\psi} \qquad (5.51)$$

where

$$\boldsymbol{\psi} = [\psi_1, \psi_2, \psi_3, \psi_4, \psi_5, \psi_6]^T \qquad (5.52)$$

represents the arbitrary nodal displacements of element e. The global nodal displacement variations $\boldsymbol{\Psi}$ are represented by

$$\boldsymbol{\Psi} = [\Psi_1, \Psi_2, \ldots, \Psi_N]^T \qquad (5.53)$$

The element internal work term in Eq. 5.49 can be expressed as

$$\int_e \boldsymbol{\epsilon}^T \mathbf{D} \boldsymbol{\epsilon}(\boldsymbol{\phi}) t \, dA = \int_e \mathbf{q}^T \mathbf{B}^T \mathbf{D} \mathbf{B} \boldsymbol{\psi} t \, dA$$

Noting that all terms of $\mathbf{B}$ and $\mathbf{D}$ are constant, and denoting t_e and A_e as thickness and area of element, respectively, we find

$$\int_\epsilon \boldsymbol{\epsilon}^T \mathbf{D} \boldsymbol{\epsilon}(\boldsymbol{\phi}) t \, dA = \mathbf{q}^T \mathbf{B}^T \mathbf{D} \mathbf{B} t_e \int_e dA \, \boldsymbol{\psi}$$

$$= \mathbf{q}^T t_e A_e \mathbf{B}^T \mathbf{D} \mathbf{B} \boldsymbol{\psi} \qquad (5.54)$$

$$= \mathbf{q}^T \mathbf{k}^e \boldsymbol{\psi}$$

where $\mathbf{k}^e$ is the element stiffness matrix given by

$$\mathbf{k}^e = t_e A_e \mathbf{B}^T \mathbf{D} \mathbf{B} \qquad (5.55)$$

The material property matrix $\mathbf{D}$ is symmetric, and, hence, the element stiffness matrix is also symmetric. The element connectivity as presented in Table 5.1 is used in adding the stiffness values of $\mathbf{k}^e$ to the global locations. Thus,

$$\sum_e \int_e \boldsymbol{\epsilon}^T \mathbf{D}\boldsymbol{\epsilon}(\boldsymbol{\phi})t \; dA = \sum_e \mathbf{q}^T \mathbf{k}^e \boldsymbol{\psi} = \sum_e \boldsymbol{\psi}^T \mathbf{k}^e \mathbf{q}$$

$$= \boldsymbol{\Psi}^T \mathbf{KQ} \tag{5.56}$$

The global stiffness matrix $\mathbf{K}$ is symmetric and banded. The treatment of external virtual work terms follows the steps involved in the treatment of force terms in the potential energy formulation, where $\mathbf{u}$ is replaced by $\boldsymbol{\phi}$. Thus,

$$\int_e \boldsymbol{\phi}^T \mathbf{f} t \; dA = \boldsymbol{\psi}^T \mathbf{f}^e \tag{5.57}$$

which follows from Eq. 5.33, with $\mathbf{f}^e$ given by Eq. 5.36. Similarly, the traction and point load treatment follows from Eqs. 5.38 and 5.43. The terms in the variational form are given by

$$\text{Internal virtual work} = \boldsymbol{\Psi}^T \mathbf{KQ} \tag{5.58a}$$

$$\text{External virtual work} = \boldsymbol{\Psi}^T \mathbf{F} \tag{5.58b}$$

The stiffness and force matrices are modified to use the full size (all degrees of freedom), using methods suggested in Chapter 3. From the Galerkin form (Eq. 5.49), the arbitrariness of $\boldsymbol{\Psi}$ gives

$$\mathbf{KQ} = \mathbf{F} \tag{5.59}$$

where $\mathbf{K}$ and $\mathbf{F}$ are modified to account for boundary conditions. Equation 5.59 turns out to be the same as Eq. 5.45, obtained in the potential energy formulation.

Stress Calculations

Since strains are constant in a constant strain triangle (CST) element, the corresponding stresses are constant. The stress values need to be calculated for each element. Using the stress–strain relations in Eq. 5.6 and element strain–displacement relations in Eq. 5.25, we have

$$\boldsymbol{\sigma} = \mathbf{DBq} \tag{5.60}$$

The connectivity in Table 5.1 is once again needed to extract the element nodal displacements $\mathbf{q}$ from the global displacements vector $\mathbf{Q}$. Equation 5.60 is used to calculate the element stresses. For interpolation purposes, the calculated stress may be used as the value at the centroid of the element.

Principal stresses and their directions are calculated using Mohr's circle relationships. The program at the end of the chapter includes the principal stress calculations.

Detailed calculations in the example below illustrate the steps involved. However, it is expected that the exercise problems at the end of the chapter will be solved using a computer.

Example 5.4

For the two-dimensional loaded plate shown in Fig. E5.4, determine the displacements of nodes 1 and 2 and the element stresses using plane stress conditions. Body force may be neglected in comparison to the external forces.

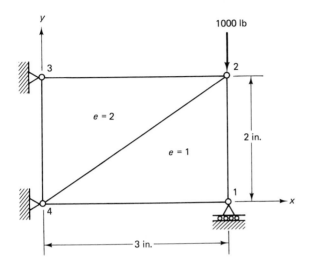

Thickness t = 0.5 in., E = 30 × 10⁶ psi, ν = 0.25   **Figure E5.4**

Solution   For plane stress conditions, the material property matrix is given by

$$\mathbf{D} = \frac{E}{1 - \nu^2} \begin{bmatrix} 1 & \nu & 0 \\ \nu & 1 & 0 \\ 0 & 0 & \dfrac{1 - \nu}{2} \end{bmatrix} = \begin{bmatrix} 3.2 \times 10^7 & 0.8 \times 10^7 & 0 \\ 0.8 \times 10^7 & 3.2 \times 10^7 & 0 \\ 0 & 0 & 1.2 \times 10^7 \end{bmatrix}$$

Using the local numbering pattern used in Fig. E5.3, we establish the connectivity as follows:

		Nodes	
Element No.	1	2	3
1	1	2	4
2	3	4	2

On performing the matrix multiplication $\mathbf{DB}^e$, we get

$$\mathbf{DB}^1 = 10^7 \begin{bmatrix} 1.067 & -0.4 & 0 & 0.4 & -1.067 & 0 \\ 0.267 & -1.6 & 0 & 1.6 & -0.267 & 0 \\ -0.6 & 0.4 & 0.6 & 0 & 0 & -0.4 \end{bmatrix}$$

and

$$\mathbf{DB}^2 = 10^7 \begin{bmatrix} -1.067 & 0.4 & 0 & -0.4 & 1.067 & 0 \\ -0.267 & 1.6 & 0 & -1.6 & 0.267 & 0 \\ 0.6 & -0.4 & -0.6 & 0 & 0 & 0.4 \end{bmatrix}$$

These two relationships will be used later in calculating stresses using $\boldsymbol{\sigma}^e = \mathbf{DB}^e\mathbf{q}$. The multiplication $t_e A_e \mathbf{B}^{e^T} \mathbf{DB}^e$ gives the element stiffness matrices,

$$\mathbf{k}^1 = 10^7 \begin{bmatrix}
\overset{1}{0.983} & \overset{2}{-0.5} & \overset{3}{-0.45} & \overset{4}{0.2} & \overset{5}{-0.533} & \overset{6}{0.3} \\
 & 1.4 & 0.3 & -1.2 & 0.2 & -0.2 \\
 & & 0.45 & 0 & 0 & -0.3 \\
 & & & 1.2 & -0.2 & 0 \\
 & \text{Symmetric} & & & 0.533 & 0 \\
 & & & & & 0.2
\end{bmatrix} \begin{array}{l} \text{Global} \\ \leftarrow \text{dof} \end{array}$$

$$\mathbf{k}^2 = 10^7 \begin{bmatrix}
\overset{5}{0.983} & \overset{6}{-0.5} & \overset{7}{-0.45} & \overset{8}{0.2} & \overset{3}{-0.533} & \overset{4}{0.3} \\
 & 1.4 & 0.3 & -1.2 & 0.2 & -0.2 \\
 & & 0.45 & 0 & 0 & -0.3 \\
 & & & 1.2 & -0.2 & 0 \\
 & \text{Symmetric} & & & 0.533 & 0 \\
 & & & & & 0.2
\end{bmatrix} \begin{array}{l} \text{Global} \\ \leftarrow \text{dof} \end{array}$$

In the above element matrices, the global dof association is shown on top. In the problem under consideration, Q_2, Q_5, Q_6, Q_7, and Q_8 are all zero. Using the elimination approach discussed in Chapter 3, it is now sufficient to consider the stiffnesses associated with the degrees of freedom Q_1, Q_3, Q_4. Since the body forces are neglected, the force vector has the component $F_4 = -1000$ lb. The set of equations is given by the matrix representation

$$10^7 \begin{bmatrix} 0.983 & -0.45 & 0.2 \\ -0.45 & 0.983 & 0 \\ 0.2 & 0 & 1.4 \end{bmatrix} \begin{Bmatrix} Q_1 \\ Q_3 \\ Q_4 \end{Bmatrix} = \begin{Bmatrix} 0 \\ 0 \\ -1000 \end{Bmatrix}$$

Solving for Q_1, Q_3, and Q_4, we get

$$Q_1 = 1.913 \times 10^{-5} \text{ in.} \qquad Q_3 = 0.875 \times 10^{-5} \text{ in.} \qquad Q_4 = -7.436 \times 10^{-5} \text{ in.}$$

For element 1, the element nodal displacement vector is given by

$$\mathbf{q}^1 = 10^{-5}[1.913, 0, 0.875, -7.436, 0, 0]^T$$

The element stresses $\boldsymbol{\sigma}^1$ are calculated from $\mathbf{DB}^1\mathbf{q}$ as

$$\boldsymbol{\sigma}^1 = [-93.3, -1138.7, -62.3]^T \text{psi}$$

Similarly

$$\mathbf{q}^2 = 10^{-5}[0, 0, 0, 0, 0.875, -7.436]^{\mathrm{T}}$$

$$\boldsymbol{\sigma}^2 = [93.4, 23.4, -297.4]^{\mathrm{T}}\mathrm{psi}$$

The computer results may differ slightly since the penalty approach for handling boundary conditions is used in the computer program.   ■

Temperature Effects

If the distribution of the change in temperature $\Delta T(x, y)$ is known, the strain due to this change in temperature can be treated as an initial strain $\boldsymbol{\epsilon}_0$. From the theory of mechanics of solids, $\boldsymbol{\epsilon}_0$ can be represented by

$$\boldsymbol{\epsilon}_0 = [\alpha\,\Delta T,\ \alpha\,\Delta T,\ 0]^{\mathrm{T}} \qquad (5.61)$$

for plane stress, and

$$\boldsymbol{\epsilon}_0 = (1 + \nu)[\alpha\,\Delta T,\ \alpha\,\Delta T,\ 0]^{\mathrm{T}} \qquad (5.62)$$

for plane strain. The stresses and strains are related by

$$\boldsymbol{\sigma} = \mathbf{D}(\boldsymbol{\epsilon} - \boldsymbol{\epsilon}_0) \qquad (5.63)$$

The effect of temperature can be accounted for by considering the strain energy term. We have

$$
\begin{aligned}
U &= \frac{1}{2}\int(\boldsymbol{\epsilon} - \boldsymbol{\epsilon}_0)^{\mathrm{T}}\mathbf{D}(\boldsymbol{\epsilon} - \boldsymbol{\epsilon}_0)t\,dA \\
&= \frac{1}{2}\int(\boldsymbol{\epsilon}^{\mathrm{T}}\mathbf{D}\boldsymbol{\epsilon} - 2\boldsymbol{\epsilon}^{\mathrm{T}}\mathbf{D}\boldsymbol{\epsilon}_0 + \boldsymbol{\epsilon}_0^{\mathrm{T}}\mathbf{D}\boldsymbol{\epsilon}_0)t\,dA
\end{aligned}
\qquad (5.64)
$$

The first term in the expansion above gives the stiffness matrix derived before. The last term is a constant, which has no effect on the minimization process. The middle term, which yields the temperature load, is now considered in detail. Using the strain–displacement relationship $\boldsymbol{\epsilon} = \mathbf{Bq}$,

$$\int_A \boldsymbol{\epsilon}^{\mathrm{T}}\mathbf{D}\boldsymbol{\epsilon}_0 t\,dA = \sum_e \mathbf{q}^{\mathrm{T}}(\mathbf{B}^{\mathrm{T}}\mathbf{D}\boldsymbol{\epsilon}_0)t_e A_e \qquad (5.65)$$

This step is directly obtained in the Galerkin approach where $\boldsymbol{\epsilon}^{\mathrm{T}}$ will be $\boldsymbol{\epsilon}^{\mathrm{T}}(\phi)$ and $\mathbf{q}^{\mathrm{T}}$ will be $\boldsymbol{\psi}^{\mathrm{T}}$.

It is convenient to designate the element temperature load as

$$\boldsymbol{\Theta}^e = t_e A_e\,\mathbf{B}^{\mathrm{T}}\mathbf{D}\boldsymbol{\epsilon}_0 \qquad (5.66)$$

where

$$\boldsymbol{\Theta}^e = [\Theta_1, \Theta_2, \Theta_3, \Theta_4, \Theta_5, \Theta_6]^{\mathrm{T}} \qquad (5.67)$$

The vector $\boldsymbol{\epsilon}_0$ is the strain in Eq. 5.61 or 5.62 due to the average temperature change in the element. $\boldsymbol{\Theta}^e$ represents the element nodal load contributions that must be added to the global force vector using the connectivity.

The stresses in an element are then obtained by using Eq. 5.63 in the form

$$\boldsymbol{\sigma} = \mathbf{D}(\mathbf{Bq} - \boldsymbol{\epsilon}_0) \tag{5.68}$$

5.4 PROBLEM MODELING AND BOUNDARY CONDITIONS

The finite element method is used for computing displacements and stresses for a wide variety of problems. The physical dimensions, loading, and boundary conditions are clearly defined in some problems, similar to what we discussed in Example 5.4. In other problems, these are not clear at the outset.

An example is the problem illustrated in Fig. 5.8a. A plate with such a loading can exist anywhere in space. Since we are interested in the deformation of the body, the symmetry of the geometry and the symmetry of the loading can be used effectively. Let x and y represent the axes of symmetry as shown in Fig. 5.8b. The points along the x axis move along x and are constrained in the y direction and points along the y axis are constrained along the x direction. This suggests that the part, which is one-quarter of the full area, with the loading and boundary conditions as shown is all that is needed to solve for the deformation and stresses.

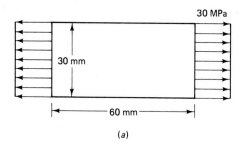

(a)

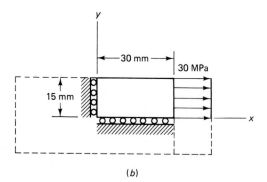

(b)

Figure 5.8 Rectangular plate.

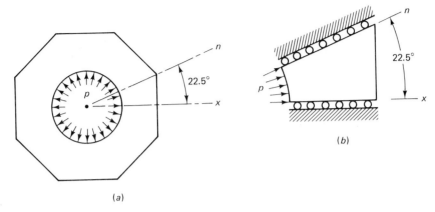

(a)

(b)

Figure 5.9   Octagonal pipe.

As another example, consider an octagonal pipe under internal pressure, shown in Fig. 5.9a. By symmetry, we observe that it is sufficient to consider the 22.5° segment shown in Fig. 5.9b. The boundary conditions require that points along x and n are constrained normal to the two lines, respectively. Note that for a *circular* pipe under internal or external pressure, by symmetry, all points move radially. In this case, any radial segment may be considered. The boundary conditions for points along the x axis in Fig. 5.9b are easily considered by using the penalty approach discussed in Chapter 3. The boundary conditions for points along the inclined direction n, which are considered perpendicular to n, are now treated in detail. If node i with degrees of freedom Q_{2i-1} and Q_{2i} moves along n as seen in Fig. 5.10, and θ is the angle of inclination of n with respect to x axis, we have

$$Q_{2i-1}\sin\theta - Q_{2i}\cos\theta = 0 \tag{5.69}$$

This boundary condition is seen to be a multipoint constraint, which is discussed in Chapter 3. Using the penalty approach presented in Chapter 3, this amounts to adding a term to the potential energy:

$$\Pi = \tfrac{1}{2}\mathbf{Q}^{\mathrm{T}}\mathbf{K}\mathbf{Q} - \mathbf{Q}^{\mathrm{T}}\mathbf{F} + \tfrac{1}{2}C(Q_{2i-1}\sin\theta - Q_{2i}\cos\theta)^2 \tag{5.70}$$

where C is a large number.

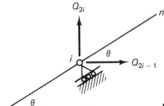

Figure 5.10   Inclined roller support.

The squared term in Eq. 5.70 can be written in the form

$$\tfrac{1}{2}C(Q_{2i-1}\sin\theta - Q_{2i}\cos\theta)^2 = \tfrac{1}{2}[Q_{2i-1}, Q_{2i}]\begin{bmatrix} C\sin^2\theta & -C\sin\theta\cos\theta \\ -C\sin\theta\cos\theta & C\cos^2\theta \end{bmatrix}\begin{Bmatrix} Q_{2i-1} \\ Q_{2i} \end{Bmatrix}$$

(5.71)

The terms $C\sin^2\theta$, $-C\sin\theta\cos\theta$, and $C\cos^2\theta$ get added to the global stiffness matrix, for every node on the incline, and the new stiffness matrix is used to solve for the displacements. Note that the above modifications can also be directly obtained from Eq. 3.82 by substituting $\beta_0 = 0$, $\beta_1 = \sin\theta$, and $\beta_2 = -\cos\theta$. The contributions to the banded stiffness matrix $\mathbf{S}$ are made in the locations $(2i - 1, 1)$, $(2i - 1, 2)$, and $(2i, 1)$ by adding $C\sin^2\theta$, $-C\sin\theta\cos\theta$, $C\cos^2\theta$, respectively.

Some General Comments on Dividing into Elements

When dividing an area into triangles, avoid large aspect ratios. Aspect ratio is defined as the ratio of maximum to minimum characteristic dimensions. Observe that the best elements are those that approach an equilateral triangular configuration. Such configurations are not usually possible. A good practice may be to choose corner angles in the range of 30° to 120°.

In problems where the stresses change widely over an area, such as in notches and fillets, it is good practice to decrease the size of elements in that area to capture the stress variations. The constant strain triangle (CST), in particular, gives constant stresses on the element. This suggests that smaller elements will better represent the distribution. Better estimates of maximum stress may be obtained even with coarser meshes by plotting and extrapolating. For this purpose, the constant element stresses may be interpreted as the values at centroids of the triangle. A method for evaluating nodal values from constant element values is presented in the postprocessing section of Chapter 12.

Coarse meshes are recommended for initial trials to check data and reasonableness of results. Errors may be fixed at this stage, before running larger numbers of elements. Increasing the number of elements in those regions where stress variations are high should give better results. This is called *convergence*. One should get a feel for convergence by successively increasing the number of elements in finite element meshes.

Example 5.5

The solution of Example 5.4 using program FE2CST is presented below, using interactive mode of input.

```
Number of Elements  =? 2
Number of Nodes =? 4
Number of Constrained DOF =? 5
Number of Component Loads =? 1
```

```
Number of Materials =? 1
Thickness =? 0.5
File Name for Output ? EG5_5.DAT
Element#, 3 Nodes, Material#
? 1,4,1,2,1
? 2,3,4,2,1
The Bandwidth is 8
Node#,  X-Coord,   Y-Coord
? 1,3,0
? 2,3,2
? 3,0,2
? 4,0,0
Material#,  E,  Poissons Ratio
? 1,30E6,.25
DOF#,  Displacement
? 2,0
? 5,0
? 6,0
? 7,0
? 8,0
DOF#,  Applied Load
? 4,-1000
```

```
NODE#   X-Displ     Y-Displ
   1    1.907E-05   -8.346E-09
   2    8.733E-06   -7.416E-05
   3    2.736E-09   -1.686E-09
   4   -2.736E-09   -1.383E-10
  DOF#        REACTION:
    2        8.2065E+02
    5       -2.6902E+02
    6        1.6575E+02
    7        2.6902E+02
    8        1.3598E+01
```

```
ELEM#    SX          SY          TXY        S1          S2         ANGLE SX->S1
  1   -9.312E+01  -1.136E+03  -6.208E+01  -8.944E+01  -1.139E+03  -3.396E+00
  2    9.312E+01   2.326E+01  -2.966E+02   3.568E+02  -2.405E+02  -4.164E+01
```

■

Example 5.6

This example shows how more detailed models can be analyzed using program FE2CST with the file mode of input. The finite element model is created using program MESHGEN.

Consider the problem shown in Fig. E5.6a. It is necessary to determine the location and magnitude of the maximum y stress in the plate.

Use of the mesh generation program MESHGEN requires mapping the region into a checkerboard and specifying the number of subdivisions for discretization. The detailed explanation of using MESHGEN is given in Chapter 12. Here, the emphasis is only on using the program to generate input data for FE2CST. Thus, using the checkerboard in Fig. E5.6b, a 36-node, 48-element mesh is created as shown in Fig. E5.6c. Program PLOT2D has been used to generate the plot after executing MESHGEN. Program DATAFEM is used to define the boundary conditions, loads, and material properties. The resulting data file is input into FE2CST. The output file is then examined for the solution, using a text processor or editor. In summary, the order in which programs are executed is MESHGEN, PLOT2D, DATAFEM, and FE2CST.

The maximum y stress is 1768.0 psi occurring in the hatched region in Fig. E5.6c.

The reader is urged to follow the above steps, which will help in the solution of complex problems with less effort. Though not used here, programs BESTFIT and CONTOUR can be used at this stage for obtaining nodal stresses and contour plots, as discussed in Chapter 12. Also, the stresses in the elements may be considered to be accurate at the centroids of the elements and can be extrapolated to obtain the maximum stresses. See Figs. E6.2c or E7.2b for examples of such extrapolation.   ∎

```
1000 REM  *************     FE2CST     *************
1010 REM  ** Two Dimensional Stress Analysis using CST **
1020 CLS
1030 PRINT TAB(22); "====================================="
1040 PRINT TAB(22); "     Program written by          "
1050 PRINT TAB(22); "  T.R.Chandrupatla and A.D.Belegundu   "
1060 PRINT TAB(22); "====================================="
1070 DEFINT I-N
1080 LOCATE 8, 28: PRINT "1. Interactive Data Input"
1090 LOCATE 9, 28: PRINT "2. Data Input from a File"
1100 LOCATE 12, 28: INPUT "   Your Choice <1 or 2> "; IFL1
1110 CLS
1120 IF IFL1 = 2 GOTO 1190
1130 INPUT "Number of Elements ="; NE
1140 INPUT "Number of Nodes ="; NN
1150 INPUT "Number of Constrained DOF ="; ND
1160 INPUT "Number of Component Loads ="; NL
1170 INPUT "Number of Materials ="; NM
1180 GOTO 1220
1190 INPUT "File Name for Input Data "; FILE1$
1200 OPEN FILE1$ FOR INPUT AS #1
1210 INPUT #1, NN, NE, ND, NL, NM, NDIM, NEN
1220 INPUT "Thickness ="; TH
1230 INPUT "File Name for Output "; FILE2$
1240 OPEN FILE2$ FOR OUTPUT AS #2
1250 REM TOTAL DOF IS "NQ"
1260 NQ = 2 * NN
1270 NBW = 0
1280 DIM X(NN, 2), NOC(NE, 3), MAT(NE), PM(NM, 3), NU(ND), U(ND)
1290 DIM F(NQ), D(3, 3), B(3, 6), DB(3, 6), SE(6, 6), Q(6), STR(3)
1300 IF IFL1 = 1 GOTO 1320
1310 GOSUB 2420: GOTO 1360
1320 REM BANDWIDTH NBW FROM CONNECTIVITY"NOC"
1330 PRINT "Element#, 3 Nodes, Material# "
1340 FOR I = 1 TO NE
1350 INPUT N, NOC(N, 1), NOC(N, 2), NOC(N, 3), MAT(N): NEXT I
1360 FOR I = 1 TO NE
1370 CMIN = NN + 1: CMAX = 0
```

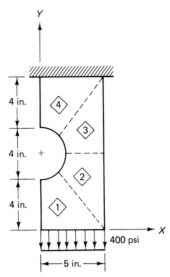

4 in.

4 in.   +

4 in.

400 psi

5 in.

Thickness = 0.4 in.
$E = 30 \times 10^6$ psi, $\nu = 0.3$

(a)

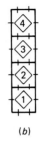

(b)

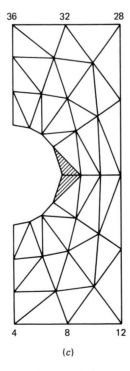

36   32   28

4   8   12

(c)

Figure E5.6

155

```
1380 FOR J = 1 TO 3
1390 IF CMIN > NOC(I, J) THEN CMIN = NOC(I, J)
1400 IF CMAX < NOC(I, J) THEN CMAX = NOC(I, J)
1410 NEXT J
1420 C = 2 * (CMAX - CMIN + 1)
1430 IF NBW < C THEN NBW = C
1440 NEXT I
1450 PRINT "The Bandwidth is"; NBW
1460 REM *** INITIALIZATION ***
1470 DIM S(NQ, NBW)
1480 FOR I = 1 TO NQ: FOR J = 1 TO NBW
1490 S(I, J) = 0: NEXT J: NEXT I
1500 IF IFL1 = 2 GOTO 1630
1510 PRINT "Node#,  X-Coord,  Y-Coord"
1520 FOR I = 1 TO NN
1530 INPUT N, X(N, 1), X(N, 2): NEXT I
1540 PRINT "Material#,  E,  Poissons Ratio"
1550 FOR I = 1 TO NM
1560 INPUT N, PM(N, 1), PM(N, 2): NEXT I
1570 PRINT "DOF#,  Displacement"
1580 FOR I = 1 TO ND
1590 INPUT NU(I), U(I): NEXT I
1600 PRINT "DOF#, Applied Load"
1610 FOR I = 1 TO NL
1620 INPUT N, F(N): NEXT I
1630 REM *** GLOBAL STIFFNESS MATRIX **
1640 FOR N = 1 TO NE
1650 PRINT "Forming Stiffness Matrix of Element "; N
1660 GOSUB 2530
1670 FOR I = 1 TO 6
1680 FOR J = 1 TO 6
1690 SE(I, J) = 0
1700 FOR K = 1 TO 3
1710 SE(I, J) = SE(I, J) + .5 * ABS(DJ) * B(K, I) * DB(K, J) * TH
1720 NEXT K: NEXT J: NEXT I
1730 PRINT ".... Placing in Global Locations"
1740 FOR II = 1 TO 3
1750 NRT = 2 * (NOC(N, II) - 1)
1760 FOR IT = 1 TO 2
1770 NR = NRT + IT
1780 I = 2 * (II - 1) + IT
1790 FOR JJ = 1 TO 3
1800 NCT = 2 * (NOC(N, JJ) - 1)
1810 FOR JT = 1 TO 2
1820 J = 2 * (JJ - 1) + JT
1830 NC = NCT + JT - NR + 1
1840 IF NC <= 0 GOTO 1860
1850 S(NR, NC) = S(NR, NC) + SE(I, J)
1860 NEXT JT: NEXT JJ: NEXT IT: NEXT II
1870 NEXT N
1880 REM *** MODIFY FOR BOUNDARY CONDITIONS ***
1890 CNST = S(1, 1) * 10000
1900 FOR I = 1 TO ND
1910 N = NU(I)
1920 S(N, 1) = S(N, 1) + CNST
1930 F(N) = F(N) + CNST * U(I): NEXT I
1940 REM *** EQUATION SOLVING ***
1950 GOSUB 5000
1960 PRINT #2,"NE,NN,ND,NL,NM ="
1970 PRINT #2,NE;  NN;  ND;  NL;  NM
1980 PRINT #2, "NODE, X-COORD., Y-COORD."
1990 FOR I=1 TO NN
2000 PRINT #2,I;  X(I,1);  X(I,2) : NEXT I
2010 PRINT #2, "ELEMENT#, 3 NODES, MATERIAL#"
2020 FOR I = 1 TO NE
```

```
2030 PRINT #2,I; NOC(I,1);  NOC(I,2);  NOC(I,3);  MAT(I) :NEXT I
2040 PRINT #2, "DOF#, SPECIFIED DISPLACEMENT"
2050 FOR I=1 TO ND: PRINT #2, NU(I): NEXT I
2060 PRINT #2, "MATERIAL#, E, POISSON'S RATIO"
2070 FOR I=1 TO NM: PRINT #2, I, PM(I,1), PM(I,2): NEXT I
2080 PRINT #2, "NODE#   X-Displ    Y-Displ"
2090 FOR I = 1 TO NN
2100 PRINT #2, USING " ###"; I;
2110 PRINT #2, USING " ##.###^^^^"; F(2 * I - 1); F(2 * I): NEXT I
2120 REM *** REACTION CALCULATION ***
2130 PRINT #2, " DOF#      REACTION:"
2140 FFF1$ = "    ###    ##.####^^^^"
2150 FOR I = 1 TO ND
2160 N = NU(I)
2170 R = CNST * (U(I) - F(N))
2180 PRINT #2, USING FFF1$; N; R: NEXT I
2190 PRINT #2,
2200 REM *** STRESS CALCULATIONS ***
2210 PRINT #2, "ELEM#   SX      SY      TXY";
2220 PRINT #2, "     S1      S2     ANGLE SX->S1"
2230 FOR N = 1 TO NE
2240 GOSUB 2530
2250 GOSUB 2830
2260 '***PRINCIPAL STRESS CALCULATIONS***
2270 IF STR(3) <> 0 GOTO 2322
2280 S1 = STR(1): S2 = STR(2): ANG = 0
2290 IF S1 > S2 GOTO 2340
2300 S1 = STR(2): S2 = STR(1): ANG = 90
2310 GOTO 2340
2322 C = .5 * (STR(1) + STR(2))
2324 R = SQR(.25 * (STR(1) - STR(2)) ^ 2 + (STR(3)) ^ 2)
2326 S1 = C + R: S2 = C - R
2328 IF C > STR(1) GOTO 2332
2330 ANG = 57.29577951# * ATN(STR(3) / (STR(1) - S2)): GOTO 2340
2332 ANG = 57.29577951# * ATN(STR(3) / (S1 - STR(1)))
2334 IF STR(3) > 0 THEN ANG = 90 - ANG
2336 IF STR(3) < 0 THEN ANG = -90 - ANG
2340 PRINT #2, USING " ###"; N;
2350 PRINT #2, USING " ##.###^^^^"; STR(1); STR(2); STR(3);
2360 PRINT #2, USING " ##.###^^^^"; S1; S2; ANG
2370 NEXT N
2380 CLOSE #2
2390 PRINT "Results are in file "; FILE2$
2400 END
2410 '=============== READ DATA ====================
2420 FOR I = 1 TO NN: FOR J = 1 TO NDIM
2430 INPUT #1, X(I, J): NEXT J: NEXT I
2440 FOR I = 1 TO NE: FOR J = 1 TO NEN
2450 INPUT #1, NOC(I, J): NEXT J: NEXT I
2460 FOR I = 1 TO NE: INPUT #1, MAT(I): NEXT I
2470 FOR I = 1 TO ND: INPUT #1, NU(I), U(I): NEXT I
2480 FOR I = 1 TO NL: INPUT #1, N, F(N): NEXT I
2490 FOR I = 1 TO NM: FOR J = 1 TO 3
2500 INPUT #1, PM(I, J): NEXT J: NEXT I
2510 CLOSE #1
2520 RETURN
2530 'ELEMENT STIFFNESS FORMATION
2540 '***FIRST THE D-MATRIX***
2550 M = MAT(N): E = PM(M, 1): PNU = PM(M, 2)
2560 C1 = E / (1 - PNU ^ 2): C2 = C1 * PNU: C3 = .5 * E / (1 + PNU)
2570 D(1, 1) = C1: D(1, 2) = C2: D(1, 3) = 0
2580 D(2, 1) = C2: D(2, 2) = C1: D(2, 3) = 0
2590 D(3, 1) = O: D(3, 2) = 0: D(3, 3) = C3
2600 '***STRAIN-DISPLACEMENT MATRIX***
2610 I1 = NOC(N, 1): I2 = NOC(N, 2): I3 = NOC(N, 3)
```

```
2620 X1 = X(I1, 1): Y1 = X(I1, 2)
2630 X2 = X(I2, 1): Y2 = X(I2, 2)
2640 X3 = X(I3, 1): Y3 = X(I3, 2)
2650 X21 = X2 – X1: X32 = X3 – X2: X13 = X1 – X3
2660 Y12 = Y1 – Y2: Y23 = Y2 – Y3: Y31 = Y3 – Y1
2670 DJ = X13 * Y23 – X32 * Y31'DETERMINANT OF JACOBIAN
2680 '***FORMATION OF B MATRIX***
2690 B(1, 1) = Y23 / DJ: B(2, 1) = 0: B(3, 1) = X32 / DJ
2700 B(1, 2) = 0: B(2, 2) = X32 / DJ: B(3, 2) = Y23 / DJ
2710 B(1, 3) = Y31 / DJ: B(2, 3) = 0: B(3, 3) = X13 / DJ
2720 B(1, 4) = 0: B(2, 4) = X13 / DJ: B(3, 4) = Y31 / DJ
2730 B(1, 5) = Y12 / DJ: B(2, 5) = 0: B(3, 5) = X21 / DJ
2740 B(1, 6) = 0: B(2, 6) = X21 / DJ: B(3, 6) = Y12 / DJ
2750 '***DB MATRIX DB=D*B***
2760 FOR I = 1 TO 3
2770 FOR J = 1 TO 6
2780 DB(I, J) = 0
2790 FOR K = 1 TO 3
2800 DB(I, J) = DB(I, J) + D(I, K) * B(K, J)
2810 NEXT K: NEXT J: NEXT I
2820 RETURN
2830 'STRESS EVALUATION
2840 Q(1) = F(2 * I1 – 1): Q(2) = F(2 * I1)
2850 Q(3) = F(2 * I2 – 1): Q(4) = F(2 * I2)
2860 Q(5) = F(2 * I3 – 1): Q(6) = F(2 * I3)
2870 FOR I = 1 TO 3
2880 STR(I) = 0
2890 FOR K = 1 TO 6
2900 STR(I) = STR(I) + DB(I, K) * Q(K): NEXT K: NEXT I
2910 RETURN
5000 N1 = NQ – 1
5010 REM  *** FORWARD ELIMINATION ***
5020 FOR K = 1 TO N1
5030 NK = NQ – K + 1
5040 IF NK > NBW THEN NK = NBW
5050 FOR I = 2 TO NK
5060 C1 = S(K, I) / S(K, 1)
5070 I1 = K + I – 1
5080 FOR J = I TO NK
5090 J1 = J – I + 1
5100 S(I1, J1) = S(I1, J1) – C1 * S(K, J): NEXT J
5110 F(I1) = F(I1) – C1 * F(K): NEXT I: NEXT K
5120 REM  *** BACK SUBSTITUTION ***
5130 F(NQ) = F(NQ) / S(NQ, 1)
5140 FOR KK = 1 TO N1
5150 K = NQ – KK
5160 C1 = 1 / S(K, 1)
5170 F(K) = C1 * F(K)
5180 NK = NQ – K + 1
5190 IF NK > NBW THEN NK = NBW
5200 FOR J = 2 TO NK
5210 F(K) = F(K) – C1 * S(K, J) * F(K + J – 1)
5220 NEXT J
5230 NEXT KK
5240 RETURN
```

PROBLEMS

5.1. The nodal coordinates of the triangular element are shown in Fig. P5.1. At the interior point P, the x coordinate is 3.3 and $N_1 = 0.3$. Determine N_2, N_3, and the y coordinate at point P.

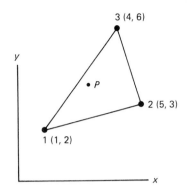

Figure P5.1

5.2. Determine the Jacobian for the $(x, y) - (\xi, \eta)$ transformation for the element shown in Fig. P5.2. Also, find the area of the triangle.

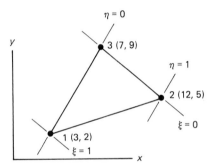

Figure P5.2

5.3. For point P located inside the triangle shown in Fig. P5.3, the shape functions N_1 and N_2 are 0.15 and 0.25, respectively. Determine the x and y coordinates of point P.

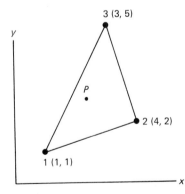

Figure P5.3

5.4. In Example 5.1, determine the shape functions using the area coordinate approach. (*Hint:* Use Area $= 0.5(x_{13}y_{23} - x_{23}y_{13})$ for triangle 1–2–3.)

5.5. Solve the plane stress problem in Fig. P5.5 using three different mesh divisions. Compare your deformation and stress results with values obtained from elementary beam theory.

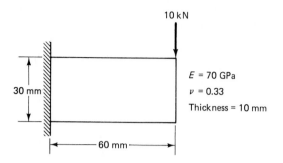

Figure P5.5

5.6. In the plate with a hole under plane stress (Fig. P5.6), find the deformed shape of the hole and determine the maximum stress distribution along *AB* by using stresses in elements adjacent to the line. (*Note:* The result in this problem is the same for any thickness. You may use $t = 1$ in.)

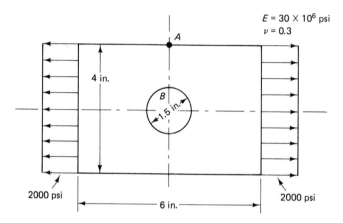

Figure P5.6

5.7. Model a half of the disk with a hole (Fig. P5.7), and find the major and minor dimensions after compression. Also, plot the distribution of maximum stress along *AB*.

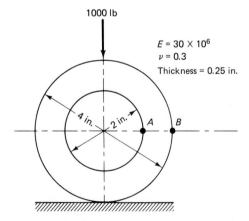

Figure P5.7

5.8. Model the 22.5° segment of the octagonal pipe shown in Fig. P5.8. Show the deformed configuration of the segment and the distribution of maximum in-plane shear stress. (*Hint:* For all points along *CD*, use stiffness modification suggested in Eq. 5.71. Also, maximum in-plane shear stress $= (\sigma_1 - \sigma_2)/2$, where σ_1 and σ_2 are the principal stresses. Assume plane strain.)

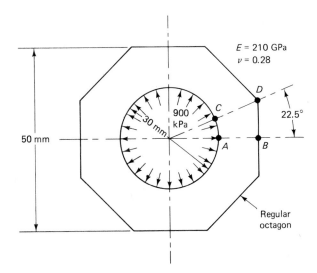

Figure P5.8

5.9. For the linearly varying distributed load on edge 1–2, shown in Fig. P5.9, determine the nodal load contributions F_1, F_2, F_3, and F_4, in terms of T_{1x}, T_{1y}, T_{2x}, T_{2y}, (x_1, y_1), and (x_2, y_2). (*Hint:* In the development in Eq. 5.38a, use $T_x = N_1 T_{1x} + N_2 T_{2x}$ and $T_y = N_1 T_{1y} + N_2 T_{2y}$, and integrate the various terms.)

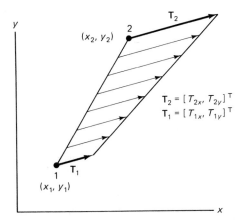

Figure P5.9

5.10. Determine the location and magnitude of maximum principal stress and maximum shearing stress in the fillet shown in Fig. P5.10.

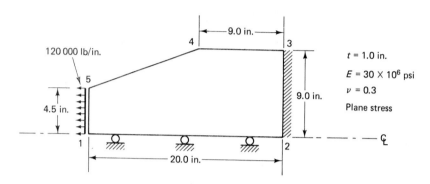

Figure P5.10

*5.11. The torque arm in Fig. P5.11 is an automotive component. Determine the location and magnitude of maximum Von Mises stress, σ_{VM}, given by

$$\sigma_{VM} = \sqrt{\sigma_x^2 - \sigma_x\sigma_y + \sigma_y^2 + 3\tau_{xy}^2}$$

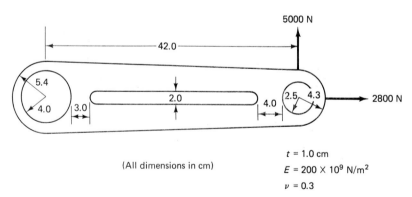

Figure P5.11

5.12. A large flat surface of a steel body is subjected to a line load of 100 lb/in. Assuming plane strain, consider an enclosure as shown in Fig. P5.12 and determine the deformation of the surface and stress distribution in the body. (*Hint:* Choose small elements close to the load.)

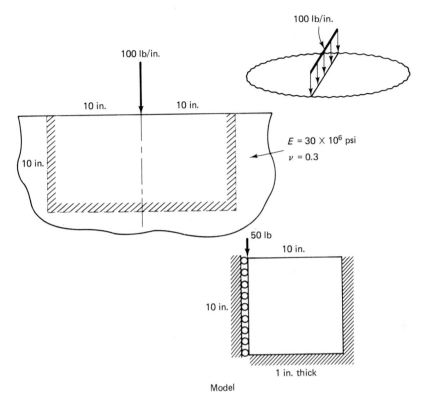

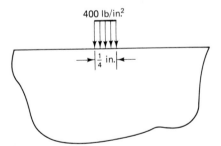

Figure P5.12

5.13. In the above problem, the load is changed to a distributed load 400 lb/in.^{2} on a 1/4-in.-wide long region, as in Fig. P5.13. Model the problem as above with this loading and find deformation of the surface and stress distribution in the body.

Figure P5.13

*5.14. A $\frac{1}{2}$ × 5-in. copper piece fits snugly into a short channel-shaped steel piece at room temperature, as shown in Fig. P5.14. The assembly is subjected to a uniform temperature increase of 80° F. Assuming that the properties are constant within this change and that the surfaces are bonded together, find the deformed shape and the stress distribution. (*Hint:* Introduce the nodal loads Eq. 5.66 due to temperature effect into the program FE2CST to solve the problem.)

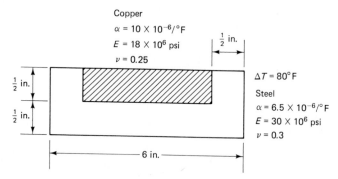

Figure P5.14

*5.15. In the slotted ring shown in Fig. P5.15, two loads of magnitude P and load R are applied such that the 3-mm gap closes. Determine the magnitude of P and show the deformed shape of the part. (*Hint:* Find the deflection of gap for say, $P = 100$, and multiply the deflections proportionately.)

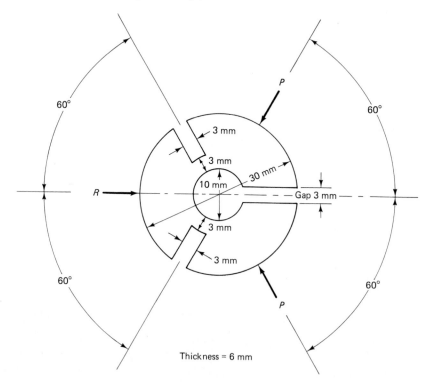

Figure P5.15

chapter six

Axisymmetric Solids Subjected to Axisymmetric Loading

6.1 INTRODUCTION

Problems involving three-dimensional axisymmetric solids or solids of revolution, subjected to axisymmetric loading, reduce to simple two-dimensional problems. Because of total symmetry about the z axis, as seen in Fig. 6.1, all deformations and stresses are independent of the rotational angle θ. Thus, the problem needs to be looked at as a two-dimensional problem in rz, defined on the revolving area (Fig. 6.1b). Gravity forces can be considered if acting in the z direction. Revolving bodies like flywheels can be analyzed by introducing centrifugal forces in the body force term. We now discuss the axisymmetric problem formulation.

6.2 AXISYMMETRIC FORMULATION

Considering the elemental volume shown in Fig. 6.2, the potential energy can be written in the form

$$\Pi = \frac{1}{2} \int_0^{2\pi} \int_A \boldsymbol{\sigma}^T \boldsymbol{\epsilon} r \, dA \, d\theta - \int_0^{2\pi} \int_A \mathbf{u}^T \mathbf{f} r \, dA \, d\theta - \int_0^{2\pi} \int_L \mathbf{u}^T \mathbf{T} r \, d\ell \, d\theta - \sum_i \mathbf{u}_i^T \mathbf{P}_i$$

(6.1)

where $r \, d\ell \, d\theta$ is the elemental surface area, and the point load $\mathbf{P}_i$ represents a line load distributed around a circle, as shown in Fig. 6.1.

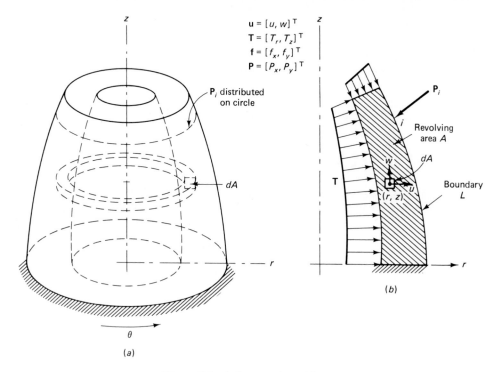

$$\mathbf{u} = [u, w]^T$$
$$\mathbf{T} = [T_r, T_z]^T$$
$$\mathbf{f} = [f_x, f_y]^T$$
$$\mathbf{P} = [P_x, P_y]^T$$

Figure 6.1 Axisymmetric problem.

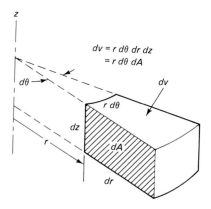

$$dv = r\, d\theta\, dr\, dz$$
$$= r\, d\theta\, dA$$

Figure 6.2 Elemental volume.

All variables in the integrals are independent of θ. Thus, Eq. 6.1 can be written as

$$\Pi = 2\pi\left(\frac{1}{2}\int_A \boldsymbol{\sigma}^T \boldsymbol{\epsilon} r\, dA - \int_A \mathbf{u}^T \mathbf{f} r\, dA - \int_L \mathbf{u}^T \mathbf{T} r\, d\ell\right) - \sum_i \mathbf{u}_i^T \mathbf{P}_i \qquad (6.2)$$

where

$$\mathbf{u} = [u, w]^{\mathrm{T}} \tag{6.3}$$

$$\mathbf{f} = [f_r, f_z]^{\mathrm{T}} \tag{6.4}$$

$$\mathbf{T} = [T_r, T_z]^{\mathrm{T}} \tag{6.5}$$

From Fig. 6.3, we can write the relationship between strains $\boldsymbol{\epsilon}$ and displacements $\mathbf{u}$ as

$$\boldsymbol{\epsilon} = [\epsilon_r, \ \epsilon_z, \ \gamma_{rz}, \ \epsilon_\theta]^{\mathrm{T}}$$

$$= \left[\frac{\partial u}{\partial r}, \ \frac{\partial w}{\partial z}, \ \frac{\partial u}{\partial z} + \frac{\partial w}{\partial r}, \ \frac{u}{r} \right]^{T} \tag{6.6}$$

The stress vector is correspondingly defined as

$$\boldsymbol{\sigma} = [\sigma_r, \ \sigma_z, \ \tau_{rz}, \ \sigma_\theta]^{\mathrm{T}} \tag{6.7}$$

The stress–strain relations are given in the usual form:

$$\boldsymbol{\sigma} = \mathbf{D}\boldsymbol{\epsilon} \tag{6.8}$$

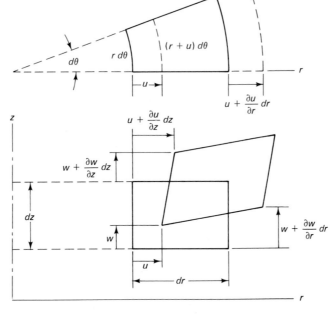

Figure 6.3 Deformation of elemental volume.

where the (4×4) matrix $\mathbf{D}$ can be written by dropping the appropriate terms from the three-dimensional matrix in Chapter 1, as

$$\mathbf{D} = \frac{E(1 - \nu)}{(1 + \nu)(1 - 2\nu)} \begin{bmatrix} 1 & \dfrac{\nu}{1 - \nu} & 0 & \dfrac{\nu}{1 - \nu} \\ \dfrac{\nu}{1 - \nu} & 1 & 0 & \dfrac{\nu}{1 - \nu} \\ 0 & 0 & \dfrac{1 - 2\nu}{2(1 - \nu)} & 0 \\ \dfrac{\nu}{1 - \nu} & \dfrac{\nu}{1 - \nu} & 0 & 1 \end{bmatrix} \quad (6.9)$$

In the Galerkin formulation, we require

$$2\pi \int \boldsymbol{\sigma}^\mathrm{T} \boldsymbol{\epsilon}(\boldsymbol{\phi}) r \, dA - \left(2\pi \int_A \boldsymbol{\phi}^\mathrm{T} \mathbf{f} r \, dA + 2\pi \int_L \boldsymbol{\phi}^\mathrm{T} \mathbf{T} r \, d\ell + \sum \boldsymbol{\phi}_i^\mathrm{T} \mathbf{P}_i \right) = 0 \quad (6.10)$$

where

$$\boldsymbol{\phi} = [\phi_r, \ \phi_z]^\mathrm{T} \quad (6.11)$$

$$\boldsymbol{\epsilon}(\boldsymbol{\phi}) = \left[\frac{\partial \phi_r}{\partial r}, \ \frac{\partial \phi_z}{\partial z}, \ \frac{\partial \phi_r}{\partial z} + \frac{\partial \phi_z}{\partial r}, \ \frac{\phi_r}{r} \right]^\mathrm{T} \quad (6.12)$$

6.3 FINITE ELEMENT MODELING: TRIANGULAR ELEMENT

The two-dimensional region defined by the revolving area is divided into triangular elements, as shown in Fig. 6.4. Though each element is completely represented by the area in the rz plane, in reality, it is a ring-shaped solid of revolution obtained by revolving the triangle about the z axis. A typical element is shown in Fig. 6.5.

The definition of connectivity of elements and the nodal coordinates follow the steps involved in the CST element discussed in Section 5.2. We note here that the r and z coordinates respectively replace x and y.

Using the three shape functions N_1, N_2, and N_3, we define

$$\mathbf{u} = \mathbf{Nq} \quad (6.13)$$

where $\mathbf{u}$ is defined in (6.3) and

$$\mathbf{N} = \begin{bmatrix} N_1 & 0 & N_2 & 0 & N_3 & 0 \\ 0 & N_1 & 0 & N_2 & 0 & N_3 \end{bmatrix} \quad (6.14)$$

$$\mathbf{q} = [q_1, q_2, q_3, q_4, q_5, q_6]^\mathrm{T} \quad (6.15)$$

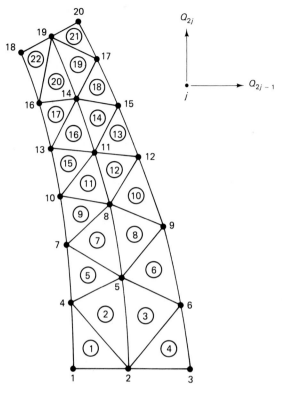

Figure 6.4   Triangulation.

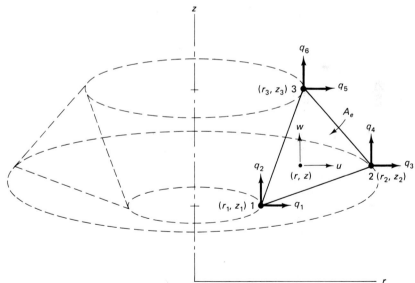

Figure 6.5   Axisymmetric triangular element.

If we denote $N_1 = \xi$ and $N_2 = \eta$, and note that $N_3 = 1 - \xi - \eta$, then Eq. 6.13 gives

$$
\begin{aligned}
u &= \xi q_1 + \eta q_3 + (1 - \xi - \eta)q_5 \\
w &= \xi q_2 + \eta q_4 + (1 - \xi - \eta)q_6
\end{aligned}
\tag{6.16}
$$

By using the isoparametric representation, we find

$$
\begin{aligned}
r &= \xi r_1 + \eta r_2 + (1 - \xi - \eta)r_3 \\
z &= \xi z_1 + \eta z_2 + (1 - \xi - \eta)z_3
\end{aligned}
\tag{6.17}
$$

The chain rule of differentiation gives

$$
\left\{ \begin{matrix} \dfrac{\partial u}{\partial \xi} \\[2mm] \dfrac{\partial u}{\partial \eta} \end{matrix} \right\} = \mathbf{J} \left\{ \begin{matrix} \dfrac{\partial u}{\partial r} \\[2mm] \dfrac{\partial u}{\partial z} \end{matrix} \right\}
\tag{6.18}
$$

and

$$
\left\{ \begin{matrix} \dfrac{\partial w}{\partial \xi} \\[2mm] \dfrac{\partial w}{\partial \eta} \end{matrix} \right\} = \mathbf{J} \left\{ \begin{matrix} \dfrac{\partial w}{\partial r} \\[2mm] \dfrac{\partial w}{\partial z} \end{matrix} \right\}
\tag{6.19}
$$

where the Jacobian $\mathbf{J}$ is given by

$$
\mathbf{J} = \begin{bmatrix} r_{13} & z_{13} \\ r_{23} & z_{23} \end{bmatrix}
\tag{6.20}
$$

In the definition of $\mathbf{J}$ above, we have used the notation $r_{ij} = r_i - r_j$ and $z_{ij} = z_i - z_j$. The determinant of $\mathbf{J}$ is

$$
\det \mathbf{J} = r_{13}z_{23} - r_{23}z_{13}
\tag{6.21}
$$

Recall that $|\det \mathbf{J}| = 2A_e$. That is, the absolute value of the determinant of $\mathbf{J}$ equals twice the area of the element. The inverse relations for Eqs. 6.18 and 6.19 are given by

$$
\left\{ \begin{matrix} \dfrac{\partial u}{\partial r} \\[2mm] \dfrac{\partial u}{\partial z} \end{matrix} \right\} = \mathbf{J}^{-1} \left\{ \begin{matrix} \dfrac{\partial u}{\partial \xi} \\[2mm] \dfrac{\partial u}{\partial \eta} \end{matrix} \right\} \quad \text{and} \quad \left\{ \begin{matrix} \dfrac{\partial w}{\partial r} \\[2mm] \dfrac{\partial w}{\partial z} \end{matrix} \right\} = \mathbf{J}^{-1} \left\{ \begin{matrix} \dfrac{\partial w}{\partial \xi} \\[2mm] \dfrac{\partial w}{\partial \eta} \end{matrix} \right\}
\tag{6.22}
$$

where

$$
\mathbf{J}^{-1} = \frac{1}{\det \mathbf{J}} \begin{bmatrix} z_{23} & -z_{13} \\ -r_{23} & r_{13} \end{bmatrix}
\tag{6.23}
$$

Introducing the above transformation relationships into the strain–displacement relations in Eq. 6.6 and using Eqs. 6.16, we get

$$\boldsymbol{\epsilon} = \left\{ \begin{array}{c} \dfrac{z_{23}(q_1 - q_5) - z_{13}(q_3 - q_5)}{\det \mathbf{J}} \\[3mm] \dfrac{-r_{23}(q_2 - q_6) + r_{13}(q_4 - q_6)}{\det \mathbf{J}} \\[3mm] \dfrac{-r_{23}(q_1 - q_5) + r_{13}(q_3 - q_5) + z_{23}(q_2 - q_6) - z_{13}(q_4 - q_6)}{\det \mathbf{J}} \\[3mm] \dfrac{N_1 q_1 + N_2 q_3 + N_3 q_5}{r} \end{array} \right\}$$

This can be written in the matrix form as

$$\boldsymbol{\epsilon} = \mathbf{Bq} \tag{6.24}$$

where the element strain–displacement matrix $\mathbf{B}$, of dimension (4×6), is given by

$$\mathbf{B} = \begin{bmatrix} \dfrac{z_{23}}{\det \mathbf{J}} & 0 & \dfrac{z_{31}}{\det \mathbf{J}} & 0 & \dfrac{z_{12}}{\det \mathbf{J}} & 0 \\[3mm] 0 & \dfrac{r_{32}}{\det \mathbf{J}} & 0 & \dfrac{r_{13}}{\det \mathbf{J}} & 0 & \dfrac{r_{21}}{\det \mathbf{J}} \\[3mm] \dfrac{r_{32}}{\det \mathbf{J}} & \dfrac{z_{23}}{\det \mathbf{J}} & \dfrac{r_{13}}{\det \mathbf{J}} & \dfrac{z_{31}}{\det \mathbf{J}} & \dfrac{r_{21}}{\det \mathbf{J}} & \dfrac{z_{12}}{\det \mathbf{J}} \\[3mm] \dfrac{N_1}{r} & 0 & \dfrac{N_2}{r} & 0 & \dfrac{N_3}{r} & 0 \end{bmatrix} \tag{6.25}$$

Potential Energy Approach

The potential energy Π on the discretized region is given by

$$\Pi = \sum_e \left[\frac{1}{2} \left(2\pi \int_e \boldsymbol{\epsilon}^T \mathbf{D} \boldsymbol{\epsilon} r \, dA \right) - 2\pi \int_e \mathbf{u}^T \mathbf{f} r \, dA - 2\pi \int_e \mathbf{u}^T \mathbf{T} r \, d\ell \right]$$
$$- \sum \mathbf{u}_i^T \mathbf{P}_i \tag{6.26}$$

The element strain energy U_e given by the first term can be written as

$$U_e = \frac{1}{2} \mathbf{q}^T \left(2\pi \int_e \mathbf{B}^T \mathbf{D} \mathbf{B} r \, dA \right) \mathbf{q} \tag{6.27}$$

The quantity inside the parentheses is the element stiffness matrix,

$$\mathbf{k}^e = 2\pi \int_e \mathbf{B}^T \mathbf{D} \mathbf{B} r \, dA \tag{6.28}$$

The fourth row in $\mathbf{B}$ has terms of the type N_i/r. Further, the integral above also has an additional r in it. As a simple approximation, $\mathbf{B}$ and r can be evaluated at the centroid of the triangle and used as representative values for the triangle. At the centroid of the triangle,

$$N_1 = N_2 = N_3 = \tfrac{1}{3} \tag{6.29}$$

and

$$\bar{r} = \frac{r_1 + r_2 + r_3}{3}$$

where $\bar{r}$ is the radius of the centroid. Denoting $\bar{\mathbf{B}}$ as the element strain–displacement matrix $\mathbf{B}$ evaluated at the centroid, we get

$$\mathbf{k}^e = 2\pi\bar{r}\,\bar{\mathbf{B}}^{\mathsf{T}}\mathbf{D}\bar{\mathbf{B}} \int_e dA$$

or

$$\mathbf{k}^e = 2\pi\bar{r}A_e\,\bar{\mathbf{B}}^{\mathsf{T}}\mathbf{D}\bar{\mathbf{B}} \tag{6.30}$$

We note here that $2\pi\bar{r}A_e$ is the volume of the ring-shaped element shown in Fig. 6.5. Also, A_e is given by

$$A_e = \tfrac{1}{2}\,|\det \mathbf{J}| \tag{6.31}$$

We also use this centroid or midpoint rule* for body forces and surface tractions as discussed below. Caution must be exerted for elements close to the axis of symmetry. For better results, smaller elements need to be chosen close to the axis of symmetry. Another approach is to introduce $r = N_1 r_1 + N_2 r_2 + N_3 r_3$ in the following equations and perform elaborate integration. More elaborate methods of numerical integration are discussed in Chapter 7.

Body Force Term

We first consider the body force term $2\pi \int_e \mathbf{u}^{\mathsf{T}}\mathbf{f}r \, dA$. We have

$$2\pi \int_e \mathbf{u}^{\mathsf{T}}\mathbf{f}r \, dA = 2\pi \int_e (uf_r + wf_z)r \, dA$$

$$= 2\pi \int_e [(N_1 q_1 + N_2 q_3 + N_3 q_5)f_r + (N_1 q_2 + N_2 q_4 + N_3 q_6)f_z]r \, dA$$

Once again, approximating the variable quantities by their values at the centroid of the triangle, we get

---

*Suggested by O. C. Zienkiewicz, *The Finite Element Method*, 3d ed. New York: McGraw-Hill, 1983.

$$2\pi \int_e \mathbf{u}^T \mathbf{f} r \, dA = \mathbf{q}^T \mathbf{f}^e \tag{6.32}$$

where the element body force vector $\mathbf{f}^e$ is given by

$$\mathbf{f}^e = \frac{2\pi \bar{r} A_e}{3} [\bar{f}_r, \bar{f}_z, \bar{f}_r, \bar{f}_z, \bar{f}_r, \bar{f}_z]^T \tag{6.33}$$

The bar on the $\mathbf{f}$ terms indicates that they are evaluated at the centroid. Where body force is the primary load, greater accuracy may be obtained by substituting $r = N_1 r_1 + N_2 r_2 + N_3 r_3$ into Eq. 6.32 and integrating to get nodal loads.

Rotating Flywheel

As an example, let us consider a rotating flywheel with its axis in the z direction. We consider the flywheel to be stationary and apply the equivalent radial centrifugal (inertial) force per unit volume of $\rho r \omega^2$, where ρ is the density (mass per unit volume), and ω the angular velocity in rad/s. In addition, if gravity acts along the negative z axis, then

$$\mathbf{f} = [f_r, f_z]^T = [\rho r \omega^2, -\rho g]^T \tag{6.34}$$

and

$$\bar{f}_r = \rho \bar{r} \omega^2, \bar{f}_z = -\rho g \tag{6.35}$$

For more precise results with coarse meshes, we need to use $r = N_1 r_1 + N_2 r_2 + N_3 r_3$ and integrate.

Surface Traction

For a uniformly distributed load with components T_r and T_z, shown in Fig. 6.6, on the edge connecting nodes 1 and 2, we get

$$2\pi \int_e \mathbf{u}^T \mathbf{T} r \, d\ell = \mathbf{q}^T \mathbf{T}^e \tag{6.36}$$

where

$$\mathbf{q} = [q_1, q_2, q_3, q_4]^T \tag{6.37}$$

$$\mathbf{T}^e = 2\pi \ell_{1-2} [aT_r, aT_z, bT_r, bT_z]^T \tag{6.38}$$

$$a = \frac{2r_1 + r_2}{6} \qquad b = \frac{r_1 + 2r_2}{6} \tag{6.39}$$

$$\ell_{1-2} = \sqrt{(r_2 - r_1)^2 + (z_2 - z_1)^2} \tag{6.40}$$

In the above derivation, r is expressed as $N_1 r_1 + N_2 r_2$ and then integrated. When the line $1-2$ is parallel to z axis, we have $r_1 = r_2$, which gives $a = b = 0.5 r_1$.

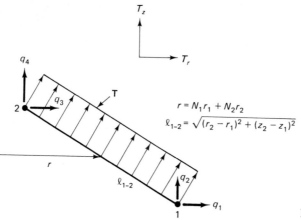

$$r = N_1 r_1 + N_2 r_2$$
$$\ell_{1-2} = \sqrt{(r_2 - r_1)^2 + (z_2 - z_1)^2}$$

Figure 6.6   Surface traction.

The load distributed along a circumference of a circle on the surface has to be applied at a point on the revolving area. We may conveniently locate a node here and add the load components.

On summing up the strain energy and force terms over all the elements and modifying for the boundary conditions while minimizing the total potential energy, we get

$$\mathbf{KQ} = \mathbf{F} \tag{6.41}$$

We note here that axisymmetric boundary conditions need be applied only on the revolving area shown in Fig. 6.1.

Galerkin Approach

In the Galerkin formulation, the consistent variation $\boldsymbol{\phi}$ in an element is expressed as

$$\boldsymbol{\phi} = \mathbf{N}\boldsymbol{\psi} \tag{6.42}$$

where

$$\boldsymbol{\psi} = [\psi_1, \psi_2, \ldots, \psi_6]^T \tag{6.43}$$

The corresponding strain $\boldsymbol{\epsilon}(\phi)$ is given by

$$\boldsymbol{\epsilon}(\phi) = \mathbf{B}\boldsymbol{\psi} \tag{6.44}$$

The global vector of variations $\boldsymbol{\Psi}$ is represented by

$$\boldsymbol{\Psi} = [\Psi_1, \Psi_2, \Psi_3, \ldots, \Psi_N]^T \tag{6.45}$$

We now introduce the interpolated displacements into the Galerkin variational form (Eq. 6.10). The first term representing the internal virtual work gives

$$\text{Internal virtual work} = 2\pi \int_A \boldsymbol{\sigma}^T \boldsymbol{\epsilon}(\boldsymbol{\phi}) r \, dA$$

$$= \sum_e 2\pi \int_e \mathbf{q}^T \mathbf{B}^T \mathbf{D} \mathbf{B} \boldsymbol{\psi} r \, dA$$

$$= \sum_e \mathbf{q}^T \mathbf{k}^e \boldsymbol{\psi} \tag{6.46}$$

where the element stiffness $\mathbf{k}^e$ is given by

$$\mathbf{k}^e = 2\pi \bar{r} A_e \bar{\mathbf{B}}^T \mathbf{D} \bar{\mathbf{B}} \tag{6.47}$$

We note that $\mathbf{k}^e$ is symmetric. Using the connectivity of the elements, the internal virtual work can be expressed in the form

$$\text{Internal virtual work} = \sum_e \mathbf{q}^T \mathbf{k}^e \boldsymbol{\psi} = \sum \boldsymbol{\psi}^T \mathbf{k}^e \mathbf{q}$$

$$= \boldsymbol{\Psi}^T \mathbf{K} \mathbf{Q} \tag{6.48}$$

where $\mathbf{K}$ is the global stiffness matrix. The external virtual work terms in Eq. 6.10 involving body forces, surface tractions, and point loads can be treated in the same way as in the potential energy approach, by replacing $\mathbf{q}$ with $\boldsymbol{\psi}$. The summation of all the force terms over the elements then yields

$$\text{External virtual work} = \boldsymbol{\Psi}^T \mathbf{F} \tag{6.49}$$

The boundary conditions are considered using the ideas discussed in Chapter 3. The stiffness matrix $\mathbf{K}$ and the force $\mathbf{F}$ are modified, resulting in the same set of equations as (6.41).

Detailed calculations in the example below are provided for illustrating the steps involved. However, it is expected that the exercise problems at the end of the chapter will be solved using program AXYSYM, which is provided.

Example 6.1

In Fig. E6.1, a long cylinder of inside diameter 80 mm and outside diameter 120 mm snugly fits in a hole over its full length. The cylinder is then subjected to an internal pressure of 2 MPa. Using two elements on the 10-mm length shown, find the displacements at the inner radius.

Solution

Element	Connectivity			Node	Coordinates	
	1	2	3		r	z
1	1	2	4	1	40	10
2	2	3	4	2	40	0
				3	60	0
				4	60	10

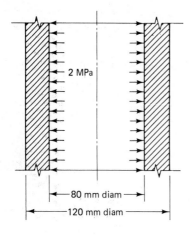 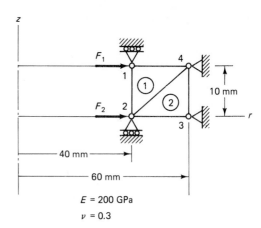

Figure E6.1

We will use the units of millimeters for length, newtons for force, and megapascals for stress and E. These units are consistent. On substituting $E = 200\,000$ MPa and $\nu = 0.3$, we have

$$
\mathbf{D} = \begin{bmatrix}
2.69 \times 10^5 & 1.15 \times 10^5 & 0 & 1.15 \times 10^5 \\
1.15 \times 10^5 & 2.69 \times 10^5 & 0 & 1.15 \times 10^5 \\
0 & 0 & 0.77 \times 10^5 & 0 \\
1.15 \times 10^5 & 1.15 \times 10^5 & 0 & 2.69 \times 10^5
\end{bmatrix}
$$

For both elements, $\det \mathbf{J} = 200$ mm^2, $A_e = 100$ mm^2. From Eq. 6.31, forces F_1 and F_3 are given by

$$
F_1 = F_3 = \frac{2\pi r_1 \ell_e p_i}{2} = \frac{2\pi (40)(10)(2)}{2} = 2514 \text{ N}
$$

The **B** matrices relating element strains to nodal displacements are obtained first.
For element 1, $\bar{r} = \frac{1}{3}(40 + 40 + 60) = 46.67$ mm,

$$
\mathbf{\bar{B}}^1 = \begin{bmatrix}
-0.05 & 0 & 0 & 0 & 0.05 & 0 \\
0 & 0.1 & 0 & -0.1 & 0 & 0 \\
0.1 & -0.05 & -0.1 & 0 & 0 & 0.05 \\
0.0071 & 0 & 0.0071 & 0 & 0.0071 & 0
\end{bmatrix}
$$

For element 2, $\bar{r} = \frac{1}{3}(40 + 60 + 60) = 53.33$ mm,

$$
\mathbf{\bar{B}}^2 = \begin{bmatrix}
-0.05 & 0 & 0.05 & 0 & 0 & 0 \\
0 & 0 & 0 & -0.1 & 0 & 0.1 \\
0 & -0.05 & -0.1 & 0.05 & 0.1 & 0 \\
0.00625 & 0 & 0.00625 & 0 & 0.00625 & 0
\end{bmatrix}
$$

The element stress–displacement matrices are obtained by multiplying **DB**:

$$\mathbf{D\bar{B}^1} = 10^4 \begin{bmatrix} -1.26 & 1.15 & 0.082 & -1.15 & 1.43 & 0 \\ -0.49 & 2.69 & 0.082 & -2.69 & 0.657 & 0.1 \\ 0.77 & -0.385 & -0.77 & 0 & 0 & 0.385 \\ -0.384 & 1.15 & 0.191 & -1.15 & 0.766 & 0 \end{bmatrix}$$

$$\mathbf{D\bar{B}^2} = 10^4 \begin{bmatrix} -1.27 & 0 & 1.42 & -1.15 & 0.072 & 1.15 \\ -0.503 & 0 & 0.647 & -2.69 & 0.072 & 2.69 \\ 0 & -0.385 & -0.77 & 0.385 & 0.77 & 0 \\ -0.407 & 0 & 0.743 & -1.15 & 0.168 & 1.15 \end{bmatrix}$$

The stiffness matrices are obtained by finding $2\pi\bar{r}A_e\,\mathbf{\bar{B}^T D\bar{B}}$ for each element.

Global dof →   1   2   3   4   7   8

$$\mathbf{k^1} = 10^7 \begin{bmatrix} 4.03 & -2.58 & -2.34 & 1.45 & -1.932 & 1.13 \\ & 8.45 & 1.37 & -7.89 & 1.93 & -0.565 \\ & & 2.30 & -0.24 & 0.16 & -1.13 \\ & & & 7.89 & -1.93 & 0 \\ & \text{Symmetric} & & & 2.25 & 0 \\ & & & & & 0.565 \end{bmatrix}$$

Global dof →   3   4   5   6   7   8

$$\mathbf{k^2} = 10^7 \begin{bmatrix} 2.05 & 0 & -2.22 & 1.69 & -0.085 & -1.69 \\ & 0.645 & 1.29 & -0.645 & -1.29 & 0 \\ & & 5.11 & -3.46 & -2.42 & 2.17 \\ & & & 9.66 & 1.05 & -9.01 \\ & \text{Symmetric} & & & 2.62 & 0.241 \\ & & & & & 9.01 \end{bmatrix}$$

Using the elimination approach, on assembling the matrices with reference to the degrees of freedom 1 and 3, we get

$$10^7 \begin{bmatrix} 4.03 & -2.34 \\ -2.34 & 4.35 \end{bmatrix} \begin{Bmatrix} Q_1 \\ Q_3 \end{Bmatrix} = \begin{Bmatrix} 2514 \\ 2514 \end{Bmatrix}$$

$$Q_1 = 0.014 \times 10^{-2} \text{ mm}$$

$$Q_2 = 0.0133 \times 10^{-2} \text{ mm}$$ ∎

Stress Calculations

From the set of nodal displacements obtained above, the element nodal displacements **q** can be found using the connectivity. Then, using stress–strain relation in Eq. 6.8 and strain–displacement relation in Eq. 6.24, we have

$$\boldsymbol{\sigma} = \mathbf{D\bar{B}q} \qquad (6.50)$$

where $\bar{\mathbf{B}}$ is $\mathbf{B}$, given in Eq. 6.25, evaluated at the centroid of the element. We also note that σ_θ is a principal stress. The two principal stresses σ_1, σ_2 corresponding to σ_r, σ_z, τ_{rz} can be calculated using Mohr's circle.

Example 6.2

Calculate the element stresses in the problem discussed in Example 6.1.

Solution   We need to find $\sigma^{eT} = [\sigma_r, \sigma_z, \tau_{rz}, \sigma_\theta]^e$ for each element. From the connectivity established in Example 6.1,

$$\mathbf{q}^1 = [0.0140, 0, 0.0133, 0, 0, 0]^T \times 10^{-2}$$

$$\mathbf{q}^2 = [0.0133, 0, 0, 0, 0, 0]^T \times 10^{-2}$$

Using the product of matrices $\mathbf{DB}^e$ and $\mathbf{q}$

$$\boldsymbol{\sigma}^e = \mathbf{D}\bar{\mathbf{B}}^e\mathbf{q}$$

we get

$$\boldsymbol{\sigma}^1 = [-166, -58.2, 5.4, -28.4]^T \times 10^{-2} \text{ MPa}$$

$$\boldsymbol{\sigma}^2 = [-169.3, -66.9, 0, -54.1]^T \times 10^{-2} \text{ MPa} \qquad \blacksquare$$

Temperature Effects

Uniform increase in temperature of ΔT introduces initial normal strains $\boldsymbol{\epsilon}_0$ given as

$$\boldsymbol{\epsilon}_0 = [\alpha\,\Delta T, \alpha\,\Delta T, 0, \alpha\,\Delta T]^T \qquad (6.51)$$

The stresses are given by

$$\boldsymbol{\sigma} = \mathbf{D}(\boldsymbol{\epsilon} - \boldsymbol{\epsilon}_0) \qquad (6.52)$$

where $\boldsymbol{\epsilon}$ is the total strain.

On substitution into the strain energy, this yields an additional term of $-\boldsymbol{\epsilon}^T\mathbf{D}\boldsymbol{\epsilon}_0$ in the potential energy Π. Using the element strain–displacement relations in Eq. 6.24, we find

$$2\pi \int_A \boldsymbol{\epsilon}^T\mathbf{D}\boldsymbol{\epsilon}_0 r\, dA = \sum_e \mathbf{q}^T(2\pi\bar{r}A_e\,\bar{\mathbf{B}}^T\mathbf{D}\bar{\boldsymbol{\epsilon}}_0) \qquad (6.53)$$

The consideration of the temperature effect in the Galerkin approach is rather simple. The term $\boldsymbol{\epsilon}^T$ above is replaced by $\boldsymbol{\epsilon}^T(\phi)$.

The expression in parentheses gives element nodal load contributions. The vector $\bar{\boldsymbol{\epsilon}}_0$ is the initial strain evaluated at the centroid, representing the average temperature rise of the element. We have

$$\boldsymbol{\Theta}^e = 2\pi\bar{r}A_e\,\bar{\mathbf{B}}^T\mathbf{D}\bar{\boldsymbol{\epsilon}}_0 \qquad (6.54)$$

where

$$\boldsymbol{\Theta}^e = [\Theta_1, \Theta_2, \Theta_3, \Theta_4, \Theta_5, \Theta_6]^T \qquad (6.55)$$

6.4 PROBLEM MODELING AND BOUNDARY CONDITIONS

We have seen that the axisymmetric problem simply reduces to consideration of the revolving area. The boundary conditions need to be enforced on this area. θ independence arrests the rotation. Axisymmetry also implies that points lying on the z axis remain radially fixed. Let us now consider some typical problems with a view to modeling them.

Cylinder Subjected to Internal Pressure

Figure 6.7 shows a hollow cylinder of length L subjected to an internal pressure. One end of the cylindrical pipe is attached to a rigid wall. In this, we need to model only the rectangular region of the length L bound between r_i and r_o. Nodes on the fixed end are constrained in the z and r directions. Stiffness and force modifications will be made for these nodes.

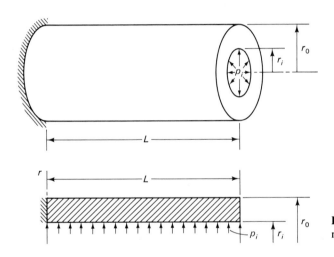

Figure 6.7   Hollow cylinder under internal pressure.

Infinite Cylinder

In Fig. 6.8, modeling of a cylinder of infinite length subjected to external pressure is shown. The length dimensions are assumed to remain constant. This plane strain condition is modeled by considering a unit length and restraining the end surfaces in the z direction.

Press Fit on a Rigid Shaft

Press fit of a ring of length L and internal radius r_i onto a rigid shaft radius $r_i + \delta$ is considered in Fig. 6.9. When symmetry is assumed about the midplane, this plane is restrained in the z direction. When we impose the condition that nodes at the inter-

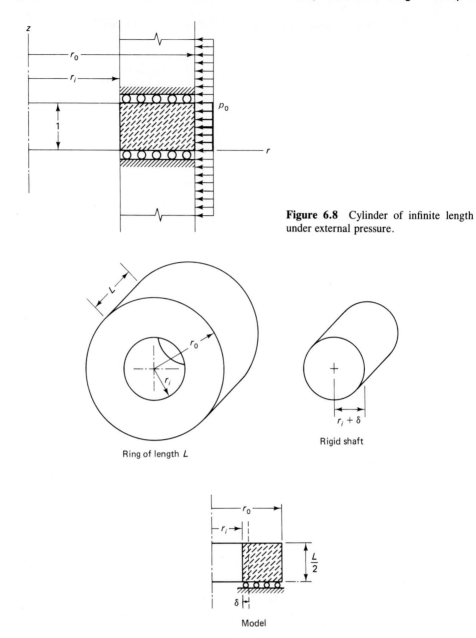

Figure 6.8 Cylinder of infinite length under external pressure.

Figure 6.9 Press fit on a rigid shaft.

nal radius have to displace radially by δ, a large stiffness C is added to the diagonal locations for the radially constrained dof's and a force $C\delta$ is added to the corresponding force components. Solution of the equations gives displacements at nodes; stresses can then be evaluated.

Press Fit on an Elastic Shaft

The condition at the contacting boundary leads to an interesting problem when an elastic sleeve is press fitted onto an elastic shaft. Take the problem of Fig. 6.9 stated above with the shaft also treated as elastic. A method to handle this is considered by referring to Fig. 6.10. We may define pairs of nodes on the contacting boundary, each pair consisting of one node on the sleeve and one on the shaft. If Q_i and Q_j are displacements of a typical pair along the radial degrees of freedom, we need to satisfy the multipoint constraint

$$Q_j - Q_i = \delta \tag{6.56}$$

When the term $\frac{1}{2}C(Q_j - Q_i - \delta)^2$ is added to the potential energy, the constraint is approximately enforced. The penalty approach for handling multipoint constraints is discussed in Chapter 3. Note that C is a large number. We have

$$\frac{1}{2}C(Q_j - Q_i - \delta)^2 = \frac{1}{2}CQ_i^2 + \frac{1}{2}CQ_j^2 - \frac{1}{2}C(Q_iQ_j + Q_jQ_i)$$
$$+ CQ_i\delta - CQ_j\delta + \frac{1}{2}C\delta^2 \tag{6.57}$$

This implies the following modifications:

$$\begin{bmatrix} K_{ii} & K_{ij} \\ K_{ji} & K_{jj} \end{bmatrix} \longrightarrow \begin{bmatrix} K_{ii} + C & K_{ij} - C \\ K_{ji} - C & K_{jj} + C \end{bmatrix} \tag{6.58}$$

and

$$\begin{bmatrix} F_i \\ F_j \end{bmatrix} \longrightarrow \begin{bmatrix} F_i - C\delta \\ F_j + C\delta \end{bmatrix} \tag{6.59}$$

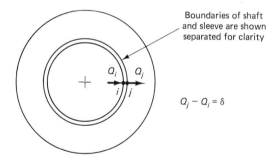

Boundaries of shaft
and sleeve are shown
separated for clarity

$Q_j - Q_i = \delta$

Figure 6.10 Elastic sleeve on an elastic shaft.

Belleville Spring

The Belleville spring, also called the Belleville washer, is a conical disk spring. The load is applied on the periphery of the circle and supported at the bottom as shown in Fig. 6.11a. As load is applied in the axial direction, the supporting edge moves out. Only the rectangular area shown shaded in Fig. 6.11c needs to be modeled. An axisymmetric load P is placed at the top corner, and the bottom supporting corner is

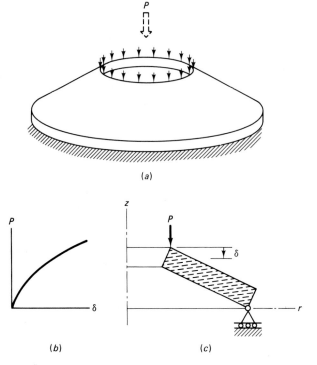

(a)

(b)   (c)   **Figure 6.11**   Belleville spring.

constrained in the z direction. Load–deflection characteristics and stress distribution can be obtained by dividing the area into elements and using a computer program. In the Belleville spring, the load–deflection curve is *nonlinear* (Fig. 6.11b). The stiffness depends on the geometry. We can find a good approximate solution by an incremental approach. We find the stiffness matrix $\mathbf{K}(\mathbf{x})$ for the given coordinate geometry. We obtain the displacements $\Delta\mathbf{Q}$ for an incremental loading of $\Delta\mathbf{F}$ from

$$\mathbf{K}(\mathbf{x})\,\Delta\mathbf{Q} = \Delta\mathbf{F} \tag{6.60}$$

The displacements $\Delta\mathbf{Q}$ are converted to the components Δu, Δw and added to $\mathbf{x}$ to update the new geometry

$$\mathbf{x} \longleftarrow \mathbf{x} + \Delta\mathbf{u} \tag{6.61}$$

$\mathbf{K}$ is recalculated for the new geometry and the new set of equations 6.60 is solved. The process is continued until the full applied load is reached.

This example illustrates the incremental approach for geometric nonlinearity.

Thermal Stress Problem

Shown in Fig. 6.12a is a steel sleeve inserted into a rigid insulated wall. The sleeve fits snugly and then the temperature is raised by ΔT. The stresses in the sleeve increase because of the constraint. The rectangular area of length $L/2$, bounded by r_i and r_0 is considered (Fig. 6.12b), with points on the outer radius constrained radially

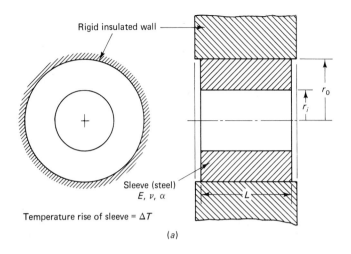

Rigid insulated wall

Sleeve (steel)
E, v, α

Temperature rise of sleeve = ΔT

(a)

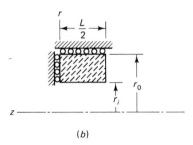

(b)

Figure 6.12   Thermal stress problem.

and points on r constrained axially. The load vector is modified using the load vector from Eq. 6.55 and the finite element equations are solved.

Modeling of simple to complex problems of engineering importance have been discussed above. In real life, each problem poses its own challenge. With a clear understanding of the loading, boundary conditions, and the material behavior, the modeling of a problem can be broken down into simple and easy steps.

Example 6.2

A steel disk flywheel rotates at 3000 rpm. The outer diameter is 24 in. and the hole diameter is 6 in. (Fig. E6.2a). Find the value of the maximum tangential stress under the following conditions: thickness = 1 in., $E = 30 \times 10^6$ psi, Poisson's ratio = 0.3, wt. density = 0.283 lb/in.^{3}.

A four-element finite element model is shown in Fig. E6.2b. The load vector is calculated from Eq. 6.34, neglecting gravity load. The result is

$$\mathbf{F} = [3449, 0, 9580, 0, 23\,380, 0, 38\,711, 0, 32\,580, 0, 18\,780, 0]^{\mathrm{T}} \text{ lb}$$

The input data for program AXYSYM is given below.

```
Number of Elements =? 4
Number of Nodes =? 6
Number of Constrained DOF =? 3
```

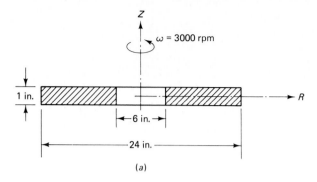

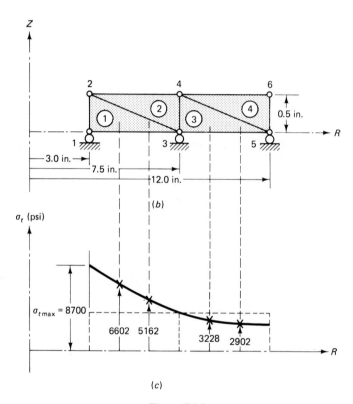

Figure E6.2

```
Number of Component Loads =? 6
Number of Materials =? 1
File Name for Output? EG6_2.DAT
Element#, 3 Nodes, Material#
? 1,1,3,2,1
? 2,2,3,4,1
? 3,4,3,5,1
? 4,4,5,6,1
```

```
THE BANDWIDTH IS 6
NODE NO,  R-COORD,  Z-COORD
? 1,3,0
? 2,3,0.5
? 3,7.5,0.0
? 4,7.5,0.5
? 5,12,0
? 6,12,0.5
Material#, E, Poissons Ratio
? 1,30E6,0.3
DOF#,  Displacement
? 2,0
? 6,0
? 10,0
DOF#,  Applied Load
? 1,3449
? 3,9580
? 5,23380
? 7,38711
? 9,32580
? 11,18780
```

The computer output gives us the tangential stresses in each of the four elements. Treating these values as centroidal values, and *extrapolating* as shown in Fig. E6.2c, the maximum tangential stress occurring at the inner boundary is obtained as $\sigma_{t\,max} = 8700$ psi. ∎

```
1000 REM ********      AXYSYM      *********
1010 REM ** Axisymmetric Stress Analysis using CST **
1020 CLS
1030 PRINT "======================================="
1040 PRINT "       Program written by         "
1050 PRINT " T.R.Chandrupatla and A.D.Belegundu   "
1060 PRINT "======================================="
1070 DEFINT I-N
1080 LOCATE 8,28: PRINT "1. Interactive Data Input"
1090 LOCATE 9,28: PRINT "2. Data Input From a File"
1100 LOCATE 10,28: INPUT "Your Choice <1 or 2>"; IFL1
1110 IF IFL1 = 2 GOTO 1180
1120 INPUT "Number of Elements ="; NE
1130 INPUT "Number of Nodes ="; NN
1140 INPUT "Number of Constrained DOF ="; ND
1150 INPUT "Number of Component Loads ="; NL
1160 INPUT "Number of Materials ="; NM
1170 GOTO 1210
1180 INPUT "File Name for Input Data"; FILE1$
1190 OPEN FILE1$ FOR INPUT AS #1
1200 INPUT #1, NN,NE,ND,NL,NM,NDIM,NEN
1210 INPUT "File Name for Output"; FILE2$
1220 OPEN FILE2$ FOR OUTPUT AS #2
1230 REM TOTAL DOF IS "NQ"
1240 NQ = 2 * NN: PI = 3.14159
1250 DIM F(NQ)
1260 DIM X(NN, 2), NOC(NE, 3), MAT(NE), PM(NM, 3), NU(ND), U(ND)
1270 DIM D(4, 4), B(4, 6), DB(4, 6), SE(6, 6), Q(6), STR(4)
```

```
1280 REM BANDWIDTH NBW FROM CONNECTIVITY"NOC"
1290 NBW = 0
1300 IF IFL1 = 1 GOTO 1320
1310 GOSUB 2410 : GOTO 1350
1320 PRINT "Element#, 3 Nodes, Material#"
1330 FOR I = 1 TO NE
1340 INPUT N, NOC(N, 1), NOC(N, 2), NOC(N, 3), MAT(N): NEXT I
1350 FOR N = 1 TO NE
1360 CMIN = NN + 1: CMAX = 0
1370 FOR J = 1 TO 3
1380 IF CMIN > NOC(N, J) THEN CMIN = NOC(N, J)
1390 IF CMAX < NOC(N, J) THEN CMAX = NOC(N, J)
1400 NEXT J
1410 C = 2 * (CMAX - CMIN + 1)
1420 IF NBW < C THEN NBW = C
1430 NEXT N
1440 PRINT "THE BANDWIDTH IS"; NBW
1450 REM *** INITIALIZATION ***
1460 DIM S(NQ, NBW)
1470 FOR I = 1 TO NQ
1480 F(NQ) = 0
1490 FOR J = 1 TO NBW
1500 S(I, J) = 0: NEXT J: NEXT I
1510 IF IFL1 = 2 GOTO 1650
1520 PRINT "NODE NO, R-COORD, Z-COORD"
1530 FOR I = 1 TO NN
1540 INPUT N, X(N, 1), X(N, 2): NEXT I
1550 PRINT "Material#, E, Poissons Ratio"
1560 FOR I = 1 TO NM
1570 INPUT N, PM(N, 1), PM(N, 2): NEXT I
1580 PRINT "DOF#, Displacement"
1590 FOR I = 1 TO ND
1600 INPUT NU(I), U(I): NEXT I
1610 PRINT "DOF#, Applied Load"
1620 FOR I = 1 TO NL
1630 INPUT N, F(N): NEXT I
1640 REM *** GLOBAL STIFFNESS MATRIX **
1650 FOR N = 1 TO NE
1660 PRINT , "FORMING STIFFNESS MATRIX OF ELEMENT "; N
1670 GOSUB 2540
1680 FOR I = 1 TO 6
1690 FOR J = 1 TO 6
1700 SE(I, J) = 0
1710 FOR K = 1 TO 4
1720 SE(I, J) = SE(I, J) + B(K, I) * DB(K, J) * PI * RBAR * ABS(DJ)
1730 NEXT K: NEXT J: NEXT I
1740 PRINT "PLACING STIFFNESS IN GLOBAL LOCATIONS"
1750 FOR II = 1 TO 3
1760 NRT = 2 * (NOC(N, II) - 1)
1770 FOR IT = 1 TO 2
1780 NR = NRT + IT
1790 I = 2 * (II - 1) + IT
1800 FOR JJ = 1 TO 3
1810 NCT = 2 * (NOC(N, JJ) - 1)
1820 FOR JT = 1 TO 2
1830 J = 2 * (JJ - 1) + JT
1840 NC = NCT + JT - NR + 1
1850 IF NC <= 0 GOTO 1870
1860 S(NR, NC) = S(NR, NC) + SE(I, J)
1870 NEXT JT: NEXT JJ: NEXT IT: NEXT II
1880 NEXT N
1890 REM *** MODIFY FOR BOUNDARY CONDITIONS ***
1900 CNST = S(1, 1) * 10000
1910 FOR I = 1 TO ND
1920 N = NU(I)
```

```
1930 S(N, 1) = S(N, 1) + CNST
1940 F(N) = F(N) + CNST * U(I): NEXT I
1950 REM  *** EQUATION SOLVING ***
1960 GOSUB 5000
1970 '** PRINT OUTPUT **
1980 PRINT #2,"NN,NE,ND,NL,NM,NDIM,NEN ="
1990 PRINT #2, NN; NE; ND; NL; NM; NDIM; NEN
2000 PRINT #2, "NODE, X-COORD., Y-COORD."
2010 FOR I=1 TO NN
2020 PRINT #2,I,X(I,1),X(I,2) :NEXT I
2030 PRINT #2, "ELEMENT#, 3 NODES, MATERIAL#"
2040 FOR I=1 TO NE
2050 PRINT #2, I;  NOC(I,1);  NOC(I,2);  NOC(I,3);  MAT(I) : NEXT I
2060 PRINT #2,"DOF#, SPECIFIED DISPLACEMENT"
2070 FOR I=1 TO ND:PRINT #2,NU(I),U(I) :NEXT I
2080 PRINT #2, "MATERIAL#, E, POISSON'S RATIO"
2090 FOR I=1 TO NM:PRINT #2, I, PM(I,1), PM(I,2): NEXT I
2100 PRINT "NODE#   X-Displ    Y-Displ"
2110 PRINT #2, "NODE, X-DISP., Y-DISP."
2120 FOR I = 1 TO NN
2130 PRINT #2, I, F(2*I-1), F(2*I)
2140 PRINT USING " ###"; I;
2150 PRINT USING " ##.###^^^^"; F(2 * I - 1); F(2 * I): NEXT I
2160 REM  *** STRESS CALCULATIONS ***
2170 PRINT "ELEM#   SR      SZ      TRZ      ST";
2180 PRINT #2, "ELEM#,  SR,  SZ,  TRZ,  ST"
2190 PRINT #2, "   S1      S2    ANGLE SX->S1  ST"
2200 PRINT "    S1      S2    ANGLE SX->S1  ST"
2210 FOR N = 1 TO NE
2220 GOSUB 2540
2230 GOSUB 2870
2240 '***PRINCIPAL STRESS CALCULATIONS***
2250 IF STR(3) <> 0 GOTO 2300
2260 S1 = STR(1): S2 = STR(2): ANG = 0
2270 IF S1 > S2 GOTO 2320
2280 S1 = STR(2): S2 = STR(1): ANG = 90
2290 GOTO 2320
2300 C = .5 * (STR(1) + STR(2))
2302 R = SQR(.25 * (STR(1) - STR(2)) ^ 2 + (STR(3)) ^ 2)
2304 S1 = C + R: S2 = C - R
2306 IF C > STR(1) GOTO 2310
2308 ANG = 57.29577951# * ATN(STR(3) / (STR(1) - S2)): GOTO 2320
2310 ANG = 57.29577951# * ATN(STR(3) / (S1 - STR(1)))
2312 IF STR(3) > 0 THEN ANG = 90 - ANG
2314 IF STR(3) < 0 THEN ANG = -90 - ANG
2320 PRINT #2,USING " ###"; N;
2330 PRINT #2,USING " ##.###^^^^"; STR(1);STR(2);STR(3);STR(4)
2340 PRINT #2,USING " ##.###^^^^"; S1;S2;ANG
2350 PRINT USING " ###"; N;
2360 PRINT USING " ##.###^^^^"; STR(1); STR(2); STR(3);STR(4);
2370 PRINT USING " ##.###^^^^"; S1; S2; ANG
2380 NEXT N
2390 CLOSE #2
2400 END
2410 '============== FILE INPUT =================
2420 FOR I=1 TO NN: FOR J=1 TO NDIM
2430 INPUT #1, X(I,J):NEXT J:NEXT I
2440 FOR I=1 TO NE: FOR J=1 TO NEN
2450 INPUT #1, NOC(I,J): NEXT J: NEXT I
2460 FOR I=1 TO NE: INPUT#1,MAT(I): NEXT I
2470 FOR I=1 TO ND: INPUT#1,NU(I),U(I): NEXT I
2480 FOR I=1 TO NL: INPUT#1, N,F(N): NEXT I
2490 FOR I=1 TO NM: FOR J=1 TO 3
2500 INPUT #1, PM(I,J):NEXT J:NEXT I
2510 CLOSE #1
```

```
2520 RETURN
2530 '===========================================
2540 'ELEMENT STIFFNESS FORMATION
2550 '***FIRST THE D–MATRIX***
2560 M = MAT(N): E = PM(M, 1): PNU = PM(M, 2)
2570 C1 = E * (1 – PNU) / ((1 + PNU) * (1 – 2 * PNU)): C2 = PNU / (1 – PNU)
2580 FOR I = 1 TO 4: FOR J = 1 TO 4: D(I, J) = 0: NEXT J: NEXT I
2590 D(1, 1) = C1: D(1, 2) = C1 * C2: D(1, 4) = C1 * C2
2600 D(2, 1) = D(1, 2): D(2, 2) = C1: D(2, 4) = C1 * C2
2610 D(3, 3) = .5 * E / (1 + PNU)
2620 D(4, 1) = D(1, 4): D(4, 2) = D(2, 4): D(4, 4) = C1
2630 '***STRAIN–DISPLACEMENT MATRIX***
2640 I1 = NOC(N, 1): I2 = NOC(N, 2): I3 = NOC(N, 3)
2650 R1 = X(I1, 1): Z1 = X(I1, 2)
2660 R2 = X(I2, 1): Z2 = X(I2, 2)
2670 R3 = X(I3, 1): Z3 = X(I3, 2)
2680 R21 = R2 – R1: R32 = R3 – R2: R13 = R1 – R3
2690 Z12 = Z1 – Z2: Z23 = Z2 – Z3: Z31 = Z3 – Z1
2700 DJ = R13 * Z23 – R32 * Z31'DETERMINANT OF JACOBIAN
2710 RBAR = (R1 + R2 + R3) / 3
2720 '***FORMATION OF B MATRIX***
2730 B(1, 1) = Z23 / DJ: B(2, 1) = 0: B(3, 1) = R32 / DJ: B(4, 1) = 1 / (3 * RBAR)
2740 B(1, 2) = 0: B(2, 2) = R32 / DJ: B(3, 2) = Z23 / DJ: B(4, 2) = 0
2750 B(1, 3) = Z31 / DJ: B(2, 3) = 0: B(3, 3) = R13 / DJ: B(4, 3) = 1 / (3 * RBAR)
2760 B(1, 4) = 0: B(2, 4) = R13 / DJ: B(3, 4) = Z31 / DJ: B(4, 4) = 0
2770 B(1, 5) = Z12 / DJ: B(2, 5) = 0: B(3, 5) = R21 / DJ: B(4, 5) = 1 / (3 * RBAR)
2780 B(1, 6) = 0: B(2, 6) = R21 / DJ: B(3, 6) = Z12 / DJ: B(4, 6) = 0
2790 '***DB MATRIX DB=D*B***
2800 FOR I = 1 TO 4
2810 FOR J = 1 TO 6
2820 DB(I, J) = 0
2830 FOR K = 1 TO 4
2840 DB(I, J) = DB(I, J) + D(I, K) * B(K, J)
2850 NEXT K: NEXT J: NEXT I
2860 RETURN
2870 'STRESS EVALUATION
2880 Q(1) = F(2 * I1 – 1): Q(2) = F(2 * I1)
2890 Q(3) = F(2 * I2 – 1): Q(4) = F(2 * I2)
2900 Q(5) = F(2 * I3 – 1): Q(6) = F(2 * I3)
2910 FOR I = 1 TO 4
2920 STR(I) = 0
2930 FOR K = 1 TO 6
2940 STR(I) = STR(I) + DB(I, K) * Q(K): NEXT K: NEXT I
2950 RETURN
5000 N1 = NQ – 1
5010 REM *** FORWARD ELIMINATION ***
5020 FOR K = 1 TO N1
5030 NK = NQ – K + 1
5040 IF NK > NBW THEN NK = NBW
5050 FOR I = 2 TO NK
5060 C1 = S(K, I) / S(K, 1)
5070 I1 = K + I – 1
5080 FOR J = I TO NK
5090 J1 = J – I + 1
5100 S(I1, J1) = S(I1, J1) – C1 * S(K, J): NEXT J
5110 F(I1) = F(I1) – C1 * F(K): NEXT I: NEXT K
5120 REM *** BACK SUBSTITUTION ***
5130 F(NQ) = F(NQ) / S(NQ, 1)
5140 FOR KK = 1 TO N1
5150 K = NQ – KK
5160 C1 = 1 / S(K, 1)
5170 F(K) = C1 * F(K)
5180 NK = NQ – K + 1
5190 IF NK > NBW THEN NK = NBW
```

```
5200 FOR J = 2 TO NK
5210 F(K) = F(K) - C1 * S(K, J) * F(K + J - 1)
5220 NEXT J
5230 NEXT KK
5240 RETURN
```

PROBLEMS

6.1. The open-ended steel cylinder shown in Fig. P6.1 is subjected to an internal pressure of 1 MPa. Find the deformed shape and the distribution of principal stresses.

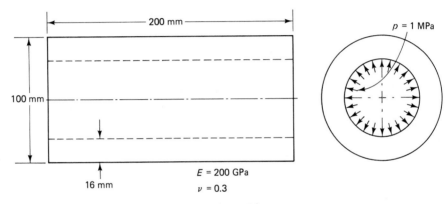

Figure P6.1

6.2. Find the deformed configuration and the stress distribution in the walls of the closed cylinder shown in Fig. P6.2.

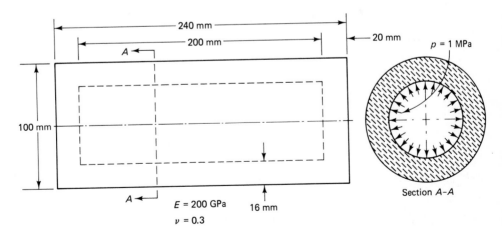

Figure P6.2

6.3. Determine the diameters after deformation and the distribution of principal stresses along the radius of the infinite cylinder subjected to internal pressure as shown in Fig. P6.3.

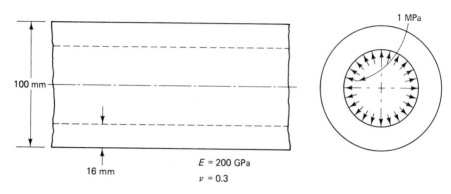

$E = 200$ GPa
$\nu = 0.3$

Figure P6.3

6.4. The steel sleeve of internal diameter 3 in. is press fitted onto a rigid shaft of diameter 3.01 in., as shown in Fig. P6.4. Determine the outer diameter of the sleeve after fitting, and the stress distribution. Estimate the contact pressure by interpolating the radial stress in the neighboring elements.

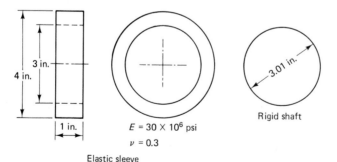

$E = 30 \times 10^6$ psi
$\nu = 0.3$

Elastic sleeve

Rigid shaft

Figure P6.4

6.5. Solve Problem 6.4 if the shaft is also made out of steel.

6.6. The steel flywheel shown in Fig. P6.6 rotates at 3000 rpm. Find the deformed shape of the flywheel and give the stress distribution.

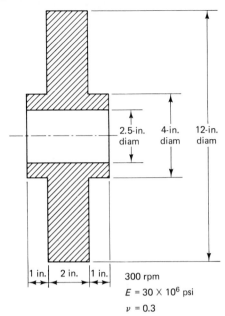

2.5-in.   4-in.   12-in.
diam   diam   diam

1 in. 2 in. 1 in.   300 rpm

$E = 30 \times 10^6$ psi

$\nu = 0.3$

Figure P6.6   Flywheel.

6.7. The circular pad hydrostatic bearing shown in Fig. P6.7 is used for supporting slides subjected to large forces. Oil under pressure is supplied through a small hole at the center and flows out through the gap. The pressure distribution in the pocket area and the gap is shown in the figure. Find the deformed configuration of the pad and determine the stress distribution. (*Note:* Neglect the dimension of the oil supply hole.)

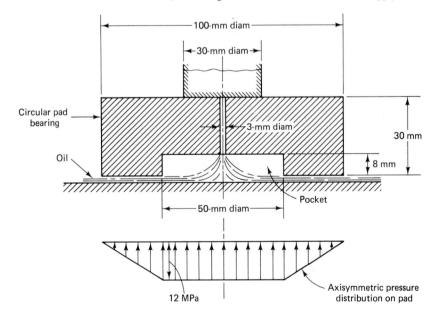

Figure P6.7   Hydrostatic bearing.

*6.8. A Belleville spring is a conical disk spring. For the spring shown in Fig. P6.8, determine the axial load required to flatten the spring. Solve the problem using the incremental approach discussed in the text and plot the load–deflection curve as the spring flattens.

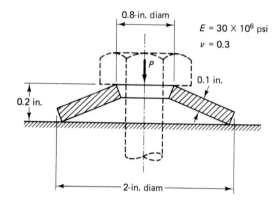

0.8-in. diam

$E = 30 \times 10^6$ psi

$\nu = 0.3$

P

0.1 in.

0.2 in.

2-in. diam

Figure P6.8   Belleville spring.

*6.9. The aluminum tube shown in Fig. P6.9 fits snugly into a rigid hole at room temperature. If the temperature of the aluminum tube is increased by 40°C, find the deformed configuration and the stress distribution.

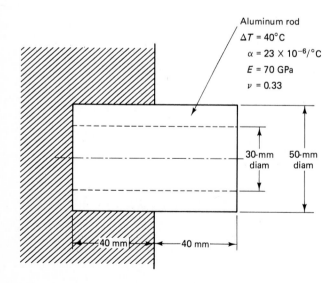

Aluminum rod

$\Delta T = 40°C$

$\alpha = 23 \times 10^{-6}/°C$

$E = 70$ GPa

$\nu = 0.33$

30-mm diam   50-mm diam

40 mm   40 mm

Figure P6.9

6.10. The steel water tank shown in Fig. P6.10 is bolted to a 5-m circular support. If the water is at a height of 3 m as shown, find the deformed shape and stress distribution. (*Note:* Pressure $= \rho gh$, water density $\rho = 1$ Mg/m^3, $g = 9.8$ m/s^2.)

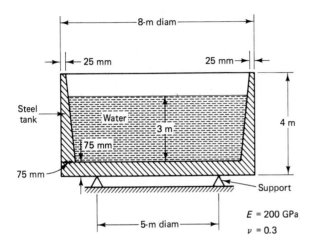

E = 200 GPa
ν = 0.3

Figure P6.10   Water tank.

chapter seven

Two-Dimensional Isoparametric Elements and Numerical Integration

7.1 INTRODUCTION

In Chapter 5 and 6, we have developed the constant strain triangular element for stress analysis. In this chapter, we develop what are popularly called _isoparametric_ elements and apply them to stress analysis. These elements have proved effective on a wide variety of two- and three-dimensional problems in engineering. We present the two-dimensional four-node quadrilateral in detail. Development of higher-order elements follow the same basic steps used in the four-node quadrilateral. The higher-order elements can capture variations in stress such as occur near fillets, holes, etc. We can view the isoparametric family of elements in a unified manner due to the simple and versatile manner in which shape functions can be derived, followed by the generation of the element stiffness matrix using numerical integration.

7.2 THE FOUR-NODE QUADRILATERAL

Consider the general quadrilateral element shown in Fig. 7.1. The local nodes are numbered as 1, 2, 3, and 4 in a _counterclockwise_ fashion as shown, and (x_i, y_i) are the coordinates of node i. The vector $\mathbf{q} = [q_1, q_2, \ldots, q_8]^T$ denotes the element displacement vector. The displacement of an interior point P located at (x, y) is represented as $\mathbf{u} = [u(x, y), v(x, y)]^T$.

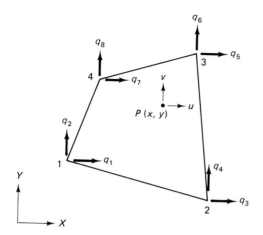

Figure 7.1 Four-node quadrilateral element.

Shape Functions

Following the steps in earlier chapters, we first develop the shape functions on a master element, shown in Fig. 7.2. The master element is defined in ξ, η coordinates (or *natural* coordinates), and is square shaped. The Lagrange shape functions N_i, where $i = 1, 2, 3, 4$, are defined such that N_i is equal to unity at node i and is zero at other nodes. In particular, consider the definition of N_i:

$$N_1 = 1 \quad \text{at node 1}$$
$$= 0 \quad \text{at nodes 2, 3, 4} \tag{7.1}$$

Now, the requirement that $N_1 = 0$ at nodes 2, 3, and 4 is equivalent to requiring that $N_1 = 0$ along edges $\xi = +1$ and $\eta = +1$ (Fig. 7.2). Thus, N_1 has to be of the form

$$N_1 = c(1 - \xi)(1 - \eta) \tag{7.2}$$

where c is some constant. The constant is determined from the condition $N_1 = 1$ at node 1. Since $\xi = -1$, $\eta = -1$ at node 1, we have

$$1 = c(2)(2) \tag{7.3}$$

which yields $c = \frac{1}{4}$. Thus,

$$N_1 = \tfrac{1}{4}(1 - \xi)(1 - \eta) \tag{7.4}$$

All the four shape functions can be written as

$$N_1 = \tfrac{1}{4}(1 - \xi)(1 - \eta)$$
$$N_2 = \tfrac{1}{4}(1 + \xi)(1 - \eta)$$
$$N_3 = \tfrac{1}{4}(1 + \xi)(1 + \eta)$$
$$N_4 = \tfrac{1}{4}(1 - \xi)(1 + \eta) \tag{7.5}$$

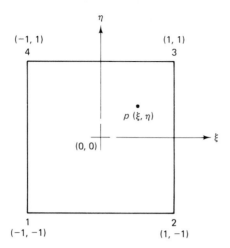

Figure 7.2   The quadrilateral element in ξ, η space (the *master* element).

While implementing in a computer program, the following compact representation of Eqs. 7.5 is useful:

$$N_i = \tfrac{1}{4}(1 + \xi\xi_i)(1 + \eta\eta_i) \qquad (7.6)$$

where (ξ_i, η_i) are the coordinates of node i.

We now express the displacement field within the element in terms of the nodal values. Thus, if $\mathbf{u} = [u, v]^T$ represents the displacement components of a point located at (ξ, η), and $\mathbf{q}$, dimension (8×1), is the element displacement vector, then

$$u = N_1q_1 + N_2q_3 + N_3q_5 + N_4q_7$$
$$v = N_1q_2 + N_2q_4 + N_3q_6 + N_4q_8 \qquad (7.7a)$$

which can be written in matrix form as

$$\mathbf{u} = \mathbf{Nq} \qquad (7.7b)$$

where

$$\mathbf{N} = \begin{bmatrix} N_1 & 0 & N_2 & 0 & N_3 & 0 & N_4 & 0 \\ 0 & N_1 & 0 & N_2 & 0 & N_3 & 0 & N_4 \end{bmatrix} \qquad (7.8)$$

In the isoparametric formulation, we use the *same* shape functions N_i to also express the coordinates of a point within the element in terms of nodal coordinates. Thus,

$$x = N_1x_1 + N_2x_2 + N_3x_3 + N_4x_4$$
$$y = N_1y_1 + N_2y_2 + N_3y_3 + N_4y_4 \qquad (7.9)$$

Subsequently, we will need to express the derivatives of a function in x, y coordinates in terms of its derivatives in ξ, η coordinates. This is done as follows. A function $f = f(x, y)$, in view of Eqs. 7.9, can be considered to be an implicit function of ξ and η as $f = f[x(\xi, \eta), y(\xi, \eta)]$. Using the chain rule of differentiation, we have

$$\frac{\partial f}{\partial \xi} = \frac{\partial f}{\partial x}\frac{\partial x}{\partial \xi} + \frac{\partial f}{\partial y}\frac{\partial y}{\partial \xi}$$

$$\frac{\partial f}{\partial \eta} = \frac{\partial f}{\partial x}\frac{\partial x}{\partial \eta} + \frac{\partial f}{\partial y}\frac{\partial y}{\partial \eta}$$

(7.10)

or

$$\left\{\begin{array}{c} \dfrac{\partial f}{\partial \xi} \\[2mm] \dfrac{\partial f}{\partial \eta} \end{array}\right\} = \mathbf{J}\left\{\begin{array}{c} \dfrac{\partial f}{\partial x} \\[2mm] \dfrac{\partial f}{\partial y} \end{array}\right\}$$

(7.11)

where $\mathbf{J}$ is the Jacobean matrix

$$\mathbf{J} = \begin{bmatrix} \dfrac{\partial x}{\partial \xi} & \dfrac{\partial y}{\partial \xi} \\[3mm] \dfrac{\partial x}{\partial \eta} & \dfrac{\partial y}{\partial \eta} \end{bmatrix}$$

(7.12)

In view of Eqs. 7.5 and 7.9, we have

$$\mathbf{J} = \frac{1}{4}\begin{bmatrix} -(1-\eta)x_1+(1-\eta)x_2+(1+\eta)x_3-(1+\eta)x_4 & -(1-\eta)y_1+(1-\eta)y_2+(1+\eta)y_3-(1+\eta)y_4 \\ -(1-\xi)x_1-(1+\xi)x_2+(1+\xi)x_3+(1-\xi)x_4 & -(1-\xi)y_1-(1+\xi)y_2+(1+\xi)y_3+(1-\xi)y_4 \end{bmatrix}$$

(7.13a)

$$\equiv \begin{bmatrix} J_{11} & J_{12} \\ J_{21} & J_{22} \end{bmatrix}$$

(7.13b)

Equation 7.11 can be inverted as

$$\left\{\begin{array}{c} \dfrac{\partial f}{\partial x} \\[2mm] \dfrac{\partial f}{\partial y} \end{array}\right\} = \mathbf{J}^{-1}\left\{\begin{array}{c} \dfrac{\partial f}{\partial \xi} \\[2mm] \dfrac{\partial f}{\partial \eta} \end{array}\right\}$$

(7.14a)

or

$$\left\{\begin{array}{c} \dfrac{\partial f}{\partial x} \\[2mm] \dfrac{\partial f}{\partial y} \end{array}\right\} = \frac{1}{\det \mathbf{J}}\begin{bmatrix} J_{22} & -J_{12} \\ -J_{21} & J_{11} \end{bmatrix}\left\{\begin{array}{c} \dfrac{\partial f}{\partial \xi} \\[2mm] \dfrac{\partial f}{\partial \eta} \end{array}\right\}$$

(7.14b)

These expressions will be used in the derivation of the element stiffness matrix.

An additional result that will be needed is the relation

$$dx \, dy = \det \mathbf{J} \, d\xi \, d\eta \tag{7.15}$$

The proof of this result, found in many textbooks on calculus, is given in the **appendix**.

Element Stiffness Matrix

The stiffness matrix for the quadrilateral element can be derived from the strain energy in the body, given by

$$U = \int_V \tfrac{1}{2} \boldsymbol{\sigma}^{\mathrm{T}} \boldsymbol{\epsilon} \, dV \tag{7.16}$$

or

$$U = \sum_e t_e \int_e \tfrac{1}{2} \boldsymbol{\sigma}^{\mathrm{T}} \boldsymbol{\epsilon} \, dA \tag{7.17}$$

where t_e is the thickness of element e.

The strain–displacement relations are

$$\boldsymbol{\epsilon} = \left\{ \begin{array}{c} \epsilon_x \\ \epsilon_y \\ \gamma_{xy} \end{array} \right\} = \left\{ \begin{array}{c} \dfrac{\partial u}{\partial x} \\[2mm] \dfrac{\partial v}{\partial y} \\[2mm] \left(\dfrac{\partial u}{\partial y} + \dfrac{\partial v}{\partial x} \right) \end{array} \right\} \tag{7.18}$$

By considering $f \equiv u$ in Eq. 7.14b, we have

$$\left\{ \begin{array}{c} \dfrac{\partial u}{\partial x} \\[2mm] \dfrac{\partial u}{\partial y} \end{array} \right\} = \frac{1}{\det \mathbf{J}} \left[\begin{array}{cc} J_{22} & -J_{12} \\ -J_{21} & J_{11} \end{array} \right] \left\{ \begin{array}{c} \dfrac{\partial u}{\partial \xi} \\[2mm] \dfrac{\partial u}{\partial \eta} \end{array} \right\} \tag{7.19a}$$

Similarly,

$$\left\{ \begin{array}{c} \dfrac{\partial v}{\partial x} \\[2mm] \dfrac{\partial v}{\partial y} \end{array} \right\} = \frac{1}{\det \mathbf{J}} \left[\begin{array}{cc} J_{22} & -J_{12} \\ -J_{21} & J_{11} \end{array} \right] \left\{ \begin{array}{c} \dfrac{\partial v}{\partial \xi} \\[2mm] \dfrac{\partial v}{\partial \eta} \end{array} \right\} \tag{7.19b}$$

Equations 7.18 and 7.19*a,b* yield

$$
\boldsymbol{\epsilon} = \mathbf{A}
\begin{Bmatrix}
\dfrac{\partial u}{\partial \xi} \\[8pt]
\dfrac{\partial u}{\partial \eta} \\[8pt]
\dfrac{\partial v}{\partial \xi} \\[8pt]
\dfrac{\partial v}{\partial \eta}
\end{Bmatrix}
\tag{7.20}
$$

where $\mathbf{A}$ is given by

$$
\mathbf{A} = \frac{1}{\det \mathbf{J}}
\begin{bmatrix}
J_{22} & -J_{12} & 0 & 0 \\
0 & 0 & -J_{21} & J_{11} \\
-J_{21} & J_{11} & J_{22} & -J_{12}
\end{bmatrix}
\tag{7.21}
$$

Now, from the interpolation equations Eqs. 7.7*a*, we have

$$
\begin{Bmatrix}
\dfrac{\partial u}{\partial \xi} \\[8pt]
\dfrac{\partial u}{\partial \eta} \\[8pt]
\dfrac{\partial v}{\partial \xi} \\[8pt]
\dfrac{\partial v}{\partial \eta}
\end{Bmatrix} = \mathbf{Gq}
\tag{7.22}
$$

where

$$
\mathbf{G} = \frac{1}{4}
\begin{bmatrix}
-(1-\eta) & 0 & (1-\eta) & 0 & (1+\eta) & 0 & -(1+\eta) & 0 \\
-(1-\xi) & 0 & -(1+\xi) & 0 & (1+\xi) & 0 & (1-\xi) & 0 \\
0 & -(1-\eta) & 0 & (1-\eta) & 0 & (1+\eta) & 0 & -(1+\eta) \\
0 & -(1-\xi) & 0 & -(1+\xi) & 0 & (1+\xi) & 0 & (1-\xi)
\end{bmatrix}
\tag{7.23}
$$

Equations 7.20 and 7.22 now yield

$$
\boxed{\boldsymbol{\epsilon} = \mathbf{Bq}}
\tag{7.24}
$$

where

$$
\mathbf{B} = \mathbf{AG}
\tag{7.25}
$$

The relation $\boldsymbol{\epsilon} = \mathbf{Bq}$ is the desired result. The strain in the element is expressed in terms of its nodal displacement. The stress is now given by

$$\boxed{\boldsymbol{\sigma} = \mathbf{DBq}}$$

(7.26)

where $\mathbf{D}$ is a (3×3) material matrix. The strain energy in Eq. 7.17 becomes

$$U = \sum_e \tfrac{1}{2} \mathbf{q}^T \left[t_e \int_{-1}^{1} \int_{-1}^{1} \mathbf{B}^T \mathbf{DB} \det \mathbf{J} \, d\xi \, d\eta \right] \mathbf{q}$$

(7.27a)

$$= \sum_e \tfrac{1}{2} \mathbf{q}^T \mathbf{k}^e \mathbf{q}$$

(7.27b)

where

$$\boxed{\mathbf{k}^e = t_e \int_{-1}^{1} \int_{-1}^{1} \mathbf{B}^T \mathbf{DB} \det \mathbf{J} \, d\xi \, d\eta}$$

(7.28)

is the element stiffness matrix of dimension (8×8).

We note here that quantities $\mathbf{B}$ and $\det \mathbf{J}$ in the above integral are involved functions of ξ and η, and so the integration has to be performed numerically. Methods of numerical integration are discussed subsequently.

Element Force Vectors

Body force. A body force which is distributed force per unit volume, contributes to the global load vector $\mathbf{F}$. This contribution can be determined by considering the body force term in the potential energy expression

$$\int_V \mathbf{u}^T \mathbf{f} \, dV$$

(7.29)

Using $\mathbf{u} = \mathbf{Nq}$, and treating the body force $\mathbf{f} = [f_x, f_y]^T$ as constant within each element, we get

$$\int_V \mathbf{u}^T \mathbf{f} \, dV = \sum_e \mathbf{q}^T \mathbf{f}^e$$

(7.30)

where the (8×1) element body force vector is given by

$$\mathbf{f}^e = t_e \left[\int_{-1}^{1} \int_{-1}^{1} \mathbf{N}^T \det \mathbf{J} \, d\xi \, d\eta \right] \begin{Bmatrix} f_x \\ f_y \end{Bmatrix}$$

(7.31)

As with the stiffness matrix derived earlier, the body force vector above has to be evaluated by numerical integration.

Traction force.   Assume that a constant traction force, $\mathbf{T} = [T_x, T_y]^T$, a force per unit area, is applied on edge 2–3 of the quadrilateral element. Along this edge, we have $\xi = 1$. If we use the shape functions given in Eq. 7.5, this becomes $N_1 = N_4 = 0$, $N_2 = (1 - \eta)/2$, $N_3 = (1 + \eta)/2$. Note that the shape functions are linear functions along the edges. Consequently, from the potential, the element traction load vector is readily given by

$$\mathbf{T}^e = \frac{t_e \ell_{2-3}}{2} [0 \quad 0 \quad T_x \quad T_y \quad T_x \quad T_y \quad 0 \quad 0]^T \tag{7.32}$$

where ℓ_{2-3} = length of edge 2–3. For varying distributed loads, we may express T_x and T_y in terms of values of nodes 2 and 3 using shape functions. Numerical integration can be used in this case.

Finally, **point** loads are considered in the usual manner by having a structural node at that point and simply adding to the global load vector $\mathbf{F}$.

7.3 NUMERICAL INTEGRATION

Consider the problem of numerically evaluating a one-dimensional integral of the form

$$I = \int_{-1}^{1} f(\xi) \, d\xi \tag{7.33}$$

The *Gaussian quadrature* approach for evaluating I is given below. This method has proved most useful in finite element work. Extension to integrals in two and three dimensions follows readily.

Consider the n-point approximation

$$I = \int_{-1}^{1} f(\xi) \, d\xi \approx w_1 f(\xi_1) + w_2 f(\xi_2) + \cdots + w_n f(\xi_n) \tag{7.34}$$

where $w_1, w_2, \ldots, w_n$ are the **weights** and $\xi_1, \xi_2, \ldots, \xi_n$ are the sampling points or **Gauss points**. The idea behind Gaussian quadrature is to select the n Gauss points and n weights such that Eq. 7.34 provides an exact answer for polynomials $f(\xi)$ of as large a degree as possible. In other words, the idea is that if the n-point integration formula is exact for all polynomials up to as high a degree as possible, then the formula will work well even if f is not a polynomial. To get some intuition for the method, the one-point and two-point approximations are discussed below.

One-point formula.   Consider the formula with $n = 1$ as

$$\int_{-1}^{1} f(\xi) \, d\xi \approx w_1 f(\xi_1) \tag{7.35}$$

Since there are two parameters, w_1 and ξ_1, we consider requiring the formula in Eq. 7.35 to be exact when $f(\xi)$ is a polynomial of order 1. Thus, if $f(\xi) = a_0 + a_1\xi$, then we require

$$\text{Error} = \int_{-1}^{1}(a_0 + a_1\xi)\,d\xi - w_1 f(\xi_1) = 0 \qquad (7.36a)$$

or

$$\text{Error} = 2a_0 - w_1(a_0 + a_1\xi_1) = 0 \qquad (7.36b)$$

or

$$\text{Error} = a_0(2 - w_1) - w_1 a_1\xi_1 = 0 \qquad (7.36c)$$

From Eq. 7.36c, we see that the error is zeroed if

$$w_1 = 2 \qquad \xi_1 = 0 \qquad (7.37)$$

For any general f, then, we have

$$I = \int_{-1}^{1} f(\xi)\,d\xi \approx 2f(0) \qquad (7.38)$$

which is seen to be the familiar *midpoint rule* (Fig. 7.3).

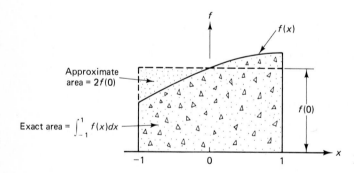

Figure 7.3   One-point gauss quadrature.

Two-point formula.   Consider the formula with $n = 2$ as

$$\int_{-1}^{1} f(\xi)\,d\xi \approx w_1 f(\xi_1) + w_2 f(\xi_2) \qquad (7.39)$$

We have four parameters to choose: w_1, w_2, ξ_1, and ξ_2. We can therefore expect the formula in Eq. 7.40 to be exact for a cubic polynomial. Thus, choosing $f(\xi) = a_0 + a_1\xi + a_2\xi^2, + a_3\xi^3$ yields

$$\text{Error} = \left[\int_{-1}^{1}(a_0 + a_1\xi + a_2\xi^2 + a_3\xi^3)\,d\xi\right] - [w_1 f(\xi_1) + w_2 f(\xi_2)] \qquad (7.40)$$

Requiring zero error yields

$$w_1 + w_2 = 2$$
$$w_1\xi_1 + w_2\xi_2 = 0$$
$$w_1\xi_1^2 + w_2\xi_2^2 = \tfrac{2}{3} \qquad (7.41)$$
$$w_1\xi_1^3 + w_2\xi_2^3 = 0$$

These nonlinear equations have the unique solution

$$w_1 = w_2 = 1 \qquad -\xi_1 = \xi_2 = 1/\sqrt{3} = 0.5773502691 \ldots \qquad (7.42)$$

From above, we can conclude that n-point Gaussian quadrature will provide an exact answer if f is a polynomial of order $(2n - 1)$ or less. Table 7.1 gives the values of w_i and ξ_i for Gauss quadrature formulas of orders $n = 1$ through $n = 6$. Note that the Gauss points are located symmetrically with respect to the origin, and that symmetrically placed points have the same weights. Moreover, the large number of digits given in Table 7.1 should be used in the calculations for accuracy; i.e., use double precision on the computer.

TABLE 7.1   GAUSS POINTS AND WEIGHTS FOR GAUSSIAN QUADRATURE

$$\int_{-1}^{1} f(\xi)\, d\xi \approx \sum_{i=1}^{n} w_i f(\xi_i)$$

Number of points, n	Location, ξ_i	Weights, w_i
1	0.0	2.0
2	$\pm 1/\sqrt{3} = \pm 0.5773502692$	1.0
3	± 0.7745966692	0.5555555556
	0.0	0.8888888889
4	± 0.8611363116	0.3478548451
	± 0.3399810436	0.6521451549
5	± 0.9061798459	0.2369268851
	± 0.5384693101	0.4786286705
	0.0	0.5688888889
6	± 0.9324695142	0.1713244924
	± 0.6612093865	0.3607615730
	± 0.2386191861	0.4679139346

Example 7.1

Evaluate

$$I = \int_{-1}^{1} \left[3e^x + x^2 + \frac{1}{(x + 2)} \right] dx$$

using one-point and two-point Gauss quadrature.

Solution   For $n = 1$, we have $w_1 = 2$, $x_1 = 0$, and

$$I \approx 2f(0)$$
$$= 7.0$$

For $n = 2$, $w_1 = w_2 = 1$, $x_1 = -0.57735\ldots$, $x_2 = +0.57735\ldots$, and $I \approx 8.7857$.

This may be compared with the exact solution

$$I_{\text{exact}} = 8.8165 \qquad \blacksquare$$

Two-Dimensional Integrals

The extension of Gaussian quadrature to two-dimensional integrals of the form

$$I = \int_{-1}^{1} \int_{-1}^{1} f(\xi, \eta) \, d\xi \, d\eta \qquad (7.43)$$

follows readily, since

$$I \approx \int_{-1}^{1} \left[\sum_{i=1}^{n} w_i f(\xi_i, \eta) \right] d\eta$$

$$\approx \sum_{j=1}^{n} w_j \left[\sum_{i=1}^{n} w_i f(\xi_i, \eta_j) \right]$$

or

$$I \approx \sum_{i=1}^{n} \sum_{j=1}^{n} w_i w_j f(\xi_i, \eta_j) \qquad (7.44)$$

Stiffness Integration

To illustrate the use of Eq. 7.44, consider the element stiffness for a quadrilateral element

$$\mathbf{k}^e = t_e \int_{-1}^{1} \int_{-1}^{1} \mathbf{B}^{\mathsf{T}} \mathbf{D} \mathbf{B} \, \det \mathbf{J} \, d\xi \, d\eta$$

where $\mathbf{B}$ and $\det \mathbf{J}$ are functions of ξ and η. Note that the integral above actually consists of the integral of each element in an (8×8) matrix. However, using the fact that $\mathbf{k}^e$ is symmetric, we do not need to integrate elements below the main diagonal.

Let ϕ represent the ijth element in the integrand. That is, let

$$\phi(\xi, \eta) = t_e(\mathbf{B}^{\mathsf{T}} \mathbf{D} \mathbf{B} \, \det \mathbf{J})_{ij} \qquad (7.45)$$

Then, if we use a *2 × 2 rule*, we get

$$k_{ij} \approx w_1^2 \phi(\xi_1, \eta_1) + w_1 w_2 \phi(\xi_1, \eta_2)$$
$$+ w_2 w_1 \phi(\xi_2, \eta_1) + w_2^2 \phi(\eta_2, \eta_2) \qquad (7.46a)$$

where $w_1 = w_2 = 1.0$, $\xi_1 = \eta_1 = -0.57735\ldots$, and $\xi_2 = \eta_2 = +0.57735\ldots$. The Gauss points for the two-point rule used above are shown in Fig. 7.4. Alternatively, if we label the Gauss points as 1, 2, 3, and 4, then k_{ij} in Eq. 7.46a can also be written as

$$k_{ij} = \sum_{IP=1}^{4} W_{IP}\phi_{IP} \qquad (7.46b)$$

where ϕ_{IP} is the value of ϕ and W_{IP} is the weight factor at integration point IP. We note that $W_{IP} = (1)(1) = 1$. Computer implementation is sometimes easier using Eq. 7.46b. We may readily follow the implementation of the above integration procedure in program FE2QUAD provided at the end of this chapter.

The evaluation of three-dimensional integrals is similar. For triangles, however, the weights and Gauss points are different, as discussed later in this chapter.

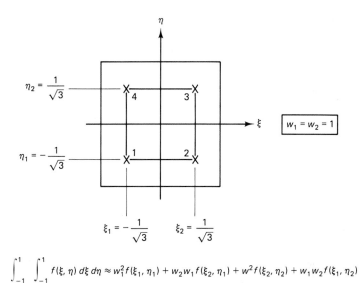

$$\int_{-1}^{1} \int_{-1}^{1} f(\xi, \eta)\, d\xi\, d\eta \approx w_1^2 f(\xi_1, \eta_1) + w_2 w_1 f(\xi_2, \eta_1) + w^2 f(\xi_2, \eta_2) + w_1 w_2 f(\xi_1, \eta_2)$$

Figure 7.4   Gaussian quadrature in two dimensions using the 2 × 2 rule.

Stress Calculations

Unlike the constant strain triangular element (Chapters 5 and 6), the stresses $\boldsymbol{\sigma} = \mathbf{DBq}$ in the quadrilateral element are not constant within the element; they are functions of ξ, η, and consequently vary within the element. In practice, the stresses are evaluated at the Gauss points, which are also the points used for numerical evaluation of $\mathbf{k}^e$, where they are found to be accurate. For a quadrilateral with 2 × 2 integration, this gives four sets of stress values. For generating less data, one may evaluate stresses at one point per element, say at $\xi = 0$, $\eta = 0$. The latter approach is used in the program FE2QUAD.

Example 7.1

Consider a rectangular element as shown in Fig. E7.1. Assume plane stress condition, $E = 30 \times 10^6$ psi, $\nu = 0.3$, and $\mathbf{q} = [0, 0, 0.002, 0.003, 0.006, 0.0032, 0, 0]^T$ in. Evaluate $\mathbf{J}$, $\mathbf{B}$, and $\boldsymbol{\sigma}$ at $\xi = 0$, $\eta = 0$.

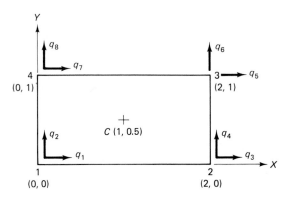

Figure E7.1

Solution   Referring to Eq. 7.13a, we have

$$\mathbf{J} = \frac{1}{4}\begin{bmatrix} 2(1 - \eta) + 2(1 + \eta) & (1 + \eta) - (1 + \eta) \\ -2(1 + \xi) + 2(1 + \xi) & (1 + \xi) + (1 - \xi) \end{bmatrix}$$

$$= \begin{bmatrix} 1 & 0 \\ 0 & \frac{1}{2} \end{bmatrix}$$

For this rectangular element, we find that $\mathbf{J}$ is a constant matrix. Now, from Eqs. 7.21,

$$\mathbf{A} = \frac{1}{1/2}\begin{bmatrix} \frac{1}{2} & 0 & 0 & 0 \\ 0 & 0 & 0 & 1 \\ 0 & 1 & \frac{1}{2} & 0 \end{bmatrix}$$

Evaluating $\mathbf{G}$ in Eq. 7.23 at $\xi = \eta = 0$, and using $\mathbf{B} = \mathbf{QG}$, we get

$$\mathbf{B}^0 = \begin{bmatrix} -\frac{1}{4} & 0 & \frac{1}{4} & 0 & \frac{1}{4} & 0 & -\frac{1}{4} & 0 \\ 0 & -\frac{1}{2} & 0 & -\frac{1}{2} & 0 & \frac{1}{2} & 0 & \frac{1}{2} \\ -\frac{1}{2} & -\frac{1}{4} & -\frac{1}{2} & \frac{1}{4} & \frac{1}{2} & \frac{1}{4} & \frac{1}{2} & -\frac{1}{4} \end{bmatrix}$$

The stresses at $\xi = \eta = 0$ are now given by the product

$$\boldsymbol{\sigma}^0 = \mathbf{D}\mathbf{B}^0\mathbf{q}$$

For the given data, we have

$$\mathbf{D} = \frac{30 \times 10^6}{(1 - 0.09)}\begin{bmatrix} 1 & 0.3 & 0 \\ 0.03 & 1 & 0 \\ 0 & 0 & 0.35 \end{bmatrix}$$

Thus

$$\boldsymbol{\sigma}^0 = [66\,920, 23\,080, 40\,960]^T \text{ psi} \qquad \blacksquare$$

Comment on degenerate quadrilaterals.   In some situations, we cannot avoid using degenerated quadrilaterals of the type shown in Fig. 7.5, where quadrilaterals degenerate into triangles. Numerical integration will permit the use of such elements, but the errors are higher than regular elements. Standard codes normally permit the use of such elements.

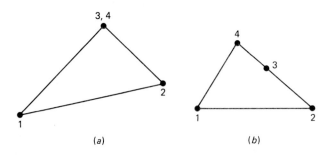

(a)   (b)   **Figure 7.5** Degenerate four-node quadrilateral elements.

7.4 HIGHER-ORDER ELEMENTS

The concepts presented above for the four-node quadrilateral element can be readily extended to other, higher-order, isoparametric elements. In the four-node quadrilateral element, the shape functions contained terms 1, ξ, η, and $\xi\eta$. In contrast, the elements to be discussed below also contain terms such as $\xi^2\eta$ and $\xi\eta^2$, which generally provide greater accuracy. Only the shape functions $\mathbf{N}$ are given below. The generation of element stiffness follows the routine steps:

$$\mathbf{u} = \mathbf{Nq} \tag{7.47a}$$

$$\boldsymbol{\epsilon} = \mathbf{Bq} \tag{7.47b}$$

$$\mathbf{k}^e = t_e \int_{-1}^{1}\int_{-1}^{1} \mathbf{B}^{\mathrm{T}}\mathbf{DB} \det \mathbf{J} \, d\xi \, d\eta \tag{7.47c}$$

where $\mathbf{k}^e$ is evaluated using Gaussian quadrature.

Nine-Node Quadrilateral

The nine-node quadrilateral has been found to be very effective in finite element practice. The local node numbers for this element are shown in Fig. 7.6a. The square master element is shown in Fig. 7.6b. The shape functions are defined as follows.

Consider, first, the ξ axis alone as shown in Fig. 7.6c. The local node numbers 1, 2, and 3 on this axis correspond to locations $\xi = -1$, 0, and $+1$, respectively. At these nodes, we define the generic shape functions L_1, L_2, and L_3 as

$$L_i(\xi) = 1 \quad \text{at node } i$$
$$= 0 \quad \text{at other two nodes} \tag{7.48}$$

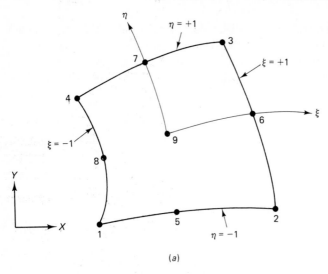

(a)

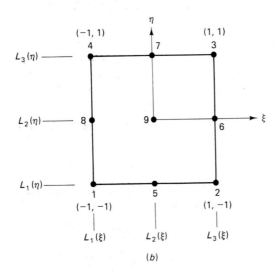

(b)

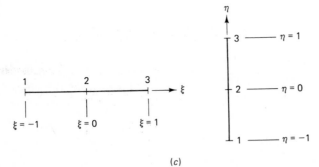

(c)

Figure 7.6 Nine-node quadrilateral (a) in x, y space and (b) in ξ, η space. (c) Definition of general shape functions.

Consider L_1. Since $L_1 = 0$ at $\xi = 0$ and at $\xi = +1$, we know that L_1 is of the form $L_1 = c\xi(1 - \xi)$. The constant c is obtained from $L_1 = 1$ at $\xi = -1$ as $c = -\frac{1}{2}$. Thus, $L_1(\xi) = -\xi(1 - \xi)/2$. L_2 and L_3 can be obtained by using similar arguments. We have

$$L_1(\xi) = -\frac{\xi(1 - \xi)}{2}$$

$$L_2(\xi) = (1 + \xi)(1 - \xi) \tag{7.49a}$$

$$L_3(\xi) = \frac{\xi(1 + \xi)}{2}$$

Similarly, generic shape functions can be defined along the η axis (Fig. 7.6c) as

$$L_1(\eta) = -\frac{\eta(1 - \eta)}{2}$$

$$L_2(\eta) = (1 + \eta)(1 - \eta) \tag{7.49b}$$

$$L_3(\eta) = \frac{\eta(1 + \eta)}{2}$$

Referring back to the master element in Fig. 7.6b, we observe that every node has the coordinates $\xi = -1$, 0, or $+1$, and $\eta = -1$, 0, or $+1$. Thus, the following *product rule* yields the shape functions $N_1, N_2, \ldots, N_9$, as

$$N_1 = L_1(\xi)L_1(\eta) \quad N_5 = L_2(\xi)L_1(\eta) \quad N_2 = L_3(\xi)L_1(\eta)$$

$$N_8 = L_1(\xi)L_2(\eta) \quad N_9 = L_2(\xi)L_2(\eta) \quad N_6 = L_3(\xi)L_2(\eta) \tag{7.50}$$

$$N_4 = L_1(\xi)L_3(\eta) \quad N_7 = L_2(\xi)L_3(\eta) \quad N_3 = L_3(\xi)L_3(\eta)$$

By the manner in which L_i are constructed, it can be readily verified that $N_i = 1$ at node i and equal to zero at other nodes, as desired.

As noted in the beginning of this section, the use of higher-order terms in **N** leads to a higher-order interpolation of the displacement field as given by $\mathbf{u} = \mathbf{Nq}$. In addition, since $x = \Sigma_i N_i x_i$ and $y = \Sigma_i N_i y_i$, it means that higher-order terms can also be used to define geometry. Thus, the elements can have curved edges if desired. However, it is possible to define a *sub-parametric* element by using nine-node shape functions to interpolate displacement and using only four-node quadrilateral shape functions to define geometry.

Eight-Node Quadrilateral

This element belongs to the **serendipity** family of elements. The element consists of eight nodes (Fig. 7.7a) all of which are located on the boundary. Our task is to define shape functions N_i such that $N_i = 1$ at node i and 0 at all other nodes. In defining N_i, we refer to the master element shown in Fig. 7.7b. First, we define

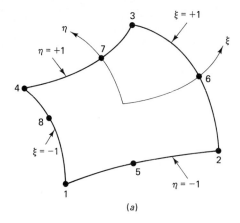

(a)

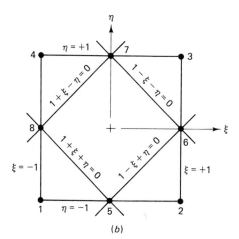

(b)

Figure 7.7 Eight-node quadrilateral (a) in x, y space and (b) in ξ, η space.

N_1–N_4. For N_1, we note that $N_1 = 1$ at node 1 and 0 at other nodes. Thus, N_1 has to vanish along the lines $\xi = +1$, $\eta = +1$, and $\xi + \eta = -1$ (Fig. 7.7a). Consequently, N_1 is of the form

$$N_1 = c(1 - \xi)(1 - \eta)(1 + \xi + \eta) \tag{7.51}$$

At node 1, $N_1 = 1$, $\xi = \eta = -1$. Thus, $c = -\frac{1}{4}$. We thus have

$$N_1 = -\frac{(1 - \xi)(1 - \eta)(1 + \xi + \eta)}{4}$$

$$N_2 = -\frac{(1 + \xi)(1 - \eta)(1 - \xi + \eta)}{4}$$

$$(7.52)$$

$$N_3 = -\frac{(1 + \xi)(1 + \eta)(1 - \xi - \eta)}{4}$$

$$N_4 = -\frac{(1 - \xi)(1 + \eta)(1 + \xi - \eta)}{4}$$

Now, we define N_5, N_6, N_7, and N_8 at the midpoints. For N_5, we know that it vanishes along edges $\xi = +1$, $\eta = +1$, and $\xi = -1$. Consequently, it has to be of the form

$$N_5 = c(1 - \xi)(1 - \eta)(1 + \xi) \qquad\qquad (7.53a)$$

$$= c(1 - \xi^2)(1 - \eta) \qquad\qquad (7.53b)$$

The constant c in the above equation is determined from the condition $N_5 = 1$ at node 5, or $N_5 = 1$ at $\xi = 0$, $\eta = -1$. Thus, $c = \frac{1}{2}$ and

$$N_5 = \frac{(1 - \xi^2)(1 - \eta)}{2} \qquad\qquad (7.53c)$$

We have

$$N_5 = \frac{(1 - \xi^2)(1 - \eta)}{2}$$

$$N_6 = \frac{(1 + \xi)(1 - \eta^2)}{2}$$

$$N_7 = \frac{(1 - \xi^2)(1 + \eta)}{2} \qquad\qquad (7.54)$$

$$N_8 = \frac{(1 - \xi)(1 - \eta^2)}{2}$$

Six-Node Triangle

The six-node triangle is shown in Figs. 7.8a, b. By referring to the master element in Fig. 7.8b, we can write the shape functions as

$$\begin{array}{ll} N_1 = \xi(2\xi - 1) & N_4 = 4\xi\eta \\ N_2 = \eta(2\eta - 1) & N_5 = 4\zeta\eta \qquad (7.55) \\ N_3 = \zeta(2\zeta - 1) & N_6 = 4\xi\zeta \end{array}$$

where $\zeta = 1 - \xi - \eta$. Because of terms ξ^2, η^2, etc. in the shape functions, this element is also called a *quadratic* triangle. The isoparametric representation is

$$\mathbf{u} = \mathbf{Nq}$$

$$(7.56)$$

$$x = \sum N_i x_i \qquad y = \sum N_i y_i$$

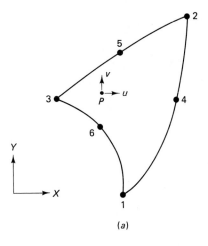

(a)

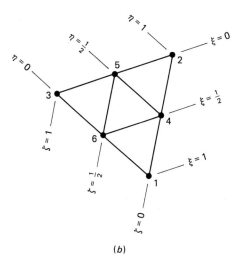

(b)   **Figure 7.8**   Six-node triangular element.

The element stiffness, which has to be integrated numerically, is given by

$$\mathbf{k}^e = t_e \iint_A \mathbf{B}^T \mathbf{D} \mathbf{B} \det \mathbf{J} \, d\xi \, d\eta \tag{7.57}$$

The Gauss points for a triangular region differ from the square region consid-ered earlier. The simplest is the one-point rule at the centroid with weight $w_1 = \frac{1}{2}$ and $\xi_1 = \eta_1 = \zeta_1 = \frac{1}{3}$. Equation 7.57 then yields

$$\mathbf{k}^e \approx \frac{1}{2} t_e \overline{\mathbf{B}}^T \overline{\mathbf{D}} \overline{\mathbf{B}} \det \overline{\mathbf{J}} \tag{7.58}$$

where $\overline{\mathbf{B}}$, $\overline{\mathbf{J}}$ are evaluated at the Gauss point. Other choices of weights and Gauss points are given in Table 7.2. The Gauss points given in Table 7.2 are arranged sym-metrically within the triangle. Because of triangular symmetry, the Gauss points oc-

TABLE 7.2   GAUSS QUADRATURE FORMULAS FOR A TRIANGLE

$$\int_0^1 \int_0^{1-\xi} f(\xi, \eta)\, d\eta\, d\xi \approx \sum_{i=1}^{n} w_i f(\xi_i, \eta_i)$$

No. of points, n	Weight, w_i	Multi-plicity	ξ_i	η_i	ζ_i
One	$\frac{1}{2}$	1	$\frac{1}{3}$	$\frac{1}{3}$	$\frac{1}{3}$
Three	$\frac{1}{6}$	3	$\frac{2}{3}$	$\frac{1}{6}$	$\frac{1}{6}$
Three	$\frac{1}{6}$	3	$\frac{1}{2}$	$\frac{1}{2}$	0
Four	$-\frac{9}{32}$	1	$\frac{1}{3}$	$\frac{1}{3}$	$\frac{1}{3}$
	$\frac{25}{96}$	3	$\frac{3}{5}$	$\frac{1}{5}$	$\frac{1}{5}$
Six	$\frac{1}{12}$	6	0.6590276223	0.2319333685	0.1090390090

cur in groups or *multiplicity* of one, three, or six. For multiplicity of three, if ξ, η, and ζ coordinates of a Gauss point are, for instance, $(\frac{2}{3}, \frac{1}{6}, \frac{1}{6})$, then the other two Gauss points are located at $(\frac{1}{6}, \frac{2}{3}, \frac{1}{6})$ and $(\frac{1}{6}, \frac{1}{6}, \frac{2}{3})$. Note that $\zeta = 1 - \xi - \eta$, as is discussed in Chapter 5. For multiplicity of six, all six possible permutations of the ξ, η, ζ coordinates are used.

Comment on midside node.   In the higher-order isoparametric elements discussed above, we note the presence of midside nodes. The midside node should be as near as possible to the center of the side. The node should not be outside of $\frac{1}{4} < s/\ell < \frac{3}{4}$, as shown in Fig. 7.9. This condition ensures that det **J** does not attain a value of zero in the element.

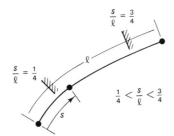

Figure 7.9 Restrictions on the location of a midside node.

Concluding Note

The concept of a master element defined in ξ, η coordinates, the definition of shape functions for interpolating displacement and geometry, and use of numerical integration are all key ingredients of the isoparametric formulation. A wide variety of elements can be formulated in a unified manner. Though only stress analysis has been considered in this chapter, the elements can be applied to nonstructural problems quite readily.

Example 7.2

The problem in Example 5.6 (Fig. E5.6) is now solved using 4-node quadrilateral elements using program FE2QUAD. The loads, boundary conditions, and node locations are same as in Fig. E5.6. The only difference is the modeling with 24 quadrilateral elements, as against 48 CST elements in Fig. E5.6. Again, MESHGEN has been used to create the mesh (Fig. E7.2a) and DATAFEM to define the loads, boundary conditions, and material properties.

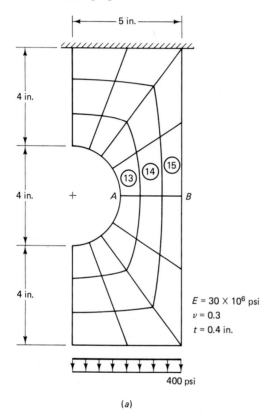

$E = 30 \times 10^6$ psi
$\nu = 0.3$
$t = 0.4$ in.

400 psi

(a)

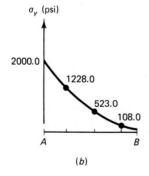

(b)

Figure E7.2

The stresses output by program FE2QUAD corresponds to the (0, 0) location in the natural coordinate system (master element). Using this fact, we *extrapolate* the *y* stresses in elements 13, 14, and 15 to obtain the maximum *y*-stress near the semicircular edge of the plate, as shown in Fig. E7.2*b*. ■

```
1000 REM **********    FE2QUAD    ************
1010 REM *** 2-D STRESS ANALYSIS USING 4-NODE ***
1020 REM ***    QUADRILATERAL ELEMENTS    ***
1030 CLS : DEFINT I-N
1040 PRINT "======================================="
1050 PRINT "    Program written by    "
1060 PRINT " T.R.Chandrupatla and A.D.Belegundu  "
1070 PRINT "======================================="
1080 PRINT " 1) Plane Stress Analysis"
1090 PRINT " 2) Plane Strain Analysis"
1100 PRINT : INPUT "Choose 1 or 2 "; LC
1110 IF LC < 1 OR LC > 2 GOTO 1080
1120 INPUT "Thickness for the Problem TH "; TH
1130 INPUT "Data File Name <path fn.ext> ", FILE1$
1140 OPEN FILE1$ FOR INPUT AS #1
1150 INPUT "Output File Name <path fn.ext> ", FILE2$
1160 OPEN FILE2$ FOR OUTPUT AS #2
1170 IF LC = 1 THEN PRINT #2, "Plane Stress Analysis"
1180 IF LC = 2 THEN PRINT #2, "Plane Strain Analysis"
1190 INPUT #1, NN, NE, ND, NL, NM, NDIM, NEN
1200 REM Total DOF is  NQ
1210 NQ = 2 * NN
1220 DIM X(NN, NDIM), NOC(NE, NEN), MAT(NE)
1230 DIM NU(ND), U(ND), F(NQ), PM(NM, 3)
1240 DIM XNI(4, 2), D(3, 3), Q(8)
1250 DIM A(3, 4), G(4, 8), B(3, 8), DB(3, 8), STR(3)
1260 GOSUB 2380: REM <-------- (((Read Data from File)))
1270 REM Bandwidth NBW from Connectivity  NOC( , )
1280 NBW = 0
1290 FOR N = 1 TO NE
1300 CMIN = NN + 1: CMAX = 0
1310 FOR J = 1 TO 4
1320 IF CMIN > NOC(N, J) THEN CMIN = NOC(N, J)
1330 IF CMAX < NOC(N, J) THEN CMAX = NOC(N, J)
1340 NEXT J
1350 C = 2 * (CMAX - CMIN + 1)
1360 IF NBW < C THEN NBW = C
1370 NEXT N
1380 PRINT "The Bandwidth is"; NBW
1390 DIM S(NQ, NBW), SE(8, 8)
1400 REM *** Initialization ***
1410 FOR I = 1 TO NQ: FOR J = 1 TO NBW
1420 S(I, J) = 0: NEXT J: NEXT I
1430 REM ******   GLOBAL STIFFNESS MATRIX  ******
1440 REM ** CORNER NODES AND ITEGRATION POINTS **
1450 GOSUB 2500: REM <--- Integ. Points of Master Element
1460 FOR N = 1 TO NE
1470 PRINT "Forming Stiffness Matrix of Element "; N
1480 GOSUB 2570: REM <---Elem. Nodal Coord. & D-Matrix
1490 GOSUB 2790: REM <---Get Element Stiffness
1500 PRINT "--- Placing in Banded Locations ---"
1510 FOR II = 1 TO 4
1520 NRT = 2 * (NOC(N, II) - 1)
1530 FOR IT = 1 TO 2
1540 NR = NRT + IT
1550 I = 2 * (II - 1) + IT
1560 FOR JJ = 1 TO 4
1570 NCT = 2 * (NOC(N, JJ) - 1)
```

```
1580 FOR JT = 1 TO 2
1590 J = 2 * (JJ - 1) + JT
1600 NC = NCT + JT - NR + 1
1610 IF NC <= 0 GOTO 1630
1620 S(NR, NC) = S(NR, NC) + SE(I, J)
1630 NEXT JT
1640 NEXT JJ: NEXT IT: NEXT II
1650 NEXT N
1660 PRINT "***** Modifying for Boundary Conditions *****"
1670 CNST = S(1, 1) * 100000!
1680 FOR I = 1 TO ND
1690 N = NU(I)
1700 S(N, 1) = S(N, 1) + CNST
1710 F(N) = F(N) + CNST * U(I): NEXT I
1720 PRINT " ****** Solving Equations ******"
1730 GOSUB 5000
1740 ' *** PRINT OUTPUT ***
1750 PRINT #2,"THICKNESS ="; TH
1760 PRINT #2,"NN, NE, ND, NL, NM, NDIM, NEN"
1770 PRINT #2,NN; NE; ND; NL; NM; NDIM; NEN
1780 PRINT #2, "NODE#  X-COORD. Y-COORD."
1790 FOR I=1 TO NN
1800 PRINT #2, I;   X(I,1);   X(I,2) : NEXT I
1810 PRINT #2, "ELEMENT#, 4 NODES, MATERIAL#"
1820 FOR I=1 TO NE
1830 PRINT #2, I;  NOC(I,1); NOC(I,2); NOC(I,3); NOC(I,4); MAT(I)
1840 NEXT I
1850 PRINT #2, "DOF#, SPECIFIED DISPLACEMENT"
1860 FOR I=1 TO ND: PRINT #2,NU(I), U(I):NEXT I
1870 PRINT #2,"MATERIAL#,  E,   POISSON'S RATIO"
1880 FOR I=1 TO NM: PRINT #2,I,PM(I,1), PM(I,2):NEXT I
1890 PRINT #2, "NODE#  X-Displ    Y-Displ"
1900 FOR I = 1 TO NN
1910 PRINT #2, USING " ###"; I;
1920 II = 2 * (I - 1)
1930 PRINT #2, USING " ##.###^^^^"; F(II + 1); F(II + 2): NEXT I
1940 PRINT #2,
1950 REM *** REACTION CALCULATION ***
1960 PRINT #2, " DOF#      REACTION:"
1970 FFF1$ = "   ###    ##.####^^^^"
1980 FOR I = 1 TO ND
1990 N = NU(I)
2000 R = CNST * (U(I) - F(N))
2010 PRINT #2, USING FFF1$; N; R: NEXT I
2020 PRINT #2,
2030 REM ***  STRESS CALCULATIONS ***
2040 REM Stress at the XI=0 and ETA=0
2050 XI = 0: ETA = 0
2060 PRINT #2, "ELEM#   SX      SY      TXY";
2070 PRINT #2, "     S1     S2     ANGLE SX->S1"
2080 FOR N = 1 TO NE
2090 GOSUB 2570: REM ** D Matrix and Nodal Coordinates etc.
2100 GOSUB 2970: REM ** Get DB Matrix
2110 REM  Stress Evaluation
2120 FOR I = 1 TO 4
2130 IN = 2 * (NOC(N, I) - 1)
2140 II = 2 * (I - 1)
2150 FOR J = 1 TO 2
2160 Q(II + J) = F(IN + J): NEXT J: NEXT I
2170 FOR I = 1 TO 3
2180 STR(I) = 0
2190 FOR K = 1 TO 8
2200 STR(I) = STR(I) + DB(I, K) * Q(K): NEXT K: NEXT I
2210 REM *** Principal Stress Calculations ***
2220 IF STR(3) <> 0 GOTO 2270
```

```
2230 S1 = STR(1): S2 = STR(2): ANG = 0
2240 IF S1 > S2 GOTO 2290
2250 S1 = STR(2): S2 = STR(1): ANG = 90
2260 GOTO 2290
2270 C = .5 * (STR(1) + STR(2))
2272 R = SQR(.25 * (STR(1) – STR(2)) ^ 2 + (STR(3)) ^ 2)
2274 S1 = C + R: S2 = C – R
2276 IF C > STR(1) GOTO 2280
2278 ANG = 57.29577951# * ATN(STR(3) / (STR(1) – S2)): GOTO 2290
2280 ANG = 57.29577951# * ATN(STR(3) / (S1 – STR(1)))
2382 IF STR(3) > 0 THEN ANG = 90 – ANG
2384 IF STR(3) < 0 THEN ANG = –90 – ANG
2290 PRINT #2, USING " ###"; N;
2300 PRINT #2, USING " ##.###^^^^"; STR(1); STR(2); STR(3);
2310 PRINT #2, USING " ##.###^^^^"; S1; S2; ANG
2320 NEXT N
2330 CLOSE #2: PRINT : PRINT
2340 PRINT "-----    All Calculations are done    -----"
2350 PRINT "The Results are available in the text file "; FILE2$
2360 PRINT "View using a text processor"
2370 END
2380 '=============== READ DATA ====================
2390 FOR I = 1 TO NN: FOR J = 1 TO NDIM
2400 INPUT #1, X(I, J): NEXT J: NEXT I
2410 FOR I = 1 TO NE: FOR J = 1 TO NEN
2420 INPUT #1, NOC(I, J): NEXT J: NEXT I
2430 FOR I = 1 TO NE: INPUT #1, MAT(I): NEXT I
2440 FOR I = 1 TO ND: INPUT #1, NU(I), U(I): NEXT I
2450 FOR I = 1 TO NL: INPUT #1, II, F(II): NEXT I
2460 FOR I = 1 TO NM: FOR J = 1 TO 3
2470 INPUT #1, PM(I, J): NEXT J: NEXT I
2480 CLOSE #1
2490 RETURN
2500 REM --- Integration Points XNI ---
2510 C = .57735026919#
2520 XNI(1, 1) = –C: XNI(1, 2) = –C
2530 XNI(2, 1) = C: XNI(2, 2) = –C
2540 XNI(3, 1) = C: XNI(3, 2) = C
2550 XNI(4, 1) = –C: XNI(4, 2) = C
2560 RETURN
2570 REM D( , ) Matrix Relates Stresses to Strains
2580 REM    ----- Formation of D Matrix -----
2590 MATN = MAT(N)
2600 E = PM(MATN, 1): PNU = PM(MATN, 2)
2610 N1 = NOC(N, 1): N2 = NOC(N, 2)
2620 N3 = NOC(N, 3): N4 = NOC(N, 4)
2630 X1 = X(N1, 1): Y1 = X(N1, 2)
2640 X2 = X(N2, 1): Y2 = X(N2, 2)
2650 X3 = X(N3, 1): Y3 = X(N3, 2)
2660 X4 = X(N4, 1): Y4 = X(N4, 2)
2670 IF LC = 2 GOTO 2710
2680 REM ***    Plane Stress    ***
2690 C1 = E / (1 – PNU ^ 2): C2 = C1 * PNU
2700 GOTO 2740
2710 REM ***    Plane Strain    ***
2720 C = E / ((1 + PNU) * (1 – 2 * PNU))
2730 C1 = C * (1 – PNU): C2 = C * PNU
2740 C3 = .5 * E / (1 + PNU)
2750 D(1, 1) = C1: D(1, 2) = C2: D(1, 3) = 0
2760 D(2, 1) = C2: D(2, 2) = C1: D(2, 3) = 0
2770 D(3, 1) = 0: D(3, 2) = 0: D(3, 3) = C3
2780 RETURN
2790 REM    Weight Factor is ONE
2800 REM Initialize Element Stiffness
2810 FOR I = 1 TO 8
```

```
2820 FOR J = 1 TO 8
2830 SE(I, J) = 0: NEXT J: NEXT I
2840 REM *** Loop on Integration Points ***
2850 FOR IP = 1 TO 4
2860 REM ***  Get DB Matrix at Integration Point IP  ***
2870 XI = XNI(IP, 1): ETA = XNI(IP, 2)
2880 GOSUB 2970
2890 REM Element Stiffness Matrix  SE
2900 FOR I = 1 TO 8
2910 FOR J = 1 TO 8
2920 FOR K = 1 TO 3
2930 SE(I, J) = SE(I, J) + B(K, I) * DB(K, J) * DJ * TH
2940 NEXT K: NEXT J: NEXT I
2950 NEXT IP
2960 RETURN
2970 REM *****  SUBROUTINE  DB MATRIX  *****
2980 REM Formation of Jacobian  TJ
2990 TJ11 = ((1 - ETA) * (X2 - X1) + (1 + ETA) * (X3 - X4)) / 4
3000 TJ12 = ((1 - ETA) * (Y2 - Y1) + (1 + ETA) * (Y3 - Y4)) / 4
3010 TJ21 = ((1 - XI) * (X4 - X1) + (1 + XI) * (X3 - X2)) / 4
3020 TJ22 = ((1 - XI) * (Y4 - Y1) + (1 + XI) * (Y3 - Y2)) / 4
3030 REM Determinant of the JACOBIAN
3040 DJ = TJ11 * TJ22 - TJ12 * TJ21
3050 REM A(3,4) Matrix relates Strains to
3060 REM Local Derivatives of u
3070 A(1, 1) = TJ22 / DJ: A(2, 1) = 0: A(3, 1) = -TJ21 / DJ
3080 A(1, 2) = -TJ12 / DJ: A(2, 2) = 0: A(3, 2) = TJ11 / DJ
3090 A(1, 3) = 0: A(2, 3) = -TJ21 / DJ: A(3, 3) = TJ22 / DJ
3100 A(1, 4) = 0: A(2, 4) = TJ11 / DJ: A(3, 4) = -TJ12 / DJ
3110 REM G(4,8) Matrix relates to Local Derivatives of u
3120 REM Local Nodal Displacements q(8)
3130 FOR I = 1 TO 4: FOR J = 1 TO 8
3140 G(I, J) = 0: NEXT J: NEXT I
3150 G(1, 1) = -(1 - ETA) / 4: G(2, 1) = -(1 - XI) / 4
3160 G(3, 2) = -(1 - ETA) / 4: G(4, 2) = -(1 - XI) / 4
3170 G(1, 3) = (1 - ETA) / 4: G(2, 3) = -(1 + XI) / 4
3180 G(3, 4) = (1 - ETA) / 4: G(4, 4) = -(1 + XI) / 4
3190 G(1, 5) = (1 + ETA) / 4: G(2, 5) = (1 + XI) / 4
3200 G(3, 6) = (1 + ETA) / 4: G(4, 6) = (1 + XI) / 4
3210 G(1, 7) = -(1 + ETA) / 4: G(2, 7) = (1 - XI) / 4
3220 G(3, 8) = -(1 + ETA) / 4: G(4, 8) = (1 - XI) / 4
3230 REM B(3,8) Matrix Relates Strains to q
3240 FOR I = 1 TO 3
3250 FOR J = 1 TO 8
3260 B(I, J) = 0
3270 FOR K = 1 TO 4
3280 B(I, J) = B(I, J) + A(I, K) * G(K, J)
3290 NEXT K: NEXT J: NEXT I
3300 REM DB(3,8) Matrix relates Stresses to q(8)
3310 FOR I = 1 TO 3
3320 FOR J = 1 TO 8
3330 DB(I, J) = 0
3340 FOR K = 1 TO 3
3350 DB(I, J) = DB(I, J) + D(I, K) * B(K, J)
3360 NEXT K: NEXT J: NEXT I
3370 RETURN
3380 REM Solution of Banded Equations
5000 N1 = NQ - 1
5010 FOR K = 1 TO N1
5020 PRINT "--- Forward Elimination in Progress ---"
5030 NK = NQ - K + 1
5040 IF NK > NBW THEN NK = NBW
5050 FOR I = 2 TO NK
```

```
5060 C1 = S(K, I) / S(K, 1)
5070 I1 = K + I - 1
5080 FOR J = I TO NK
5090 J1 = J - I + 1
5100 S(I1, J1) = S(I1, J1) - C1 * S(K, J): NEXT J
5110 F(I1) = F(I1) - C1 * F(K): NEXT I: NEXT K
5120 F(NQ) = F(NQ) / S(NQ, 1)
5130 FOR KK = 1 TO N1
5140 PRINT "--- Back Substitution in Progress ---"
5150 K = NQ - KK
5160 C1 = 1 / S(K, 1)
5170 F(K) = C1 * F(K)
5180 NK = NQ - K + 1
5190 IF NK > NBW THEN NK = NBW
5200 FOR J = 2 TO NK
5210 F(K) = F(K) - C1 * S(K, J) * F(K + J - 1)
5220 NEXT J
5230 NEXT KK
5240 RETURN
```

PROBLEMS

7.1. Figure P7.1 shows a four-node quadrilateral. The (x, y) coordinates of each node are given in the figure. The element displacement vector **q** is given as

$$\mathbf{q} = [0, 0, 0.20, 0, 0.15, 0.10, 0, 0.05]^T$$

Find:
  (a) The x, y coordinates of a point P whose location in the master element is given by $\xi = 0.5$, $\eta = 0.5$.
  (b) The u, v displacements of the point P.

7.2. Using a 2×2 rule, evaluate the integral

$$\iint\limits_{A} (x^2 + xy^2) \, dx \, dy$$

by Gaussian quadrature, where A denotes the region shown in Fig. P7.1.

7.3. State whether the following statements are true or false:
  (a) The shape functions are linear along an edge of a four-node quadrilateral element.
  (b) For isoparametric elements, such as four-, eight-, and nine-node quadrilaterals, the point $\xi = 0$, $\eta = 0$ in the master element corresponds to the centroid of the element in x, y coordinates.
  (c) The maximum stresses within an element occur at the Gauss points.
  (d) The integral of a cubic polynomial can be performed exactly using two-point Gauss quadrature.

7.4. Solve Problem P5.5 with four-node quadrilaterals. Use program FE2QUAD.

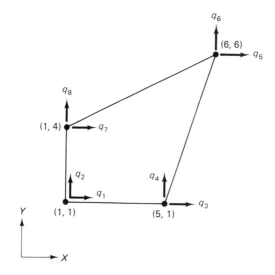

Figure P7.1

7.5. A half-symmetry model of a culvert is shown in Fig. P7.5. The pavement load is a uniformly distributed load of 5000 N/m². Using the MESHGEN program (discussed in Chapter 12), develop a finite element mesh with four-node quadrilateral elements. Using program FE2QUAD determine the location and magnitude of maximum principal stress. First, try a mesh with about 6 elements and then compare results using about 18 elements.

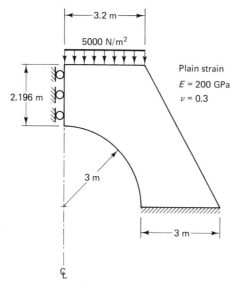

Figure P7.5

7.6. Solve Problem P5.6 using four-node quadrilateral elements (program FE2QUAD). Compare your results with the solution obtained with CST elements. Use comparable size meshes.

7.7. Solve Problem P5.7 using four-node quadrilaterals (program FE2QUAD).

7.8. Solve Problem P5.10 using four-node quadrilaterals (program FE2QUAD).

*7.9. Develop a program for axisymmetric stress analysis with four-node quadrilateral elements. Use your program to solve Example 6.1. Compare results. *Hint:* The first three rows of the **B** matrix are the same as for the plane stress problem in Eq. 7.25, and the last row can be obtained from $\epsilon_\theta = u/r$.

7.10. This problem focuses on a concept used in the MESHGEN program discussed in Chapter 12. An eight-node region is shown in Fig. P7.10a. The corresponding master element or *block* is shown in Fig. P7.10b. The block is divided into a grid of $3 \times 3 = 9$ smaller blocks of equal size, as indicated by dotted lines. Determine the corresponding x, y coordinates of all the 16 nodal points and plot the 9 subregions in Fig. P7.10a. Use the shape functions given in Eqs. 7.52 and 7.54.

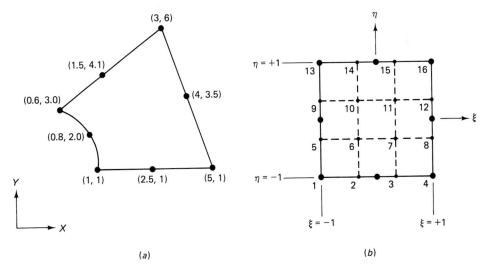

(a)   (b)

Figure P7.10

*7.11 Develop a computer program for the eight-node quadrilateral. Analyze the cantilever beam shown in Fig. P7.11 with 3 finite elements. Compare results of x stress and center-line deflections with
(a) 6-element CST model.
(b) Elementary beam theory.

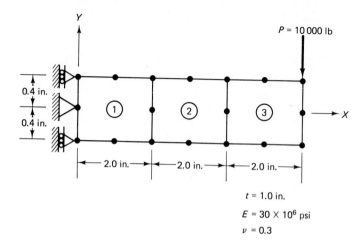

Figure P7.11

chapter eight

Beams and Frames

8.1 INTRODUCTION

Beams are slender members that are used for supporting transverse loading. Long horizontal members used in buildings and bridges, and shafts supported in bearings are some examples of beams. Complex structures with rigidly connected members are called frames and may be found in automobile and aeroplane structures and motion- and force-transmitting machines and mechanisms. In this chapter, we first present the finite element formulation for beams and extend these ideas to formulate and solve two-dimensional frame problems.

Beams with cross sections that are symmetric with respect to plane of loading are considered here. A general horizontal beam is shown in Fig. 8.1. Figure 8.2 shows the cross section and the bending stress distribution. For small deflections, we recall from elementary beam theory that

$$\sigma = -\frac{M}{I} y \tag{8.1}$$

$$\epsilon = \frac{\sigma}{E} \tag{8.2}$$

$$\frac{d^2 v}{dx^2} = \frac{M}{EI} \tag{8.3}$$

where σ is the normal stress, ϵ is the normal strain, M is the bending moment at the section, v is the deflection of the centroidal axis at x, and I is the moment of inertia of the section about the neutral axis (z axis passing through the centroid).

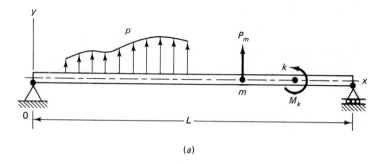

(a)

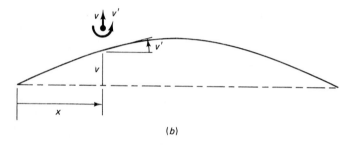

(b)

Figure 8.1 (a) Beam loading and (b) deformation of the neutral axis.

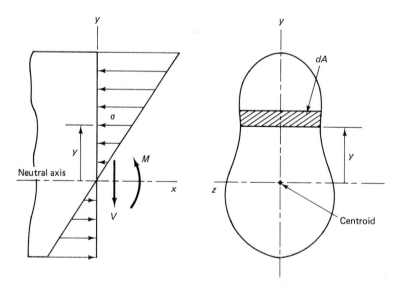

Figure 8.2 Beam section and stress distribution.

Potential Energy Approach

The strain energy dU in an element of length dx is

$$dU = \frac{1}{2} \int_A \sigma \epsilon \, dA \, dx$$

$$= \frac{1}{2} \left(\frac{M^2}{EI^2} \int_A y^2 dA \right) dx$$

Noting that $\int_A y^2 dA$ is the moment of inertia I, we have

$$dU = \frac{1}{2} \frac{M^2}{EI} \, dx \tag{8.4}$$

When Eq. 8.3 is used, then the total strain energy in the beam is given by

$$U = \frac{1}{2} \int_0^L EI \left(\frac{d^2v}{dx^2} \right)^2 dx \tag{8.5}$$

The potential energy Π of the beam is then given by

$$\Pi = \frac{1}{2} \int_0^L EI \left(\frac{d^2v}{dx^2} \right)^2 dx - \int_0^L pv \, dx - \sum_m P_m v_m - \sum_k M_k v_k' \tag{8.6}$$

where p is the distributed load per unit length, P_m is the point load at point m, M_k is the moment of the couple applied at point k, v_m is the deflection at point m and v_k' is the slope at point k.

Galerkin Approach

For the Galerkin formulation, we start from equilibrium of an elemental length. From Fig. 8.3, we recall that

$$\frac{dV}{dx} = p \tag{8.7}$$

$$\frac{dM}{dx} = V \tag{8.8}$$

When Eqs. 8.3, 8.7, and 8.8 are combined, the equilibrium equation is given by

$$\frac{d^2}{dx^2} \left(EI \frac{d^2v}{dx^2} \right) - p = 0 \tag{8.9}$$

For approximate solution by the Galerkin approach, we look for the approximate solution v constructed of finite element shape functions such that

$$\int_0^L \left[\frac{d^2}{dx^2} \left(EI \frac{d^2v}{dx^2} \right) - p \right] \phi \, dx = 0 \tag{8.10}$$

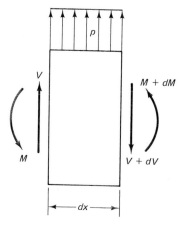

Figure 8.3 Free body diagram of an elemental length dx.

where ϕ is an arbitrary function using same basis functions as v. Note that ϕ is zero where v has a specified value. We integrate the first term of Eq. 8.10 by parts. The integral from 0 to L is split into intervals 0 to x_m, x_m to x_k, and x_k to L. We obtain

$$\int_0^L EI\frac{d^2v}{dx^2}\frac{d^2\phi}{dx^2}\,dx - \int_0^L p\phi\,dx + \frac{d}{dx}\left(EI\frac{d^2v}{dx^2}\right)\phi\bigg|_0^{x_m} + \frac{d}{dx}\left(EI\frac{d^2v}{dx^2}\right)\phi\bigg|_{x_m}^{L}$$

$$-EI\frac{d^2v}{dx^2}\frac{d\phi}{dx}\bigg|_0^{x_k} - EI\frac{d^2v}{dx^2}\frac{d\phi}{dx}\bigg|_{x_k}^{L} = 0 \qquad (8.11)$$

We note that $EI(d^2v/dx^2)$ equals the bending moment M from Eq. 8.3 and $(d/dx)[EI(d^2v/dx^2)]$ equals the shear force V from (8.8). Also, ϕ and M are zero at the supports. At x_m, the jump in shear force is P_m and at x_k, the jump in bending moment is $-M_k$. Thus, we get

$$\int_0^L EI\frac{d^2v}{dx^2}\frac{d^2\phi}{dx^2}\,dx - \int_0^L p\phi\,dx - \sum_m P_m\phi_m - \sum_k M_k\phi_k' = 0 \qquad (8.12)$$

For the finite element formulation based on Galerkin's approach, v and ϕ are constructed using the same shape functions. Equation 8.12 is precisely the statement of the principle of virtual work.

8.2 FINITE ELEMENT FORMULATION

The beam is divided into elements, as shown in Fig. 8.4. Each node has two degrees of freedom. Typically the degrees of freedom of node I are Q_{2i-1} and Q_{2i}. The degree of freedom Q_{2i-1} is transverse displacement and Q_{2i} is slope or rotation. The vector

$$\mathbf{Q} = [Q_1, Q_2, \ldots, Q_{10}]^T \qquad (8.13)$$

represents the global displacement vector. For a single element, the local degrees of freedom are represented by

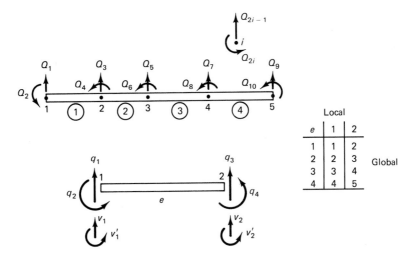

Figure 8.4 Finite element discretization.

$$\mathbf{q} = [q_1, q_2, q_3, q_4]^T \tag{8.14}$$

The local–global correspondence is easy to see from the table given in Figure 8.4. $\mathbf{q}$ is same as $[v_1, v_1', v_2, v_2']^T$. The shape functions for interpolating v on an element are defined in terms of ξ on -1 to $+1$, as shown in Fig. 8.5. *The shape functions for beam elements differ from those discussed before.* Since nodal values and nodal slopes are involved, we define Hermite shape functions, which satisfy nodal value and slope continuity requirements. Each of the shape functions is of cubic order represented by

$$H_i = a_i + b_i\xi + c_i\xi^2 + d_i\xi^3 \qquad i = 1, 2, 3, 4 \tag{8.15}$$

The conditions given in the following table must be satisfied.

	H_1	H_1'	H_2	H_2'	H_3	H_3'	H_4	H_4'
$\xi = -1$	1	0	0	1	0	0	0	0
$\xi = 1$	0	0	0	0	1	0	0	1

The coefficients a_i, b_i, c_i, d_i can be easily obtained by imposing the above conditions. Thus,

$$\begin{aligned}
H_1 &= \tfrac{1}{4}(1 - \xi)^2(2 + \xi) \quad \text{or} \quad \tfrac{1}{4}(2 - 3\xi + \xi^3) \\
H_2 &= \tfrac{1}{4}(1 - \xi)^2(\xi + 1) \quad \text{or} \quad \tfrac{1}{4}(1 - \xi - \xi^2 + \xi^3) \\
H_3 &= \tfrac{1}{4}(1 + \xi)^2(2 - \xi) \quad \text{or} \quad \tfrac{1}{4}(2 + 3\xi - \xi^3) \\
H_4 &= \tfrac{1}{4}(1 + \xi)^2(\xi - 1) \quad \text{or} \quad \tfrac{1}{4}(-1 - \xi + \xi^2 + \xi^3)
\end{aligned} \tag{8.16}$$

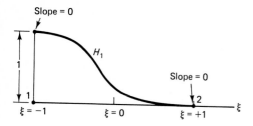

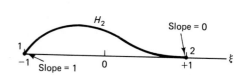

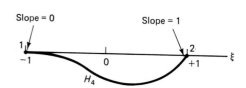

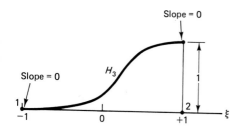

Figure 8.5   Hermite shape functions.

The Hermite shape functions can be used to write v in the form

$$v(\xi) = H_1 v_1 + H_2\left(\frac{dv}{d\xi}\right)_1 + H_3 v_2 + H_4\left(\frac{dv}{d\xi}\right)_2 \tag{8.17}$$

The coordinates transform by the relationship

$$x = \frac{1-\xi}{2} x_1 + \frac{1+\xi}{2} x_2$$

$$= \frac{x_1 + x_2}{2} + \frac{x_2 - x_1}{2}\xi \tag{8.18}$$

Since $\ell_e = x_2 - x_1$ is the length of the element, we have

$$dx = \frac{\ell_e}{2} d\xi \tag{8.19}$$

The chain rule $dv/d\xi = (dv/dx)(dx/d\xi)$ gives us

$$\frac{dv}{d\xi} = \frac{\ell_e}{2}\frac{dv}{dx} \tag{8.20}$$

Noting that dv/dx evaluated at nodes 1 and 2 is q_2 and q_4, respectively, we have

$$v(\xi) = H_1 q_1 + \frac{\ell_e}{2}H_2 q_2 + H_3 q_3 + \frac{\ell_e}{2}H_4 q_4 \tag{8.21}$$

which may be denoted as

$$v = \mathbf{Hq} \tag{8.22}$$

where

$$\mathbf{H} = \left[H_1, \frac{\ell_e}{2} H_2, H_3, \frac{\ell_e}{2} H_4 \right] \tag{8.23}$$

In the total potential energy of the system, we consider the integrals as summations over the integrals over the elements. The element strain energy is given by

$$U_e = \tfrac{1}{2} EI \int_e \left(\frac{d^2 v}{dx^2} \right)^2 dx \tag{8.24}$$

From Eq. 8.20,

$$\frac{dv}{dx} = \frac{2}{\ell_e} \frac{dv}{d\xi} \quad \text{and} \quad \frac{d^2 v}{dx^2} = \frac{4}{\ell_e^2} \frac{d^2 v}{d\xi^2}$$

Then, substituting $v = \mathbf{Hq}$, we obtain

$$\left(\frac{d^2 v}{dx^2} \right)^2 = \mathbf{q}^{\mathrm{T}} \frac{16}{\ell_e^4} \left(\frac{d^2 \mathbf{H}}{d\xi^2} \right)^{\mathrm{T}} \left(\frac{d^2 \mathbf{H}}{d\xi^2} \right) \mathbf{q} \tag{8.25}$$

$$\left(\frac{d^2 \mathbf{H}}{d\xi^2} \right) = \left[\frac{3}{2}\xi, \frac{-1 + 3\xi}{2} \frac{\ell_e}{2}, -\frac{3}{2}\xi, \frac{1 + 3\xi}{2} \frac{\ell_e}{2} \right] \tag{8.26}$$

On substituting $dx = (\ell_e/2)\, d\xi$ and the above in Eq. 8.24 we get

$$U_e = \frac{1}{2} \mathbf{q}^{\mathrm{T}} \frac{8EI}{\ell_e^3} \int_{-1}^{+1} \begin{bmatrix} \frac{9}{4}\xi^2 & \frac{3}{8}\xi(-1 + 3\xi)\ell_e & -\frac{9}{4}\xi^2 & \frac{3}{8}\xi(1 + 3\xi)\ell_e \\ & \left(\frac{-1 + 3\xi}{4}\right)^2 \ell_e^2 & -\frac{3}{8}\xi(-1 + 3\xi)\ell_e & \frac{-1 + 9\xi^2}{16}\ell_e^2 \\ & \text{Symmetric} & \frac{9}{4}\xi^2 & -\frac{3}{8}\xi(1 + 3\xi)\ell_e \\ & & & \left(\frac{1 + 3\xi}{4}\right)^2 \ell_e^2 \end{bmatrix} d\xi\, \mathbf{q}^{\mathrm{T}} \tag{8.27}$$

Each term in the matrix needs to be integrated. Note that

$$\int_{-1}^{+1} \xi^2\, d\xi = \frac{2}{3} \qquad \int_{-1}^{+1} \xi\, d\xi = 0 \qquad \int_{-1}^{+1} d\xi = 2$$

This results in the element strain energy given by

$$U_e = \tfrac{1}{2} \mathbf{q}^{\mathrm{T}} \mathbf{k}^e \mathbf{q} \tag{8.28}$$

where the element stiffness matrix $\mathbf{k}^e$ is

$$\mathbf{k}^e = \frac{EI}{\ell_e^3} \begin{bmatrix} 12 & 6\ell_e & -12 & 6\ell_e \\ 6\ell_e & 4\ell_e^2 & -6\ell_e & 2\ell_e^2 \\ -12 & -6\ell_e & 12 & -6\ell_e \\ 6\ell_e & 2\ell_e^2 & -6\ell_e & 4\ell_e^2 \end{bmatrix} \tag{8.29}$$

which is symmetric.

In the development based on Galerkin's approach (see Eq. 8.12), we note that

$$EI\frac{d^2\phi}{dx^2}\frac{d^2v}{dx^2} = \boldsymbol{\psi}^{\mathrm{T}}EI\frac{16}{\ell_e^4}\left(\frac{d^2\mathbf{H}}{d\xi^2}\right)^{\mathrm{T}}\left(\frac{d^2\mathbf{H}}{d\xi^2}\right)\mathbf{q} \tag{8.30}$$

where

$$\boldsymbol{\psi} = [\psi_1 \quad \psi_2 \quad \psi_3 \quad \psi_4]^{\mathrm{T}} \tag{8.31}$$

is the set of generalized virtual displacements on the element, $v = \mathbf{Hq}$, and $\phi = \mathbf{H}\boldsymbol{\psi}$. Equation 8.30 yields the same element stiffness as Eq. 8.28 on integration, with $\boldsymbol{\psi}^{\mathrm{T}}\mathbf{k}^e\mathbf{q}$ being the internal virtual work in an element.

8.3 LOAD VECTOR

The load contributions from the distributed load p in the element is first considered. We assume that the distributed load is uniform over the element:

$$\int_{\ell_e} pv\, dx = \left(\frac{p\ell_e}{2}\int_{-1}^{1}\mathbf{H}\, d\xi\right)\mathbf{q} \tag{8.32}$$

On substituting for $\mathbf{H}$ from Eqs. 8.16 and 8.23 and integrating, we obtain

$$\int_{\ell_e} pv\, dx = \mathbf{f}^{e^{\mathrm{T}}}\mathbf{q} \tag{8.33}$$

where

$$\mathbf{f}^e = \left[\frac{p\ell_e}{2}, \frac{p\ell_e^2}{12}, \frac{p\ell_e}{2}, \frac{-p\ell_e^2}{12}\right]^{\mathrm{T}} \tag{8.34}$$

This equivalent load on an element is shown in Fig. 8.6. The same result is obtained by considering the term $\int_e p\phi\, dx$ in Eq. 8.12 for the Galerkin formulation. The point loads P_m and M_k are readily taken care of by introducing nodes at the points of application. On introducing the local–global correspondence, from the potential energy approach, we get

$$\Pi = \tfrac{1}{2}\mathbf{Q}^{\mathrm{T}}\mathbf{K}\mathbf{Q} - \mathbf{Q}^{\mathrm{T}}\mathbf{F} \tag{8.35}$$

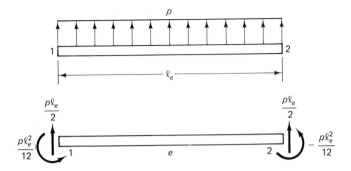

Figure 8.6   Distributed load on an element.

and from Galerkin's approach, we get

$$\mathbf{\Psi}^T\mathbf{KQ} - \mathbf{\Psi}^T\mathbf{F} = 0 \tag{8.36}$$

where $\mathbf{\Psi}$ = arbitrary admissible global virtual displacement vector.

8.4 BOUNDARY CONSIDERATIONS

When the generalized displacement value is specified as a for the degree of freedom (dof) r, we follow the penalty approach and add $\frac{1}{2}C(Q_r - a)^2$ to Π and $\Psi_j C(Q_r - a)$ to the left side of the Galerkin formulation and place no restrictions on the degrees of freedom. The number C represents stiffness and is large in comparison to beam stiffness terms. This amounts to adding stiffness C to K_{rr} and load Ca to F_r (see Fig. 8.7). Both Eqs. 8.35 and 8.36 independently yield

$$\mathbf{KQ} = \mathbf{F} \tag{8.37}$$

These equations are now solved to get the nodal displacements.

Reactions at constrained degrees of freedom may be calculated using Eq. 3.71 or 3.75.

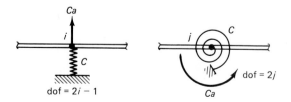

a = known generalized displacement

Figure 8.7   Boundary conditions for a beam.

8.5 SHEAR FORCE AND BENDING MOMENT

Using the bending moment and shear force equations

$$M = EI\frac{d^2v}{dx^2} \qquad V = \frac{dM}{dx} \quad \text{and} \quad v = \mathbf{Hq}$$

we get the element bending moment and shear force:

$$M = \frac{EI}{\ell_e^2}[6\xi q_1 + (3\xi - 1)\ell_e q_2 - 6\xi q_3 + (3\xi + 1)\ell_e q_4] \qquad (8.38)$$

$$V = \frac{6EI}{\ell_e^3}(2q_1 + \ell_e q_2 - 2q_3 + \ell_e q_4) \qquad (8.39)$$

Above bending moment and shear force values are for the loading as modeled using equivalent point loads. Denoting element end equilibrium loads as R_1, R_2, R_3, and R_4, we note that

$$\begin{Bmatrix} R_1 \\ R_2 \\ R_3 \\ R_4 \end{Bmatrix}^e = \frac{EI}{\ell_e^3} \begin{bmatrix} 12 & 6\ell_e & -12 & 6\ell_e \\ 6\ell_e & 4\ell_e^2 & -6\ell_e & 2\ell_e^2 \\ -12 & -6\ell_e & 12 & -6\ell_e \\ 6\ell_e & 2\ell_e^2 & -6\ell_e & 4\ell_e^2 \end{bmatrix} \begin{Bmatrix} q_1 \\ q_2 \\ q_3 \\ q_4 \end{Bmatrix} + \begin{Bmatrix} \dfrac{-p\ell_e}{2} \\ \dfrac{-p\ell_e^2}{12} \\ \dfrac{-p\ell_e}{2} \\ \dfrac{p\ell_e^2}{12} \end{Bmatrix} \qquad (8.40)$$

It is easily seen that the first term on the right is $\mathbf{k}^e\mathbf{q}$. Also note that the second term needs to be added only on elements with distributed load. In books on matrix structural analysis, the above equations are written directly from element equilibrium. Also, the last vector on the right side of the equation consists of terms that are called *fixed-end reactions*. The shear forces at the two ends of the element are $V_1 = R_1$ and $V_2 = -R_3$. The end bending moments are $M_1 = -R_2$ and $M_2 = R_4$.

Example 8.1

For the beam and loading shown in Fig. E8.1, determine the slopes at 2 and 3, and the vertical deflection at the midpoint of the distributed load.

Solution   We consider the two elements formed by the three nodes. Displacements Q_1, Q_2, Q_3, and Q_5 are constrained to be zero, and Q_4 and Q_6 need to be found. Since the lengths and sections are equal, the element matrices are calculated from Eq. 8.29 as

$$\frac{EI}{\ell^3} = \frac{(200 \times 10^9)(4 \times 10^{-6})}{1^3} = 8 \times 10^5 \text{ N/m}$$

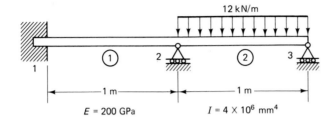

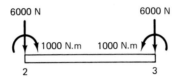

Figure E8.1

$$\mathbf{k}^1 = \mathbf{k}^2 = 8 \times 10^5 \begin{bmatrix} 12 & 6 & -12 & 6 \\ 6 & 4 & -6 & 2 \\ -12 & -6 & 12 & -6 \\ 6 & 2 & -6 & 4 \end{bmatrix}$$

$$e = 1 \qquad Q_1 \quad Q_2 \quad Q_3 \quad Q_4$$

$$e = 2 \qquad Q_3 \quad Q_4 \quad Q_5 \quad Q_6$$

We note that global applied loads are $F_4 = -1000$ N.m, $F_6 = +1000$ N.m obtained from $p\ell^2/12$, as seen in Fig. 8.6. We use here the elimination approach presented in Chapter 3. Using the connectivity, we obtain the global stiffness after elimination:

$$\mathbf{K} = \begin{bmatrix} (k_{44}^{(1)} + k_{22}^{(2)}) & k_{24}^{(2)} \\ k_{42}^{(2)} & k_{44}^{(2)} \end{bmatrix}$$

$$= 8 \times 10^5 \begin{bmatrix} 8 & 2 \\ 2 & 4 \end{bmatrix}$$

The set of equations is given by

$$8 \times 10^5 \begin{bmatrix} 8 & 2 \\ 2 & 4 \end{bmatrix} \begin{Bmatrix} Q_4 \\ Q_6 \end{Bmatrix} = \begin{Bmatrix} -1000 \\ +1000 \end{Bmatrix}$$

The solution is

$$\begin{Bmatrix} Q_4 \\ Q_6 \end{Bmatrix} = \begin{Bmatrix} -2.679 \times 10^{-4} \\ 4.464 \times 10^{-4} \end{Bmatrix}$$

For element 2, $q_1 = 0$, $q_2 = Q_4$, $q_3 = 0$, $q_4 = Q_6$. To get vertical deflection at the midpoint of the element, use $v = \mathbf{Hq}$ at $\xi = 0$:

$$v = 0 + \frac{\ell_e}{2}H_2Q_4 + 0 + \frac{\ell_e}{2}H_4Q_6$$

$$= (\tfrac{1}{2})(\tfrac{1}{4})(-2.679 \times 10^{-4}) + (\tfrac{1}{2})(-\tfrac{1}{4})(4.464 \times 10^{-4})$$

$$= -8.93 \times 10^{-5} \text{ m}$$

$$= -0.0893 \text{ mm} \qquad\blacksquare$$

8.6 BEAMS ON ELASTIC SUPPORTS

In many engineering applications, beams are supported on elastic members. Shafts are supported on ball, roller, or journal bearings. Large beams are supported on elastic walls. Beams supported on soil form a class of applications known as Winkler foundations.

Single-row ball bearings can be considered by having a node at each bearing location and adding the bearing stiffness k_B to the diagonal location of vertical degree of freedom (Fig. 8.8a). Rotational (moment) stiffness has to be considered for roller bearings and journal bearings.

In wide journal bearings and Winkler foundations, we use stiffness per unit length, s, of the supporting medium (Fig. 8.8b). Over the length of the support, this adds the following term to the total potential energy

$$\frac{1}{2}\int_0^\ell sv^2\, dx \qquad (8.41)$$

In Galerkin's approach, this term is $\int_0^\ell sv\phi\, dx$. When we substitute for $v = \mathbf{H}\mathbf{q}$ for the discretized model, the above term becomes

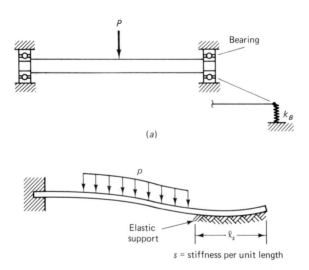

(a)

(b)   **Figure 8.8** Elastic support.

$$\frac{1}{2} \sum_e \mathbf{q}^\mathsf{T} s \int_e \mathbf{H}^\mathsf{T}\mathbf{H} \, dx \, \mathbf{q} \tag{8.42}$$

We recognize the stiffness term in the above summation,

$$\mathbf{k}_s^e = s \int_e \mathbf{H}^\mathsf{T}\mathbf{H} \, dx = \frac{s\ell_e}{2} \int_{-1}^{+1} \mathbf{H}^\mathsf{T}\mathbf{H} \, d\xi \tag{8.43}$$

On integration we have

$$\mathbf{k}_s^e = \frac{s\ell_e}{420} \begin{bmatrix} 156 & 22\ell_e & 54 & -13\ell_e \\ 22\ell_e & 4\ell_e^2 & 13\ell_e & -3\ell_e^2 \\ 54 & 13\ell_e & 156 & -22\ell_e \\ -13\ell_e & -3\ell_e^2 & -22\ell_e & 4\ell_e^2 \end{bmatrix} \tag{8.44}$$

For elements supported on an elastic foundation, this stiffness has to be added to the element stiffness given by Eq. 8.29. Matrix $\mathbf{k}_s^e$ is the consistent stiffness matrix for the elastic foundation.

8.7 PLANE FRAMES

We consider here plane structures with rigidly connected members. These members will be similar to the beams except that axial loads and axial deformations are present. The elements also have different orientations. Fig. 8.9 shows a frame ele-

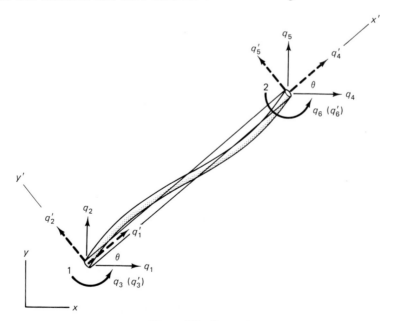

Figure 8.9   Frame element.

ment. We have two displacements and a rotational deformation for each node. The nodal displacement vector is given by

$$\mathbf{q} = [q_1, q_2, q_3, q_4, q_5, q_6]^T \tag{8.45}$$

We also define the local or body coordinate system x', y' such that x' is oriented along 1–2, with direction cosines ℓ, m (where $\ell = \cos\theta$, $m = \sin\theta$). These are evaluated using relationships given for the truss element, shown in Fig. 4.4. The nodal displacement vector in the local system is

$$\mathbf{q}' = [q_1', q_2', q_3', q_4', q_5', q_6']^T \tag{8.46}$$

Recognizing that $q_3' = q_3$ and $q_6' = q_6$, which are rotations with respect to the body, we obtain the local–global transformation

$$\mathbf{q}' = \mathbf{Lq} \tag{8.47}$$

where

$$\mathbf{L} = \begin{bmatrix} \ell & m & 0 & 0 & 0 & 0 \\ -m & \ell & 0 & 0 & 0 & 0 \\ 0 & 0 & 1 & 0 & 0 & 0 \\ 0 & 0 & 0 & \ell & m & 0 \\ 0 & 0 & 0 & -m & \ell & 0 \\ 0 & 0 & 0 & 0 & 0 & 1 \end{bmatrix} \tag{8.48}$$

It is now observed that q_2', q_3', q_5', and q_6' are like the beam degrees of freedom, while q_1' and q_4' are similar to the displacements of a rod element, as discussed in Chapter 3. Combining the two stiffnesses and arranging in proper locations, we get the element stiffness for a frame element as

$$\mathbf{k}'^e = \begin{bmatrix} \dfrac{EA}{\ell} & 0 & 0 & \dfrac{-EA}{\ell} & 0 & 0 \\ 0 & \dfrac{12EI}{\ell^3} & \dfrac{6EI}{\ell^2} & 0 & \dfrac{12EI}{\ell^3} & \dfrac{6EI}{\ell^2} \\ 0 & \dfrac{6EI}{\ell^2} & \dfrac{4EI}{\ell} & 0 & \dfrac{-6EI}{\ell^2} & \dfrac{2EI}{\ell} \\ \dfrac{-EA}{\ell} & 0 & 0 & \dfrac{EA}{\ell} & 0 & 0 \\ 0 & \dfrac{-12EI}{\ell^3} & \dfrac{-6EI}{\ell^2} & 0 & \dfrac{12EI}{\ell^3} & \dfrac{-6EI}{\ell^2} \\ 0 & \dfrac{6EI}{\ell^2} & \dfrac{2EI}{\ell} & 0 & \dfrac{-6EI}{\ell^2} & \dfrac{4EI}{\ell} \end{bmatrix} \tag{8.49}$$

As discussed in the development of a truss element in Chapter 4, we recognize that the element strain energy is given by

$$U_e = \tfrac{1}{2}\mathbf{q}'^{\mathrm{T}}\mathbf{k}'^e\mathbf{q}' = \tfrac{1}{2}\mathbf{q}^{\mathrm{T}}\mathbf{L}^{\mathrm{T}}\mathbf{k}'^e\mathbf{L}\mathbf{q} \qquad (8.50)$$

or in Galerkin's approach, the internal virtual work of an element is

$$W_e = \boldsymbol{\psi}'^{\mathrm{T}}\mathbf{k}'^e\mathbf{q}' = \boldsymbol{\psi}^{\mathrm{T}}\mathbf{L}^{\mathrm{T}}\mathbf{k}'^e\mathbf{L}\mathbf{q} \qquad (8.51)$$

where $\boldsymbol{\psi}'$ and $\boldsymbol{\psi}$ are virtual nodal displacements in local and global coordinate systems, respectively. From Eq. 8.50 or 8.51 we recognize the element stiffness matrix in global coordinates to be

$$\boxed{\mathbf{k}^e = \mathbf{L}^{\mathrm{T}}\mathbf{k}'^e\mathbf{L}} \qquad (8.52)$$

In the finite element program implementation, $\mathbf{k}'^e$ can first be defined, and then the above matrix multiplication can be carried out.

If there is distributed load on a member, as shown in Fig. 8.10, we have

$$\mathbf{q}'^{\mathrm{T}}\mathbf{f}' = \mathbf{q}^{\mathrm{T}}\mathbf{L}^{\mathrm{T}}\mathbf{f}' \qquad (8.53)$$

where

$$\mathbf{f}' = \left[0, \ \frac{p\ell_e}{2}, \ \frac{p\ell_e^2}{12}, \ 0, \ \frac{p\ell_e}{2}, \ -\frac{p\ell_e^2}{12}\right]^{\mathrm{T}} \qquad (8.54)$$

The nodal loads due to the distributed load p are given by

$$\mathbf{f} = \mathbf{L}^{\mathrm{T}}\mathbf{f}' \qquad (8.55)$$

The values of $\mathbf{f}$ are added to the global load vector. Note here that positive p is in the y' direction.

The point loads and couples are simply added to the global load vector. On gathering stiffnesses and loads, we get the system of equations

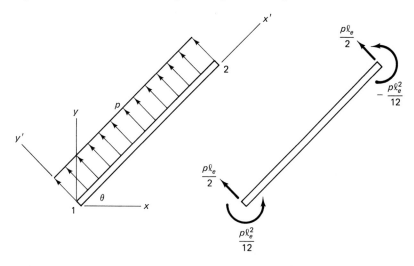

Figure 8.10   Distributed load on a frame element.

$$KQ = F$$

where the boundary conditions are considered by applying the penalty terms in the energy or Galerkin formulations.

Example 8.2

Determine the displacements and rotations of the joints for the portal frame shown in Fig. E8.2.

Solution

Step 1. Connectivity

	Node	
Element No.	1	2
1	1	2
2	3	1
3	4	2

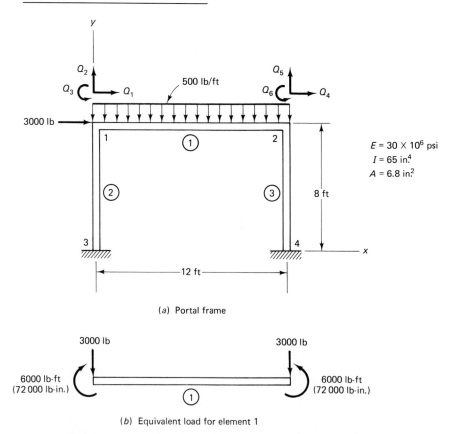

(a) Portal frame

(b) Equivalent load for element 1

Figure E8.2   (a) Portal frame. (b) Equivalent load for Element 1

Step 2. Element Stiffnesses

Element 1. Using the matrix given in Eq. 8.45 and noting that $\mathbf{k}^1 = \mathbf{k}'^1$, we find

$$
\mathbf{k}^1 = 10^4 \times
\begin{array}{c}
\begin{array}{cccccc}
Q_1 & Q_2 & Q_3 & Q_4 & Q_5 & Q_6
\end{array} \\
\begin{bmatrix}
141.7 & 0 & 0 & -141.7 & 0 & 0 \\
0 & 0.784 & 56.4 & 0 & -0.784 & 56.4 \\
0 & 56.4 & 5417 & 0 & -56.4 & 2708 \\
-141.7 & 0 & 0 & 141.7 & 0 & 0 \\
0 & -0.784 & -56.4 & 0 & 0.784 & -56.4 \\
0 & 56.4 & 2708 & 0 & -56.4 & 5417
\end{bmatrix}
\end{array}
$$

Elements 2 and 3. Local element stiffnesses for elements 2 and 3 are obtained by substituting for E, A, I and ℓ_2 in matrix $\mathbf{k}'$ of Eq. 8.49:

$$
\mathbf{k}'^2 = 10^4 \times
\begin{bmatrix}
212.5 & 0 & 0 & -212.5 & 0 & 0 \\
0 & 2.65 & 127 & 0 & -2.65 & 127 \\
0 & 127 & 8125 & 0 & -127 & 4063 \\
-212.5 & 0 & 0 & 212.5 & 0 & 0 \\
0 & -2.65 & -127 & 0 & 2.65 & -127 \\
0 & 127 & 4063 & 0 & -127 & 8125
\end{bmatrix}
$$

Transformation matrix $\mathbf{L}$. We have noted that for element 1, $\mathbf{k} = \mathbf{k}'$. For elements 2 and 3, which are oriented similarly with respect to the x, y axes, we have $\ell = 0$, $m = 1$. Then,

$$
\mathbf{L} =
\begin{bmatrix}
0 & 1 & 0 & 0 & 0 & 0 \\
-1 & 0 & 0 & 0 & 0 & 0 \\
0 & 0 & 1 & 0 & 0 & 0 \\
0 & 0 & 0 & 0 & 1 & 0 \\
0 & 0 & 0 & -1 & 0 & 0 \\
0 & 0 & 0 & 0 & 0 & 1
\end{bmatrix}
$$

Noting that $\mathbf{k}^2 = \mathbf{L}^T \mathbf{k}'^2 \mathbf{L}$, we get

$$
\mathbf{k} = 10^4 \times
\begin{array}{c}
\begin{array}{cccccc}
 & & e = 3 & Q_4 & Q_5 & Q_6 \\
 & & e = 2 \rightarrow Q_1 & Q_2 & Q_3 &
\end{array} \\
\begin{bmatrix}
2.65 & 0 & -127 & -2.65 & 0 & -127 \\
0 & 212.5 & 0 & 0 & -212.5 & 0 \\
-127 & 0 & 8125 & 127 & 0 & 4063 \\
-2.65 & 0 & 127 & 2.65 & 0 & 127 \\
0 & -212.5 & 0 & 0 & 212.5 & 0 \\
-127 & 0 & 4063 & 127 & 0 & 8125
\end{bmatrix}
\end{array}
$$

Stiffness $\mathbf{k}^1$ has all its elements in the global locations. For elements 2 and 3, the shaded part of the stiffness matrix shown above is added to the appropriate global locations of $\mathbf{K}$. The global stiffness matrix is given by

$$\mathbf{K} = 10^4 \times \begin{bmatrix} 144.3 & 0 & 127 & -141.7 & 0 & 0 \\ 0 & 213.3 & 56.4 & 0 & -0.784 & 56.4 \\ 127 & 56.4 & 13542 & 0 & -56.4 & 2708 \\ -141.7 & 0 & 0 & 144.3 & 0 & 127 \\ 0 & -0.784 & -56.4 & 0 & 213.3 & -56.4 \\ 0 & 56.4 & 2708 & 127 & -56.4 & 13542 \end{bmatrix}$$

From Fig. E8.2, the load vector can easily be written

$$\mathbf{F} = \begin{Bmatrix} 3\,000 \\ -3\,000 \\ -72\,000 \\ 0 \\ -3\,000 \\ +72\,000 \end{Bmatrix}$$

The set of equations is given by

$$\mathbf{KQ} = \mathbf{F}$$

On solving, we get

$$\mathbf{Q} = \begin{Bmatrix} 0.092 \text{ in.} \\ -0.00104 \text{ in.} \\ -0.00139 \text{ rad} \\ 0.0901 \text{ in.} \\ -0.0018 \text{ in.} \\ -3.88 \times 10^{-5} \text{ rad} \end{Bmatrix}$$

$\blacksquare$

8.8. SOME COMMENTS

Symmetric beams and plane frames have been discussed in this chapter. In engineering applications, there are several challenging problems, such as frames and mechanisms with pin-jointed members, unsymmetric beams, space frames, buckling of members due to axial loads, shear considerations, and structures with large deformations. For help in formulating and analyzing such problems, the reader may refer to some advanced publications in mechanics of solids, structural analysis, elasticity and plasticity, and finite element analysis.

Example 8.3

The solution of Example 8.1 using program BEAM is shown below.

```
Number of Elements =? 2
Number of Constrained DOF =? 4
Number of Combined Loads and Moments =? 4
Young's Modulus =? 200E3
```

```
NODE#, Coordinate
? 1,0
? 2,1000
? 3,2000
Element#, Moment of Inertia
? 1,4E6
? 2,4E6
DOF#,  Displacement
? 1,0
? 2,0
? 3,0
? 5,0
DOF#, Applied Load
? 3,-6000
? 4,-1E6
? 5,-6000
? 6,1E6
```

```
Node#     Displ.        Rotation
    1     4.0177E-11    1.3392E-08
    2    -2.5446E-10   -2.6786E-04
    3    -1.6071E-10    4.4643E-04
DOF#     Reaction
    1    -1.2857E+03
    2    -4.2853E+05
    3     8.1428E+03
    5     5.1429E+03
```

■

```
1000 REM *******       BEAM      ********
1010 REM *****   Beam Bending Analysis  *****
1012 CLS : DEFINT I-N
1020 PRINT "======================================="
1030 PRINT "      Program written by         "
1040 PRINT " T.R.Chandrupatla and A.D.Belegundu   "
1050 PRINT "======================================="
1060 INPUT "Number of Elements ="; NE
1070 INPUT "Number of Constrained DOF ="; ND
1080 INPUT "Number of Combined Loads and Moments ="; NL
1090 REM ** SINGLE MATERIAL FOR ALL ELEMENTS **
1100 INPUT "Young's Modulus ="; YM
1110 NN = NE + 1
1120 REM ** THE TOTAL DOF IS "NQ" **
1130 NQ = 2 * NN
1140 NBW = 4
1150 DIM X(NN), NU(ND), U(ND), SMI(NE)
1160 DIM S(NQ, 4), F(NQ)
1170 PRINT "NODE#, Coordinate"
1180 FOR I = 1 TO NN
1190 INPUT N, X(N): NEXT I
1200 PRINT "Element#, Moment of Inertia"
1210 FOR I = 1 TO NE
```

```
1220 INPUT N, SMI(N): NEXT I
1230 FOR I = 1 TO NQ
1240 F(I) = 0
1250 FOR J = 1 TO NBW
1260 S(I, J) = 0: NEXT J: NEXT I
1270 PRINT "DOF#,  Displacement"
1280 FOR I = 1 TO ND
1290 INPUT NU(I), U(I): NEXT I
1300 PRINT "DOF#, Applied Load"
1310 FOR I = 1 TO NL
1320 INPUT N, F(N): NEXT I
1330 REM *** GLOBAL STIFFNESS MATRIX ***
1340 FOR I = 1 TO NE
1350 I1 = 2 * I - 1
1360 EL = ABS(X(I + 1) – X(I))
1370 EIL = YM * SMI(I) / EL ^ 3
1380 S(I1, 1) = S(I1, 1) + 12 * EIL
1390 S(I1, 2) = S(I1, 2) + EIL * 6 * EL
1400 S(I1, 3) = S(I1, 3) – 12 * EIL
1410 S(I1, 4) = S(I1, 4) + EIL * 6 * EL
1420 S(I1 + 1, 1) = S(I1 + 1, 1) + EIL * 4 * EL * EL
1430 S(I1 + 1, 2) = S(I1 + 1, 2) – EIL * 6 * EL
1440 S(I1 + 1, 3) = S(I1 + 1, 3) + EIL * 2 * EL * EL
1450 S(I1 + 2, 1) = S(I1 + 2, 1) + EIL * 12
1460 S(I1 + 2, 2) = S(I1 + 2, 2) – EIL * 6 * EL
1470 S(I1 + 3, 1) = S(I1 + 3, 1) + EIL * 4 * EL * EL
1480 NEXT I
1490 REM *** MODIFY FOR BOUNDARY CONDITIONS ***
1500 CNST = (S(1, 1) + S(2, 1)) * 10000!
1510 FOR I = 1 TO ND
1520 N = NU(I)
1530 S(N, 1) = S(N, 1) + CNST
1540 F(N) = F(N) + CNST * U(I): NEXT I
1550 REM *** EQUATION SOLVING ***
1560 GOSUB 5000
1570 PRINT "Node#   Displ.      Rotation"
1580 FOR I = 1 TO NN
1590 PRINT USING "  ###"; I;
1600 PRINT USING "  ##.####^^^^"; F(2 * I - 1); F(2 * I)
1610 NEXT I
1620 REM  *** REACTION CALCULATION ***
1630 PRINT "DOF#   Reaction"
1640 FOR I = 1 TO ND
1650 N = NU(I)
1660 R = CNST * (U(I) – F(N))
1670 PRINT USING " ###"; N; : PRINT USING "  ##.####^^^^  "; R: NEXT I
1680 END
5000 N1 = NQ – 1
5010 REM *** FORWARD ELIMINATION ***
5020 FOR K = 1 TO N1
5030 NK = NQ – K + 1
5040 IF NK > NBW THEN NK = NBW
5050 FOR I = 2 TO NK
5060 C1 = S(K, I) / S(K, 1)
5070 I1 = K + I – 1
5080 FOR J = I TO NK
5090 J1 = J – I + 1
5100 S(I1, J1) = S(I1, J1) – C1 * S(K, J): NEXT J
5110 F(I1) = F(I1) – C1 * F(K): NEXT I: NEXT K
5120 REM ** BACK SUBSTITUTION **
5130 F(NQ) = F(NQ) / S(NQ, 1)
5140 FOR KK = 1 TO N1
5150 K = NQ – KK
5160 C1 = 1 / S(K, 1)
```

```
5170 F(K) = C1 * F(K)
5180 NK = NQ - K + 1
5190 IF NK > NBW THEN NK = NBW
5200 FOR J = 2 TO NK
5210 F(K) = F(K) - C1 * S(K, J) * F(K + J - 1)
5220 NEXT J
5230 NEXT KK
5240 RETURN

1000 REM ********      FRAME      ********
1010 REM *****    FRAME ANALYSIS BY FEM    *****
1020 DEFINT I-N: CLS
1030 PRINT "===================================="
1040 PRINT "       Program written by         "
1050 PRINT "  T.R.Chandrupatla and A.D.Belegundu   "
1060 PRINT "===================================="
1070 INPUT "Number of Elements ="; NE
1080 INPUT "Number of Nodes ="; NN
1090 INPUT "Number of Constrained DOF ="; ND
1100 INPUT "Number of Combined Loads and Moments ="; NL
1110 INPUT "Number of Materials ="; NM
1120 INPUT "Number of Different Sections ="; NA
1130 REM ** THE TOTAL DOF IS "NQ" **
1140 NQ = 3 * NN
1150 DIM X(NN, 2), NOC(NE, 4), PM(NM), ARIN(NA, 2), NU(ND), U(ND)
1160 REM BANDWIDTH NBW FROM CONNECTIVITY "NOC"
1170 NBW = 0
1180 PRINT "Element#, 2 Nodes, Material#, Section#"
1190 FOR I = 1 TO NE
1200 INPUT N, NOC(N, 1), NOC(N, 2), NOC(N, 3), NOC(N, 4)
1210 C = 3 * (ABS(NOC(N, 2) - NOC(N, 1)) + 1)
1220 IF NBW < C THEN NBW = C
1230 NEXT I
1240 PRINT "THE BANDWIDTH IS"; NBW
1250 REM *** INITIALIZATION ***
1260 DIM S(NQ, NBW), F(NQ), SE(6, 6)
1270 FOR I = 1 TO NQ
1280 F(NQ) = 0
1290 FOR J = 1 TO NBW
1300 S(I, J) = 0: NEXT J: NEXT I
1310 PRINT "Node#, X-Coord, Y-Coord"
1320 FOR I = 1 TO NN
1330 INPUT N, X(N, 1), X(N, 2): NEXT I
1340 PRINT "Material#, E"
1350 FOR I = 1 TO NM
1360 INPUT N, PM(N): NEXT I
1370 PRINT "Section#, Area, Mom_Inertia"
1380 FOR I = 1 TO NA
1390 INPUT N, ARIN(N, 1), ARIN(N, 2): NEXT I
1400 PRINT "DOF#, Displacement"
1410 FOR I = 1 TO ND
1420 INPUT NU(I), U(I): NEXT I
1430 PRINT "DOF#, Applied Load"
1440 FOR I = 1 TO NL
1450 INPUT N, F(N): NEXT I
1460 REM *** GLOBAL STIFFNESS MATRIX ***
1470 FOR N = 1 TO NE
1480 '+++Call Subroutine Element Stiffness+++
1490 GOSUB 1870
```

```
1500 PRINT "PLACING STIFFNESS IN GLOBAL LOCATIONS"
1510 FOR II = 1 TO 2
1520 NRT = 3 * (NOC(N, II) - 1)
1530 FOR IT = 1 TO 3
1540 NR = NRT + IT
1550 I = 3 * (II - 1) + IT
1560 FOR JJ = 1 TO 2
1570 NCT = 3 * (NOC(N, JJ) - 1)
1580 FOR JT = 1 TO 3
1590 J = 3 * (JJ - 1) + JT
1600 NC = NCT + JT - NR + 1
1610 IF NC <= 0 GOTO 1630
1620 S(NR, NC) = S(NR, NC) + SE(I, J)
1630 NEXT JT: NEXT JJ: NEXT IT: NEXT II
1640 NEXT N
1650 REM *** MODIFY FOR BOUNDARY CONDITIONS ***
1660 CNST = (S(1, 1) + S(2, 1) + S(3, 1)) * 100000!
1670 FOR I = 1 TO ND
1680 N = NU(I)
1690 S(N, 1) = S(N, 1) + CNST
1700 F(N) = F(N) + CNST * U(I): NEXT I
1710 REM *** EQUATION SOLVING ***
1720 GOSUB 5010
1730 PRINT "NODE#  X-Displ.    Y-Displ.    Rotation"
1740 FOR I = 1 TO NN
1750 PRINT USING " ###"; I;
1760 I1 = 3 * I - 2: I2 = I1 + 1: I3 = I1 + 2
1770 PRINT USING "  ##.###^^^^"; F(I1); F(I2); F(I3)
1780 NEXT I
1790 REM  *** REACTION CALCULATION ***
1800 PRINT "DOF#   Reaction"
1810 FOR I = 1 TO ND
1820 N = NU(I)
1830 R = CNST * (U(I) - F(N))
1840 PRINT USING "###"; N; : PRINT USING "  ##.###^^^^"; R: NEXT I
1850 END
1860 '===== SUBROUTINE ELEMENT STIFFNESS =====
1870 I1 = NOC(N, 1): I2 = NOC(N, 2)
1880 I3 = NOC(N, 3): I4 = NOC(N, 4)
1890 X21 = X(I2, 1) - X(I1, 1)
1900 Y21 = X(I2, 2) - X(I1, 2)
1910 EL = SQR(X21 * X21 + Y21 * Y21)
1920 EAL = PM(I3) * ARIN(I4, 1) / EL
1930 EIL = PM(I3) * ARIN(I4, 2) / EL
1940 CS = X21 / EL: SN = Y21 / EL
1950 SE(1, 1) = EAL * CS * CS + 12 * EIL * SN * SN / EL ^ 2
1960 SE(1, 2) = EAL * CS * SN - 12 * EIL * CS * SN / EL ^ 2
1970 SE(2, 1) = SE(1, 2)
1980 SE(1, 3) = -6 * EIL * SN / EL: SE(3, 1) = SE(1, 3)
1990 SE(1, 4) = -SE(1, 1): SE(4, 1) = SE(1, 4)
2000 SE(1, 5) = -SE(1, 2): SE(5, 1) = SE(1, 5)
2010 SE(1, 6) = SE(1, 3): SE(6, 1) = SE(1, 6)
2020 SE(2, 2) = EAL * SN * SN + 12 * EIL * CS * CS / EL ^ 2
2030 SE(2, 3) = 6 * EIL * CS / EL: SE(3, 2) = SE(2, 3)
2040 SE(2, 4) = -SE(1, 2): SE(4, 2) = SE(2, 4)
2050 SE(2, 5) = -SE(2, 2): SE(5, 2) = SE(2, 5)
2060 SE(2, 6) = SE(2, 3): SE(6, 2) = SE(2, 6)
```

2070 SE(3, 3) = 4 * EIL
2080 SE(3, 4) = –SE(1, 3): SE(4, 3) = SE(3, 4)
2090 SE(3, 5) = –SE(2, 3): SE(5, 3) = SE(3, 5)
2100 SE(3, 6) = 2 * EIL: SE(6, 3) = SE(3, 6)
2110 SE(4, 4) = SE(1, 1)
2120 SE(4, 5) = SE(1, 2): SE(5, 4) = SE(4, 5)
2130 SE(4, 6) = –SE(1, 3): SE(6, 4) = SE(4, 6)
2140 SE(5, 5) = SE(2, 2)
2150 SE(5, 6) = –SE(2, 3): SE(6, 5) = SE(5, 6)
2160 SE(6, 6) = 4 * EIL
2170 RETURN
5000 '===== EQUATION SOLVING =====
5010 N1 = NQ – 1
5020 REM *** FORWARD ELIMINATION ***
5030 FOR K = 1 TO N1
5040 NK = NQ – K + 1
5050 IF NK > NBW THEN NK = NBW
5060 FOR I = 2 TO NK
5070 C1 = S(K, I) / S(K, 1)
5080 I1 = K + I – 1
5090 FOR J = I TO NK
5100 J1 = J – I + 1
5110 S(I1, J1) = S(I1, J1) – C1 * S(K, J): NEXT J
5120 F(I1) = F(I1) – C1 * F(K): NEXT I: NEXT K
5130 REM ** BACK SUBSTITUTION **
5140 F(NQ) = F(NQ) / S(NQ, 1)
5150 FOR KK = 1 TO N1
5160 K = NQ – KK
5170 C1 = 1 / S(K, 1)
5180 F(K) = C1 * F(K)
5190 NK = NQ – K + 1
5200 IF NK > NBW THEN NK = NBW
5210 FOR J = 2 TO NK
5220 F(K) = F(K) – C1 * S(K, J) * F(K + J – 1)
5230 NEXT J
5240 NEXT KK
5250 RETURN

PROBLEMS

8.1. Find the deflection at the load and the slopes at the ends for the steel shaft shown in Fig. P8.1. Consider the shaft to be simply supported at bearings A and B.

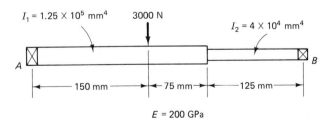

E = 200 GPa

Figure P8.1   Problems 8.1 and 8.4.

8.2. A three-span beam is shown in Fig. P8.2. Determine the deflection curve of the beam and evaluate the reactions at the supports.

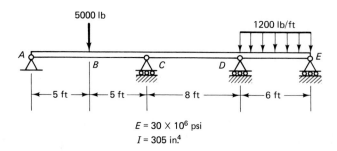

$E = 30 \times 10^6$ psi

$I = 305$ in.^{4}

Figure P8.2

8.3. A reinforced concrete slab floor is shown in Fig. P8.3. Using a unit width of the slab in the z direction, determine the deflection curve of the neutral surface under its own weight.

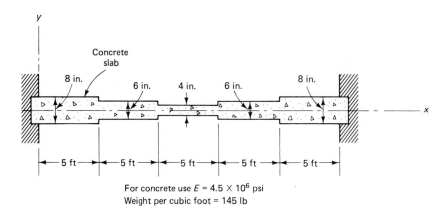

For concrete use $E = 4.5 \times 10^6$ psi
Weight per cubic foot = 145 lb

Figure P8.3

8.4. In the shaft shown in Fig. P8.1, determine the deflection at the loads and the slopes at the ends if the bearings at A and B have radial stiffnesses of 20 and 12 kN/mm, respectively.

8.5. Fig. P8.5 shows a beam AD pinned at A and welded at B and C to long and slender rods BE and CF. A load of 3000 lb is applied at D as shown. Model the beam AD using beam elements and determine deflections at B, C, and D, and stresses in rods BE and CF.

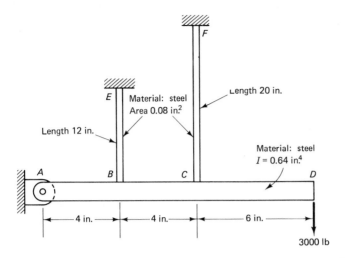

E for steel = 30 × 10⁶ psi

Figure P8.5

8.6. Figure P8.6 shows a cantilever beam with three rectangular openings. Find the deflections for the beam shown and compare the deflections with a beam without openings.

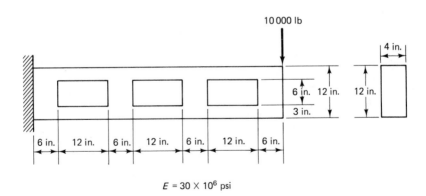

$E = 30 \times 10^6$ psi

Figure P8.6

8.7. A simplified section of a machine tool spindle is shown in Fig. P8.7. Bearing *B* has a radial stiffness of 60 N/μm and a rotational stiffness (against moment) of 8×10^5 N.m/rad. Bearing *C* has a radial stiffness of 20 N/μm and its rotational stiffness can be neglected. For a load of 1000 N, as shown, determine the deflection and slope at *A*. Also give the deflected shape of the spindle center line.

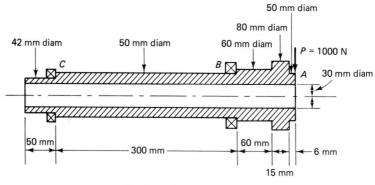

Machine tool spindle

Figure P8.7

8.8. Determine the deflection at the center of *BC* for the frame shown in Fig. P8.8, using program FRAME. Also determine the reactions at *A* and *D*.

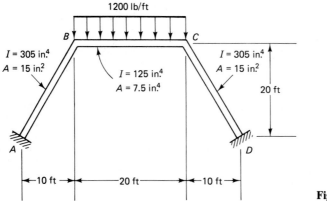

Figure P8.8

8.9. Figure P8.9 shows a hollow square section with two loading conditions. Using a 1-in. width perpendicular to the section, determine the deflection at the load for each of the two cases.

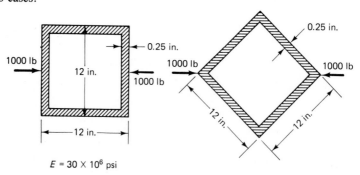

Figure P8.9

8.10. Figure P8.10 shows a five-member steel frame subjected to loads at the free end. The cross section of each member is a tube of wall thickness $t = 1$ cm and mean radius $R = 6$ cm. Determine:

(a) The displacement of node 3.

(b) The maximum axial compressive stress in a member.

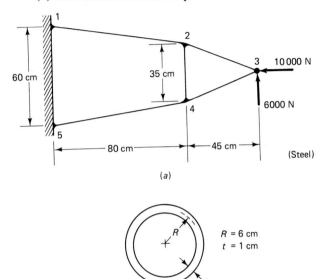

(a)

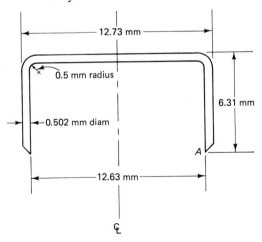

(b)

Figure P8.10

8.11. Dimensions of a common paper staple are shown in Fig. P8.11. While the staple is penetrating into the paper, a force of about 120 N is applied. Find the deformed shape for the following cases:

(a) Load uniformly distributed on the horizontal member and pinned condition at A at entry.

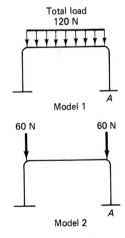

Figure P8.11

(b) Load as above with fixed condition at *A* after some penetration.

(c) Load divided into two point loads, with *A* pinned.

(d) Load as in (c) with *A* fixed.

8.12. A commonly used street light arrangement is shown in Fig. P8.12. Assuming fixed condition at *A*, compare the deformed shapes for the following two cases:

(a) Without the rod *BC* (that is, only member *ACD* supports the light).

(b) With tie rod *BC*.

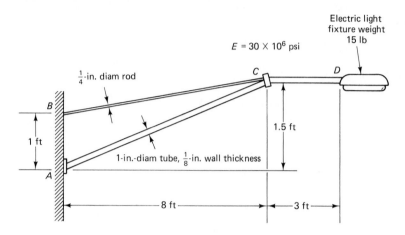

Figure P8.12

Three-Dimensional Problems in Stress Analysis

9.1 INTRODUCTION

Most engineering problems are three dimensional. So far, we have studied the possibilities of finite element analysis of simplified models, where rod elements, constant strain triangles, axisymmetric elements, beams, and so on give reasonable results. In this chapter, we deal with the formulation of three-dimensional stress analysis problems. The four-node tetrahedral element is presented in detail. Problem modeling and brick elements are also discussed.

We recall from the formulation given in Chapter 1 that

$$\mathbf{u} = [u, v, w]^{\mathrm{T}} \tag{9.1}$$

where u, v, and w are displacements in the x, y, and z directions, respectively. The stresses and strains are given by

$$\boldsymbol{\sigma} = [\sigma_x, \sigma_y, \sigma_z, \tau_{yz}, \tau_{xz}, \tau_{xy}]^{\mathrm{T}} \tag{9.2}$$

$$\boldsymbol{\epsilon} = [\epsilon_x, \epsilon_y, \epsilon_z, \gamma_{yz}, \gamma_{xz}, \gamma_{xy}]^{\mathrm{T}} \tag{9.3}$$

The stress–strain relations are given by

$$\boldsymbol{\sigma} = \mathbf{D}\boldsymbol{\epsilon} \tag{9.4}$$

where $\mathbf{D}$ is a (6×6) symmetric matrix. For isotropic materials $\mathbf{D}$ is given by Eq. 1.15.

The strain–displacement relations are given by

$$\boldsymbol{\epsilon} = \left[\frac{\partial u}{\partial x}, \frac{\partial v}{\partial y}, \frac{\partial w}{\partial z}, \frac{\partial v}{\partial z} + \frac{\partial w}{\partial y}, \frac{\partial u}{\partial z} + \frac{\partial w}{\partial x}, \frac{\partial u}{\partial y} + \frac{\partial v}{\partial x} \right]^{\mathrm{T}} \tag{9.5}$$

The body force and traction vectors are given by

$$\mathbf{f} = [\,f_x, f_y, f_z]^\mathrm{T} \tag{9.6}$$

$$\mathbf{T} = [T_x, T_y, T_z]^\mathrm{T} \tag{9.7}$$

The total potential and the Galerkin/virtual work form for three dimensions are given in Chapter 1.

9.2 FINITE ELEMENT FORMULATION

We divide the volume into four-node tetrahedra. Each node is assigned a number and the x, y, z coordinates are read in. A typical element e is shown in Fig. 9.1. The connectivity may be defined as shown in Table 9.1.

For each local node i we assign the three degrees of freedom $q_{3i-2}, q_{3i-1}, q_{3i}$ and for the corresponding global node I, we assign $Q_{3I-2}, Q_{3I-1}, Q_{3I}$. Thus, the element and global displacement vectors are

$$\mathbf{q} = [q_1, q_2, q_3, \ldots, q_{12}]^\mathrm{T} \tag{9.8}$$

$$\mathbf{Q} = [Q_1, Q_2, Q_3, \ldots, Q_N]^\mathrm{T} \tag{9.9}$$

where N is the total number of degrees of freedom for the structure, three per node.

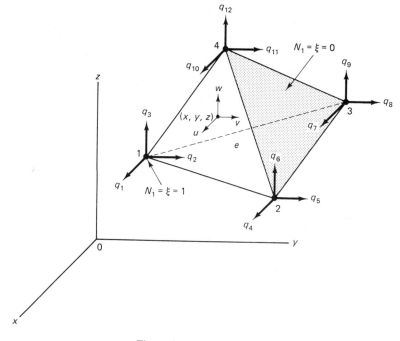

Figure 9.1 Tetrahedral element.

TABLE 9.1 CONNECTIVITY

	Nodes			
Element No	1	2	3	4
e	I	J	K	L

We define four Lagrange-type shape functions N_1, N_2, N_3, and N_4, where shape function N_i has a value of 1 at node i and is zero at the other three nodes. Specifically, N_1 is 0 at nodes 2, 3, and 4, and linearly increases to 1 at node 1. Using the master element shown in Fig. 9.2, we can define the shape functions as

$$N_1 = \xi \qquad N_2 = \eta \qquad N_3 = \zeta \qquad N_4 = 1 - \xi - \eta - \zeta \qquad (9.10)$$

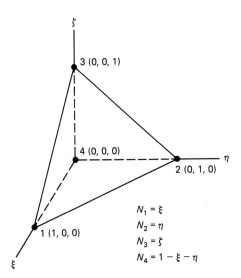

$N_1 = \xi$
$N_2 = \eta$
$N_3 = \zeta$
$N_4 = 1 - \xi - \eta$

Figure 9.2 Master element for shape functions.

The displacements u, v, w at $\mathbf{x}$ can be written in terms of the unknown nodal values as

$$\mathbf{u} = \mathbf{Nq} \qquad (9.11)$$

where

$$\mathbf{N} = \begin{bmatrix} N_1 & 0 & 0 & N_2 & 0 & 0 & N_3 & 0 & 0 & N_4 & 0 & 0 \\ 0 & N_1 & 0 & 0 & N_2 & 0 & 0 & N_3 & 0 & 0 & N_4 & 0 \\ 0 & 0 & N_1 & 0 & 0 & N_2 & 0 & 0 & N_3 & 0 & 0 & N_4 \end{bmatrix} \qquad (9.12)$$

It is easy to see that the shape functions given by Eq. 9.10 can be used to define the coordinates x, y, z of the point at which the displacements u, v, w are interpolated.

The isoparametric transformation is given by

$$x = N_1 x_1 + N_2 x_2 + N_3 x_3 + N_4 x_4$$

$$y = N_1 y_1 + N_2 y_2 + N_3 y_3 + N_4 y_4 \qquad (9.13)$$

$$z = N_1 z_1 + N_2 z_2 + N_3 z_3 + N_4 z_4$$

which, on substituting for N_i from Eq. 9.10 and using the notation $x_{ij} = x_i - x_j$, $y_{ij} = y_i - y_j$, $z_{ij} = z_i - z_j$, yields

$$x = x_4 + x_{14}\xi + x_{24}\eta + x_{34}\zeta$$

$$y = y_4 + y_{14}\xi + y_{24}\eta + y_{34}\zeta \qquad (9.14)$$

$$z = z_4 + z_{14}\xi + z_{24}\eta + z_{34}\zeta$$

Using the chain rule for partial derivatives, say of u, we have

$$\left\{ \begin{matrix} \dfrac{\partial u}{\partial \xi} \\[6pt] \dfrac{\partial u}{\partial \eta} \\[6pt] \dfrac{\partial u}{\partial \zeta} \end{matrix} \right\} = \mathbf{J} \left\{ \begin{matrix} \dfrac{\partial u}{\partial x} \\[6pt] \dfrac{\partial u}{\partial y} \\[6pt] \dfrac{\partial u}{\partial z} \end{matrix} \right\} \qquad (9.15)$$

Thus, the partial derivatives with respect to ξ, η, ζ are related to x, y, z derivatives by the above relationship. The Jacobean $\mathbf{J}$ of the transformation is given by

$$\mathbf{J} = \begin{bmatrix} \dfrac{\partial x}{\partial \xi} & \dfrac{\partial y}{\partial \xi} & \dfrac{\partial z}{\partial \xi} \\[6pt] \dfrac{\partial x}{\partial \eta} & \dfrac{\partial y}{\partial \eta} & \dfrac{\partial z}{\partial \eta} \\[6pt] \dfrac{\partial x}{\partial \zeta} & \dfrac{\partial y}{\partial \zeta} & \dfrac{\partial z}{\partial \zeta} \end{bmatrix} = \begin{bmatrix} x_{14} & y_{14} & z_{14} \\[6pt] x_{24} & y_{24} & z_{24} \\[6pt] x_{34} & y_{34} & z_{34} \end{bmatrix} \qquad (9.16)$$

We note here that

$$\det \mathbf{J} = x_{14}(y_{24}z_{34} - y_{34}z_{24}) + y_{14}(z_{24}x_{34} - z_{34}x_{24}) + z_{14}(x_{24}y_{34} - x_{34}y_{24}) \qquad (9.17)$$

The volume of the element is given by

$$V_e = \left| \int_0^1 \int_0^{1-\xi} \int_0^{1-\xi-\eta} \det \mathbf{J} \, d\xi \, d\eta \, d\zeta \right| \qquad (9.18)$$

Since $\det \mathbf{J}$ is constant,

$$V_e = |\det \mathbf{J}| \int_0^1 \int_0^{1-\xi} \int_0^{1-\xi-\eta} d\xi \, d\eta \, d\zeta \qquad (9.19)$$

Using the polynomial integral formula

$$\int_0^1 \int_0^{1-\xi} \int_0^{1-\xi-\eta} \xi^m \eta^n \zeta^p \, d\xi \, d\eta \, d\zeta = \frac{m! \, n! \, p!}{(m + n + p + 3)!} \qquad (9.20)$$

we get

$$V_e = \tfrac{1}{6} |\det \mathbf{J}| \qquad (9.21)$$

The inverse relation corresponding to Eq. 9.15 is given by

$$\begin{Bmatrix} \dfrac{\partial u}{\partial x} \\[2mm] \dfrac{\partial u}{\partial y} \\[2mm] \dfrac{\partial u}{\partial z} \end{Bmatrix} = \mathbf{A} \begin{Bmatrix} \dfrac{\partial u}{\partial \xi} \\[2mm] \dfrac{\partial u}{\partial \eta} \\[2mm] \dfrac{\partial u}{\partial \zeta} \end{Bmatrix} \qquad (9.22)$$

where $\mathbf{A}$ is the inverse of the Jacobian matrix $\mathbf{J}$ given in Eq. 9.16:

$$\mathbf{A} = \mathbf{J}^{-1} = \frac{1}{\det \mathbf{J}} \begin{bmatrix} y_{24}z_{34} - y_{34}z_{24} & y_{34}z_{14} - y_{14}z_{34} & y_{14}z_{24} - y_{24}z_{14} \\ z_{24}x_{34} - z_{34}x_{24} & z_{34}x_{14} - z_{14}x_{34} & z_{14}x_{24} - z_{24}x_{14} \\ x_{24}y_{34} - x_{34}z_{24} & x_{34}y_{14} - x_{14}y_{34} & x_{14}y_{24} - x_{24}y_{14} \end{bmatrix} \qquad (9.23)$$

Using the strain–displacement relations in Eq. 9.5, the relation between derivatives in x, y, z and ξ, η, ζ in Eq. 9.22, and the assumed displacement field $\mathbf{u} = \mathbf{Nq}$ in Eq. 9.11, we get

$$\boldsymbol{\epsilon} = \mathbf{Bq} \qquad (9.24)$$

where $\mathbf{B}$ is a (6×12) matrix given by

$$\mathbf{B} = \begin{bmatrix} A_{11} & 0 & 0 & A_{12} & 0 & 0 & A_{13} & 0 & 0 & -\tilde{A}_1 & 0 & 0 \\ 0 & A_{21} & 0 & 0 & A_{22} & 0 & 0 & A_{23} & 0 & 0 & -\tilde{A}_2 & 0 \\ 0 & 0 & A_{31} & 0 & 0 & A_{32} & 0 & 0 & A_{33} & 0 & 0 & -\tilde{A}_3 \\ 0 & A_{31} & A_{21} & 0 & A_{32} & A_{22} & 0 & A_{33} & A_{23} & 0 & -\tilde{A}_3 & -\tilde{A}_2 \\ A_{31} & 0 & A_{11} & A_{32} & 0 & A_{12} & A_{33} & 0 & A_{13} & -\tilde{A}_3 & 0 & -\tilde{A}_1 \\ A_{21} & A_{11} & 0 & A_{22} & A_{12} & 0 & A_{23} & A_{13} & 0 & -\tilde{A}_2 & -\tilde{A}_1 & 0 \end{bmatrix} \qquad (9.25)$$

where $\tilde{A}_1 = A_{11} + A_{12} + A_{13}$, $\tilde{A}_2 = A_{21} + A_{22} + A_{23}$, and $\tilde{A}_3 = A_{31} + A_{32} + A_{33}$. All the terms of $\mathbf{B}$ are constants. Thus, Eq. 9.24 gives constant strains after the nodal displacements are calculated.

Element Stiffness

The element strain energy in the total potential is given by

$$U_e = \frac{1}{2} \int_e \boldsymbol{\epsilon}^\mathrm{T} \mathbf{D} \boldsymbol{\epsilon} \, dV$$

$$= \tfrac{1}{2} \mathbf{q}^\mathrm{T} \mathbf{B}^\mathrm{T} \mathbf{D} \mathbf{B} \mathbf{q} \int_e dV$$

$$= \tfrac{1}{2} \mathbf{q}^\mathrm{T} V_e \mathbf{B}^\mathrm{T} \mathbf{D} \mathbf{B} \mathbf{q} \tag{9.26}$$

$$= \tfrac{1}{2} \mathbf{q}^\mathrm{T} \mathbf{k}^e \mathbf{q}$$

where the element stiffness matrix $\mathbf{k}^e$ is given by

$$\mathbf{k}^e = V_e \mathbf{B}^\mathrm{T} \mathbf{D} \mathbf{B} \tag{9.27}$$

where V_e is the volume of the element given by $\frac{1}{6} |\det \mathbf{J}|$. In the Galerkin approach, the internal virtual work of the element comes out to be

$$\int_e \boldsymbol{\sigma}^\mathrm{T} \boldsymbol{\epsilon}(\phi) \, dV = \boldsymbol{\psi}^\mathrm{T} V_e \mathbf{B}^\mathrm{T} \mathbf{D} \mathbf{B} \mathbf{q} \tag{9.28}$$

which gives the element stiffness in Eq. 9.27.

Force Terms

The potential term associated with body force is

$$\int_e \mathbf{u}^\mathrm{T} \mathbf{f} \, dV = \mathbf{q}^\mathrm{T} \iiint \mathbf{N}^\mathrm{T} \mathbf{f} \det \mathbf{J} \, d\xi \, d\eta \, d\zeta$$

$$= \mathbf{q}^\mathrm{T} \mathbf{f}^e \tag{9.29}$$

Using the integration formula in Eq. 9.20, we have

$$\mathbf{f}^e = \frac{V_e}{4} [f_x, f_y, f_z, f_x, f_y, f_z, \dots, f_z]^\mathrm{T} \tag{9.30}$$

Above, the element body force vector $\mathbf{f}^e$ is of dimension 12×1. Note that $V_e f_x$ is the x component of the body force which is distributed to the degrees of freedom q_1, q_4, q_7, and q_{10}.

Let us now consider uniformly distributed traction on the boundary surface. The boundary surface of a tetrahedron is a triangle. Without loss of generality, if A_e is the boundary surface on which traction is applied, formed by local nodes 1, 2, and 3, then

$$\int_{A_e} \mathbf{u}^\mathrm{T} \mathbf{T} \, dA = \mathbf{q}^\mathrm{T} \int_{A_e} \mathbf{N}^\mathrm{T} \mathbf{T} \, dA = \mathbf{q}^\mathrm{T} \mathbf{T}^e \tag{9.31}$$

The element traction load vector $\mathbf{T}^e$ is given by

$$\mathbf{T}^e = \frac{A_e}{3}[T_x, T_y, T_z, T_x, T_y, T_z, T_x, T_y, T_z, 0, 0, 0] \tag{9.32}$$

The stiffnesses and forces are gathered into global locations using element connectivity. Point loads are added into proper locations of the force vector. Boundary conditions are considered using penalty or other approaches. The energy and Galerkin approaches yield the set of equations

$$\mathbf{KQ} = \mathbf{F} \tag{9.33}$$

9.3 STRESS CALCULATIONS

After the above equations are solved, the element nodal displacements $\mathbf{q}$ can be obtained. Since $\boldsymbol{\sigma} = \mathbf{D}\boldsymbol{\epsilon}$ and $\boldsymbol{\epsilon} = \mathbf{Bq}$, the element stresses are given by

$$\boldsymbol{\sigma} = \mathbf{DBq} \tag{9.34}$$

The three principal stresses can be calculated by using the relationships given below. The three invariants of the (3×3) stress tensor are

$$
\begin{aligned}
I_1 &= \sigma_x + \sigma_y + \sigma_z \\
I_2 &= \sigma_x\sigma_y + \sigma_y\sigma_z + \sigma_x\sigma_z - \tau_{yz}^2 - \tau_{xz}^2 - \tau_{xy} \\
I_3 &= \sigma_x\sigma_y\sigma_z + 2\tau_{yz}\tau_{xz}\tau_{xy} - \sigma_x\tau_{yz}^2 - \sigma_y\tau_{xz}^2 - \sigma_z\tau_{xy}^2
\end{aligned}
\tag{9.35}
$$

We define

$$
\begin{aligned}
a &= \frac{I_1^2}{3} - I_2 \\
b &= -2\left(\frac{I_1}{3}\right)^3 + \frac{I_1 I_2}{3} - I_3 \\
c &= 2\sqrt{\frac{a}{3}} \\
\theta &= \frac{1}{3}\cos^{-1}\left(-\frac{3b}{ac}\right)
\end{aligned}
\tag{9.36}
$$

The principal stresses are given by

$$
\begin{aligned}
\sigma_1 &= \frac{I_1}{3} + c\cos\theta \\
\sigma_2 &= \frac{I_1}{3} + c\cos\left(\theta + \frac{2\pi}{3}\right) \\
\sigma_3 &= \frac{I_1}{3} + c\cos\left(\theta + \frac{4\pi}{3}\right)
\end{aligned}
\tag{9.37}
$$

9.4 MESH PREPARATION

While complex three-dimensional regions can be effectively filled by tetrahedral elements, similar to triangular elements filling a two-dimensional region, it is a tedious affair to carry out manual data preparation. To overcome this, for simple regions, it is easier to divide the regions into eight-node blocks. Consider the master cube shown in Fig. 9.3. The cube can be divided into five tetrahedra, as shown in Fig. 9.4, with the connectivity as given in Table 9.2.

In the above division, the first four elements are of equal volume and element 5 has

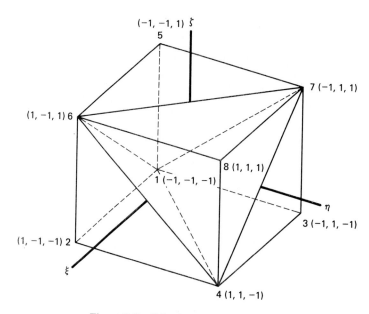

Figure 9.3 Cube for tetrahedral division.

TABLE 9.2   FIVE TETRAHEDRA

Element No.	Nodes			
	1	2	3	4
1	1	4	2	6
2	1	4	3	7
3	6	7	5	1
4	6	7	8	4
5	1	4	6	7

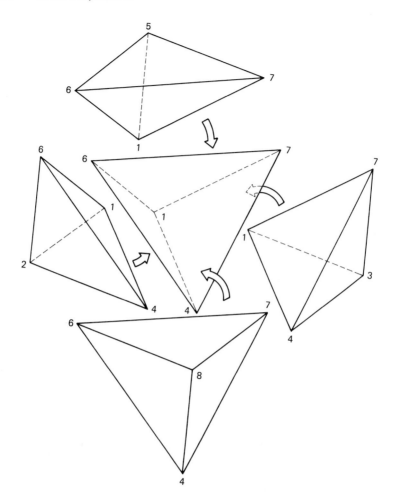

Figure 9.4   Division of a cube into five tetrahedra.

twice the volume of other elements. In this case, care must be taken to match element edges on adjacent blocks.

The master cube can also be divided into 6 elements with equal volume. A typical division is given in Table 9.3. The element division of one-half of the cube is shown in Fig. 9.5. For the division shown in Table 9.3, the same division pattern repeats for adjacent elements.

Use of det **J** in the calculation of **B** in Eq. 9.24 and use of $|\det \mathbf{J}|$ in the estimation of element volume V_e enables us to use element node numbers in any order. Among solid elements, this holds for four-node tetrahedra, since every node is connected to the other three. Some codes may still require consistent numbering schemes.

TABLE 9.3   SIX TETRAHEDRA

Element No.	Nodes			
	1	2	3	4
1	1	2	4	8
2	1	2	8	5
3	2	8	5	6
4	1	3	4	7
5	1	7	8	5
6	1	8	4	7

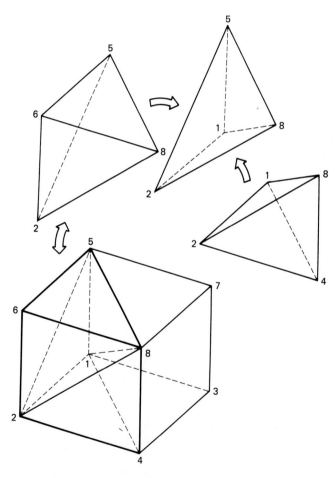

Figure 9.5   Division of a cube into six tetrahedra.

9.5 HEXAHEDRAL ELEMENTS AND HIGHER-ORDER ELEMENTS

In the hexahedral elements, a consistent node numbering scheme must be followed for defining the connectivity. For an eight-node hexahedral or brick element, we consider the mapping onto a cube of 2 unit sides placed symmetrically with ξ, η, ζ coordinates as shown in Fig. 9.6. The corresponding element in two dimensions is the four-node quadrilateral discussed in Chapter 7.

On the master cube, the Lagrange shape functions can be written as

$$N_i = \tfrac{1}{8}(1 + \xi_i\xi)(1 + \eta_i\eta)(1 + \zeta_i\zeta) \qquad i = 1 \text{ to } 8 \tag{9.38}$$

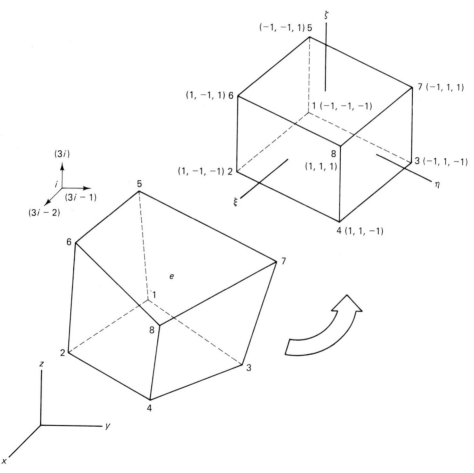

Figure 9.6   Hexahedral element.

where (ξ_i, η_i, ζ_i) represents the coordinates of node i of the element in the (ξ, η, ζ) system. The element nodal displacements are represented by the vector

$$\mathbf{q} = [q_1, q_2, \ldots, q_{24}]^T \qquad (9.39)$$

We use the shape functions N_i to define displacements at any point inside the element in terms of its nodal values:

$$u = N_1 q_1 + N_2 q_4 + \cdots + N_8 q_{22}$$
$$v = N_1 q_2 + N_2 q_5 + \cdots + N_8 q_{23} \qquad (9.40)$$
$$w = N_1 q_3 + N_2 q_6 + \cdots + N_8 q_{24}$$

Also,

$$x = N_1 x_1 + N_2 x_2 + \cdots + N_8 x_8$$
$$y = N_1 y_1 + N_2 y_2 + \cdots + N_8 y_8 \qquad (9.41)$$
$$z = N_1 z_1 + N_2 z_2 + \cdots + N_8 z_8$$

Following the steps used in the development of the quadrilateral element in Chapter 7, we can get the strains in the form

$$\boldsymbol{\epsilon} = \mathbf{Bq} \qquad (9.42)$$

The element stiffness matrix is given by

$$\mathbf{k}^e = \int_{-1}^{+1} \int_{-1}^{+1} \int_{-1}^{+1} \mathbf{B}^T \mathbf{D} \mathbf{B} \, |\det \mathbf{J}| \, d\xi \, d\eta \, d\zeta \qquad (9.43)$$

where we have used $dV = |\det \mathbf{J}| \, d\xi \, d\eta \, d\zeta$, and $\mathbf{J}$ is the (3×3) Jacobian matrix. The integration in Eq. 9.43 is performed numerically using Gauss quadrature.

Higher-order elements, for example, 10-node tetrahedral elements or 20-node and 27-node hexahedral elements, can be developed using the ideas discussed in Chapter 7.

9.6 PROBLEM MODELING

In solving a problem, the first step is to start with a coarse model. The data needed will be nodal coordinates, element nodal connectivity, material properties, constraint conditions, and nodal loads. In the three-dimensional cantilever shown in Fig. 9.7, the geometry and loading conditions demand a three-dimensional model. Element and connectivities can easily be established by defining the four 8-cornered blocks. We can model the first block, near the base of the cantilever, as a hexahedral element with connectivity 2-1-6-5-3-4-7-8. For each subsequent block, the connectivity can be generated by increasing each number in the current set by 4. Coordinates of nodes can be generated using the shape functions of Eq. 9.38 for geometry definition. These aspects will be discussed in Chapter 12. Alternatively, each block

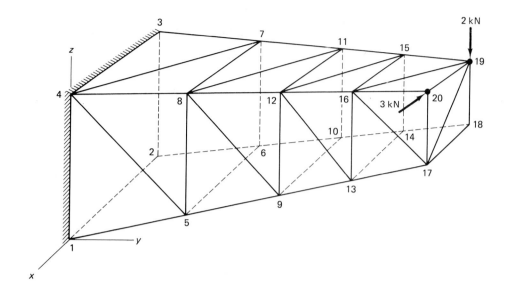

Figure 9.7   Three-dimensional elastic body.

in the 3-D cantilever can be modeled using tetrahedral elements. For the repeating block pattern shown in Fig. 9.6, the six element division given in Table 9.3 may be used.

The consideration of boundary conditions follows those presented for one- and two-dimensional problems. However, to give a general idea of constraints and their consideration in finite element analysis we refer to Fig. 9.8. A point fully restrained is a point constraint. This is considered by adding a large stiffness C to the diagonal locations corresponding to degrees of freedom of node I. When the node is constrained to move along a line, say $\mathbf{t}$, with direction cosines (ℓ, m, n), the penalty term comes from setting $\mathbf{u} \times \mathbf{t} = \mathbf{0}$. This results in the addition of following stiffness terms when the node is constrained along a line.

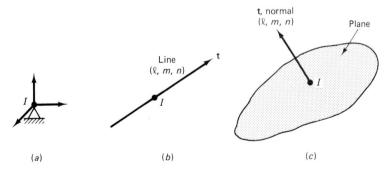

Figure 9.8   Nodal constraints: (*a*) point constraint, (*b*) line constraint, (*c*) plane constraint.

$$\begin{array}{c} \begin{matrix} \; 3l-2 & 3l-1 & 3l \end{matrix} \\ \begin{matrix} 3I-2 \\ 3I-1 \\ 3I \end{matrix} \begin{bmatrix} C\ell^2 & C\ell m & C\ell n \\ & Cm^2 & Cmn \\ \text{Symmetric} & & Cn^2 \end{bmatrix} \end{array}$$

When a node is forced to lie on a plane with normal direction t, shown in Fig. 9.8c, the penalty terms come from $\mathbf{u} \cdot \mathbf{t} = 0$. This requires that the following terms be added to the stiffness matrix:

$$\begin{array}{c} \begin{matrix} \;\; 3l-2 & \;\; 3l-1 & \;\; 3l \end{matrix} \\ \begin{matrix} 3I-2 \\ 3I-1 \\ 3I \end{matrix} \begin{bmatrix} C(1-\ell^2) & -C\ell m & -C\ell n \\ & C(1-m^2) & -Cmn \\ \text{Symmetric} & & C(1-n^2) \end{bmatrix} \end{array}$$

Fig. 9.9 shows a pyramid-shaped metal part and its finite element model. We observe here that nodes at A and B are line-constrained and nodes along C and D are plane-constrained. This discussion should help one to handle the modeling of three-dimensional problems with relative ease.

Example 9.1

The cantilever beam shown in Fig. E9.1 is analyzed using program FE3TETRA. The beam is first divided into three hexahedra. Each hexahedra is then divided into five tetrahedra, as was shown in Fig. 9.4. The element connectivity follows readily from

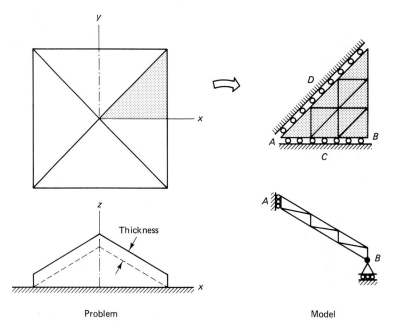

Problem   Model

Figure 9.9   Metal part with a pyramid surface.

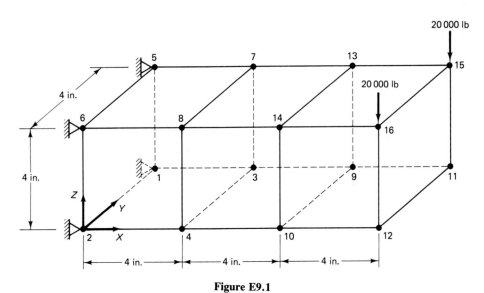

Figure E9.1

Table 9.2. The program DATAFEM has been used to create the input data file Tetra.Dat, whose listing is given below.

```
Listing of Tetra.Dat
  16    15   12  2  1  3  4
  0   4   0
  0   0   0
  10   4   0
  10   0   0
  0   4   4
  0   0   4
  10   4   4
  10   0   4
  20   4   0
  20   0   0
  30   4   0
  30   0   0
  20   4   4
  20   0   4
  30   4   4
  30   0   4
  1   4   2   6
  1   4   3   7
  6   7   5   1
  6   7   8   4
  1   4   6   7
  3   10   4   8
  3   10   9   13
```

```
 8   13    7    3
 8   13   14   10
 3   10    8   13
 9   12   10   14
 9   12   11   15
14   15   13    9
14   15   16   12
 9   12   14   15
 1    1    1    1    1    1    1    1    1    1    1    1    1    1    1
 1    0
 2    0
 3    0
 4    0
 5    0
 6    0
13    0
14    0
15    0
16    0
17    0
18    0
45  -20000
48  -20000
3E+07    .3    0
```

■

```
1000 REM *******        FE3TETRA        *******
1010 REM ***  Three Dimensional Stress Analysis  ***
1020 REM ***        Tetrahedral Elements        ***
1030 CLS : DEFINT I-N
1040 PRINT "======================================="
1050 PRINT "       Program written by          "
1060 PRINT "  T.R.Chandrupatla and A.D.Belegundu   "
1070 PRINT "======================================="
1072 INPUT "Data file name <path fn.ext> ", FILE1$
1080 OPEN FILE1$ FOR INPUT AS #1
1082 INPUT "Output File Name <path fn.ext> ", FILE2$
1084 OPEN FILE2$ FOR OUTPUT AS #2
1090 INPUT #1, NN, NE, ND, NL, NM, NDIM, NEN
1092 REM Prob dim. NDIM and Num. of Elem Nodes NEN
1094 REM are included for consistency with DATAFEM
1100 REM TOTAL DOF IS "NQ"
1110 NQ = 3 * NN
1120 DIM X(NN, 3), F(NQ), NOC(NE, 4), MAT(NE), PM(NM, 3), NU(ND), U(ND)
1130 DIM D(6, 6), B(6, 12), DB(6, 12), SE(12, 12), Q(12), STR(6)
1132 REM Read all Data from Data File
1134 GOSUB 4500
1140 REM Bandwidth NBW from Connectivity  NOC( , )
1150 NBW = 0
1160 FOR N = 1 TO NE
1180 CMIN = NQ + 1: CMAX = 0
1190 FOR J = 1 TO 4
1200 IF CMIN > NOC(N, J) THEN CMIN = NOC(N, J)
1210 IF CMAX < NOC(N, J) THEN CMAX = NOC(N, J)
```

```
1220 NEXT J
1230 C = 3 * (CMAX - CMIN + 1)
1240 IF NBW < C THEN NBW = C
1250 NEXT N
1260 PRINT "The Bandwidth is"; NBW
1270 REM *** INITIALIZATION ***
1280 DIM S(NQ, NBW)
1290 FOR I = 1 TO NQ
1310 FOR J = 1 TO NBW
1320 S(I, J) = 0: NEXT J: NEXT I
1420 REM *** GLOBAL STIFFNESS MATRIX **
1430 FOR N = 1 TO NE
1440 PRINT "FORMING STIFFNESS MATRIX OF ELEMENT  "; N
1450 GOSUB 3050
1460 FOR I = 1 TO 12
1470 FOR J = 1 TO 12
1480 SE(I, J) = 0
1490 FOR K = 1 TO 6
1500 SE(I, J) = SE(I, J) + B(K, I) * DB(K, J) * ABS(DJ) / 6
1510 NEXT K: NEXT J: NEXT I
1520 PRINT "PLACING STIFFNESS IN GLOBAL LOCATIONS"
1530 FOR II = 1 TO 4
1540 NRT = 3 * (NOC(N, II) - 1)
1550 FOR IT = 1 TO 3
1560 NR = NRT + IT
1570 I = 3 * (II - 1) + IT
1580 FOR JJ = 1 TO 4
1590 NCT = 3 * (NOC(N, JJ) - 1)
1600 FOR JT = 1 TO 3
1610 J = 3 * (JJ - 1) + JT
1620 NC = NCT + JT - NR + 1
1630 IF NC <= 0 GOTO 1650
1640 S(NR, NC) = S(NR, NC) + SE(I, J)
1650 NEXT JT: NEXT JJ: NEXT IT: NEXT II
1660 NEXT N
1670 REM  *** MODIFY FOR BOUNDARY CONDITIONS  ***
1680 CNST = S(1, 1) * 100000!
1690 FOR I = 1 TO ND
1700 N = NU(I)
1710 S(N, 1) = S(N, 1) + CNST
1720 F(N) = F(N) + CNST * U(I): NEXT I
1730 REM  *** EQUATION SOLVING ***
1740 GOSUB 5000
1750 PRINT #2, "NODE#  X-Displ    Y-Displ    Z-Displ"
1760 FOR I = 1 TO NN
1770 PRINT #2, USING " ####"; I;
1780 PRINT #2, USING " #.####^^^^"; F(3 * I - 2); F(3 * I - 1); F(3 * I)
1790 NEXT I
1800 REM  *** STRESS CALCULATIONS  ***
1810 FOR N = 1 TO NE
1820 GOSUB 3050
1830 GOSUB 4000
1840 '***PRINCIPAL STRESS CALCULATIONS***
1850 ' *** CALCULATION OF PRINCIPAL STRESSES ***
1860 AI1 = STR(1) + STR(2) + STR(3)
1870 AI21 = STR(1) * STR(2) + STR(2) * STR(3) + STR(3) * STR(1)
1880 AI22 = STR(4) * STR(4) + STR(5) * STR(5) + STR(6) * STR(6)
1890 AI2 = AI21 - AI22
1900 AI31 = STR(1) * STR(2) * STR(3) + 2 * STR(4) * STR(5) * STR(6)
```

```
1910 AI32 = STR(1) * STR(4) ^ 2 + STR(2) * STR(5) ^ 2 + STR(3) * STR(6) ^ 2
1920 AI3 = AI31 - AI32
1930 PI = 3.141593
1940 C1 = AI2 - AI1 ^ 2 / 3
1950 C2 = -2 * (AI1 / 3) ^ 3 + AI1 * AI2 / 3 - AI3
1960 C3 = 2 * SQR(-C1 / 3)
1970 TH = -3 * C2 / (C1 * C3): TH2 = ABS(1 - TH * TH)
1972 IF TH = 0 THEN TH = PI / 2
1980 IF TH > 0 THEN TH = ATN(SQR(TH2) / TH)
1982 IF TH < 0 THEN TH = PI - ATN(SQR(TH2) / TH)
1990 TH = TH / 3
2000 '*** PRINCIPAL STRESSES ***
2010 P1 = AI1 / 3 + C3 * COS(TH)
2020 P2 = AI1 / 3 + C3 * COS(TH + 2 * PI / 3)
2030 P3 = AI1 / 3 + C3 * COS(TH + 4 * PI / 3)
2040 PRINT #2,
2050 PRINT #2, "STRESSES IN ELEMENT NO.  "; N
2060 PRINT #2, "  Normal Stresses SX,SY,SZ"
2070 PRINT #2, USING "  #.####^^^^"; STR(1); STR(2); STR(3)
2080 PRINT #2, "  Shear Stresses TYZ,TXZ,TXY"
2090 PRINT #2, USING "  #.####^^^^"; STR(4); STR(5); STR(6)
2100 PRINT #2,
3000 PRINT #2, "  Principal Stresses"
3010 PRINT #2, USING "  #.####^^^^"; P1; P2; P3
3020 NEXT N
3030 CLOSE #2
3032 PRINT "Results are in the file  "; FILE2$
3040 END
3050 'ELEMENT STIFFNESS FORMATION
3060 '***FIRST THE D-MATRIX***
3070 M = MAT(N): E = PM(M, 1): PNU = PM(M, 2)
3080 C4 = E / ((1 + PNU) * (1 - 2 * PNU))
3090 C1 = C4 * (1 - PNU): C2 = C4 * PNU: C3 = .5 * E / (1 + PNU)
3100 FOR I = 1 TO 6: FOR J = 1 TO 6: D(I, J) = 0: NEXT J: NEXT I
3110 D(1, 1) = C1: D(1, 2) = C2: D(1, 3) = C2
3120 D(2, 1) = C2: D(2, 2) = C1: D(2, 3) = C2
3130 D(3, 1) = C2: D(3, 2) = C2: D(3, 3) = C1
3140 D(4, 4) = C3: D(5, 5) = C3: D(6, 6) = C3
3150 '***STRAIN-DISPLACEMENT MATRIX***
3160 I1 = NOC(N, 1): I2 = NOC(N, 2): I3 = NOC(N, 3): I4 = NOC(N, 4)
3170 X14 = X(I1, 1) - X(I4, 1): X24 = X(I2, 1) - X(I4, 1): X34 = X(I3, 1) - X(I4, 1)
3180 Y14 = X(I1, 2) - X(I4, 2): Y24 = X(I2, 2) - X(I4, 2): Y34 = X(I3, 2) - X(I4, 2)
3190 Z14 = X(I1, 3) - X(I4, 3): Z24 = X(I2, 3) - X(I4, 3): Z34 = X(I3, 3) - X(I4, 3)
3200 DJ1 = X14 * (Y24 * Z34 - Z24 * Y34)
3210 DJ2 = Y14 * (Z24 * X34 - X24 * Z34)
3220 DJ3 = Z14 * (X24 * Y34 - Y24 * X34)
3230 DJ = DJ1 + DJ2 + DJ3
3240 A11 = (Y24 * Z34 - Z24 * Y34) / DJ
3250 A21 = (Z24 * X34 - X24 * Z34) / DJ
3260 A31 = (X24 * Y34 - Y24 * X34) / DJ
3270 A12 = (Y34 * Z14 - Z34 * Y14) / DJ
3280 A22 = (Z34 * X14 - X34 * Z14) / DJ
3290 A32 = (X34 * Y14 - Y34 * X14) / DJ
3300 A13 = (Y14 * Z24 - Z14 * Y24) / DJ
3310 A23 = (Z14 * X24 - X14 * Z24) / DJ
3320 A33 = (X14 * Y24 - Y14 * X24) / DJ
3330 '***FORMATION OF B MATRIX***
3340 FOR I = 1 TO 6: FOR J = 1 TO 12: B(I, J) = 0: NEXT J: NEXT I
```

```
3350 B(1, 1) = A11: B(1,4) = A12: B(1, 7) = A13: B(1, 10) = -A11 - A12 - A13
3360 B(2, 2) = A21: B(2, 5) = A22: B(2, 8) = A23: B(2, 11) = -A21 - A22 - A23
3370 B(3, 3) = A31: B(3, 6) = A32: B(3, 9) = A33: B(3, 12) = -A31 - A32 - A33
3380 B(4, 2) = A31: B(4, 3) = A21: B(4, 5) = A32: B(4, 6) = A22: B(4, 8) = A33
3390        B(4, 9) = A23: B(4, 11) = B(3, 12): B(4, 12) = B(2, 11)
3400 B(5, 1) = A31: B(5, 3) = A11: B(5, 4) = A32: B(5, 6) = A12: B(5, 7) = A33
3410        B(5, 9) = A13: B(5, 10) = B(3, 12): B(5, 12) = B(1, 10)
3420 B(6, 1) = A21: B(6, 2) = A12: B(6, 4) = A22: B(6, 5) = A12: B(6, 7) = A23
3430        B(6, 8) = A13: B(6, 10) = B(2, 11): B(6, 11) = B(1, 10)
3440 '***DB MATRIX DB=D*B***
3450 FOR I = 1 TO 6
3460 FOR J = 1 TO 12
3470 DB(I, J) = 0
3480 FOR K = 1 TO 6
3490 DB(I, J) = DB(I, J) + D(I, K) * B(K, J)
3500 NEXT K: NEXT J: NEXT I
3510 RETURN
4000 'STRESS EVALUATION
4010 FOR I = 1 TO 4
4020 IN = 3 * (NOC(N, I) - 1): II = 3 * (I - 1)
4030 FOR J = 1 TO 3
4040 Q(II + J) = F(IN + J): NEXT J: NEXT I
4050 FOR I = 1 TO 6
4060 STR(I) = 0
4070 FOR K = 1 TO 12
4080 STR(I) = STR(I) + DB(I, K) * Q(K): NEXT K: NEXT I
4090 RETURN
4500 '=============== READ DATA ====================
4510 FOR I = 1 TO NN: FOR J = 1 TO NDIM
4520 INPUT #1, X(I, J): NEXT J: NEXT I
4530 FOR I = 1 TO NE: FOR J = 1 TO NEN
4540 INPUT #1, NOC(I, J): NEXT J: NEXT I
4550 FOR I = 1 TO NE: INPUT #1, MAT(I): NEXT I
4560 FOR I = 1 TO ND: INPUT #1, NU(I), U(I): NEXT I
4580 FOR I = 1 TO NL: INPUT #1, II, F(II): NEXT I
4590 FOR I = 1 TO NM: FOR J = 1 TO 3
4600 INPUT #1, PM(I, J): NEXT J: NEXT I
4610 CLOSE #1
4620 RETURN
5000 N1 = NQ - 1
5010 REM *** FORWARD ELIMINATION ***
5020 FOR K = 1 TO N1
5030 NK = NQ - K + 1
5040 IF NK > NBW THEN NK = NBW
5050 FOR I = 2 TO NK
5060 C1 = S(K, I) / S(K, 1)
5070 I1 = K + I - 1
5080 FOR J = I TO NK
5090 J1 = J - I + 1
5100 S(I1, J1) = S(I1, J1) - C1 * S(K, J): NEXT J
5110 F(I1) = F(I1) - C1 * F(K): NEXT I: NEXT K
5120 REM *** BACK SUBSTITUTION ***
5130 F(NQ) = F(NQ) / S(NQ, 1)
5140 FOR KK = 1 TO N1
5150 K = NQ - KK
5160 C1 = 1 / S(K, 1)
5170 F(K) = C1 * F(K)
5180 NK = NQ - K + 1
```

5190 IF NK > NBW THEN NK = NBW
5200 FOR J = 2 TO NK
5210 F(K) = F(K) – C1 * S(K, J) * F(K + J – 1)
5220 NEXT J
5230 NEXT KK
5240 RETURN

PROBLEMS

9.1. Determine the deflections at the corner points of the steel cantilever beam shown in Fig. P9.1.

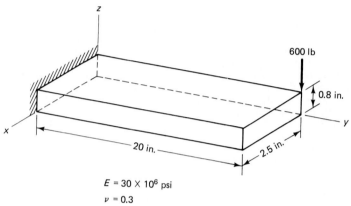

$E = 30 \times 10^6$ psi
$\nu = 0.3$

Figure P9.1

9.2. A cast iron hollow member used in a machine tool structure is fixed at one end and loaded at the other, as shown in Fig. P9.2. Find the deflection at the load and maximum principal stresses. Compare the values with the structure without an opening.

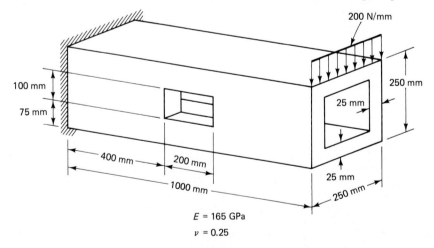

$E = 165$ GPa
$\nu = 0.25$

Figure P9.2

9.3. An S-shaped block used in force measurement is subjected to a load as shown in Fig. P9.3. Determine the amount by which the sensor is compressed. Take $E = 70\,000$ N/mm², $\nu = 0.3$.

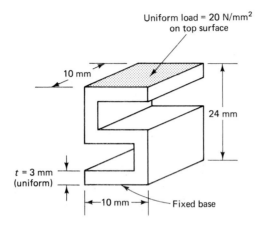

Uniform load = 20 N/mm² on top surface

10 mm

24 mm

$t = 3$ mm (uniform)

10 mm

Fixed base

Figure P9.3

9.4. A device is hydraulically loaded as shown in Fig. P9.4. Plot the deformed configuration and determine the magnitude and location of the maximum principal stresses.

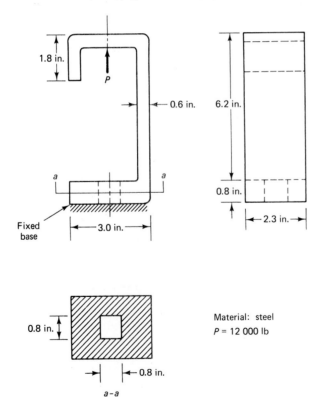

1.8 in.

P

0.6 in.   6.2 in.

a   a

Fixed base   3.0 in.   0.8 in.   2.3 in.

0.8 in.   0.8 in.   a–a

Material: steel
P = 12 000 lb

Figure P9.4

9.5. A portion of the brake pedal in an automobile is modeled as shown in Fig. P9.5. Determine the deflection at the pedal for a 500-N load.

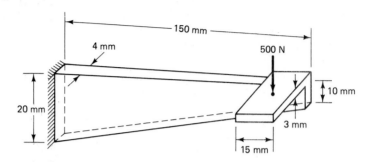

Figure P9.5

*9.6. Determine the axial elongation and location and magnitude of maximum Von Mises stress in the connecting rod shown in Fig. P9.6.

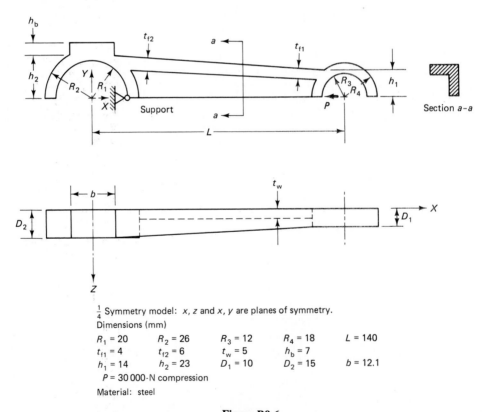

$\frac{1}{4}$ Symmetry model: x, z and x, y are planes of symmetry.
Dimensions (mm)

$R_1 = 20$	$R_2 = 26$	$R_3 = 12$	$R_4 = 18$	$L = 140$
$t_{f1} = 4$	$t_{f2} = 6$	$t_w = 5$	$h_b = 7$	
$h_1 = 14$	$h_2 = 23$	$D_1 = 10$	$D_2 = 15$	$b = 12.1$
$P = 30\,000$-N compression				

Material: steel

Figure P9.6

chapter ten

Scalar Field Problems

10.1 INTRODUCTION

In previous chapters, the unknowns in the problem represented components of a *vector* field. In a two-dimensional plate, for example, the unknown quantity is the vector field $\mathbf{u}(x, y)$, where $\mathbf{u}$ is a (2×1) displacement vector. On the other hand, quantities such as temperature, pressure, and stream potentials are *scalar* in nature. In two-dimensional steady-state heat conduction, for example, the temperature field $T(x, y)$ is the unknown to be determined.

In this chapter, the finite element method for solving such problems is discussed. In Section 10.2, one-dimensional and two-dimensional steady-state heat conduction is considered, as well as heat flow in one-dimensional fins. Section 10.3 deals with torsion of solid shafts. Scalar field problems related to fluid flow, seepage, electric/magnetic fields, and flow in ducts are defined in Section 10.4.

The striking feature of scalar field problems is that they are to be found in almost all branches of engineering and physics. Most of them can be viewed as special forms of the general Helmholtz equation, given by

$$\frac{\partial}{\partial x}\left(k_x \frac{\partial \phi}{\partial x}\right) + \frac{\partial}{\partial y}\left(k_y \frac{\partial \phi}{\partial y}\right) + \frac{\partial}{\partial z}\left(k_z \frac{\partial \phi}{\partial z}\right) + \lambda\phi + Q = 0 \qquad (10.1)$$

together with boundary conditions on ϕ and its derivatives. In the above equation, $\phi = \phi(x, y, z)$ is the field variable that is to be determined. Table 10.1 lists some of the engineering problems described by Eq. 10.1. For example, if we set $\phi = T$, $k_x = k_y = k$, $\lambda = 0$, and consider only x and y, we get $\partial^2 T/\partial x^2 + \partial^2 T/\partial y^2 + Q = 0$, which describes the heat conduction problem for temperature T, where k is

TABLE 10.1 EXAMPLES OF SCALAR FIELD PROBLEMS IN ENGINEERING

Problem	Equation	Field variable	Parameter	Boundary conditions
Heat conduction	$k\left(\dfrac{\partial^2 T}{\partial x^2} + \dfrac{\partial^2 T}{\partial y^2}\right) + Q = 0$	Temperature, T	Thermal conductivity, k	$T = T_0$ $-k\dfrac{\partial T}{\partial n} = q_0$ $-k\dfrac{\partial T}{\partial n} = h(T - T_\infty)$
Torsion	$\left(\dfrac{\partial^2 \theta}{\partial x^2} + \dfrac{\partial^2 \theta}{\partial y^2}\right) + 2 = 0$	Stress function, θ		$\theta = 0$
Potential flow	$\left(\dfrac{\partial^2 \psi}{\partial x^2} + \dfrac{\partial^2 \psi}{\partial y^2}\right) = 0$	Stream function, ψ		$\psi = \psi_0$
Seepage and groundwater flow	$k\left(\dfrac{\partial^2 \phi}{\partial x^2} + \dfrac{\partial^2 \phi}{\partial y^2}\right) + Q = 0$	Hydraulic potential, ϕ	Hydraulic conductivity, k	$\phi = \phi_0$ $\dfrac{\partial \phi}{\partial n} = 0$ $\phi = y$
Electric potential	$\epsilon\left(\dfrac{\partial^2 u}{\partial x^2} + \dfrac{\partial^2 u}{\partial y^2}\right) = -\rho$	Electric potential, u	Permittivity, ϵ	$u = u_0$ $\dfrac{\partial u}{\partial n} = 0$
Fluid flow in ducts	$\left(\dfrac{\partial^2 W}{\partial X^2} + \dfrac{\partial^2 W}{\partial Y^2}\right) + 1 = 0$	Nondimensional velocity, W		$W = 0$

the thermal conductivity and Q is the heat source/sink. Mathematically, we can develop the finite element method for various field problems in a general manner by considering Eq. 10.1. The solution to specific problems can then be obtained by suitable definition of variables. We discuss here the heat transfer and torsion problems in some detail. These are important in themselves, because they provide us an opportunity to understand the physical problem and how to handle different boundary conditions needed for modeling. Once the steps are understood, extension to other areas in engineering should present no difficulty.

10.2 STEADY-STATE HEAT TRANSFER

Introduction

We now discuss the finite element formulation for the solution of steady-state heat transfer problems. Heat transfer occurs when there is a temperature difference within a body or between a body and its surrounding medium. Heat is transferred in the form of conduction, convection, and thermal radiation. Only conduction and convection modes are treated here.

The heat flow through the wall of a heated room on a winter day is an example of conduction. The conduction process is quantified by Fourier's law. In a thermally isotropic medium, Fourier's law for two-dimensional heat flow is given by

$$q_x = -k\frac{\partial T}{\partial x} \qquad q_y = -k\frac{\partial T}{\partial y} \tag{10.2}$$

where $T = T(x, y)$ is a temperature field in the medium, q_x and q_y are the components of the heat flux (W/m^2), k is the thermal conductivity (W/m.°C), and $\partial T/\partial x$, $\partial T/\partial y$ are the temperature gradients along x and y, respectively. The resultant heat flux $\mathbf{q} = q_x\mathbf{i} + q_y\mathbf{j}$ is at right angles to an isotherm or a line of constant temperature (Fig. 10.1). Note that 1 W = 1 J/s = 1 N.m/s. The minus sign in Eq. 10.2 reflects the fact that heat is transferred in the direction of decreasing temperature. Thermal conductivity k is a material property.

In convection heat transfer, there is transfer of energy between a fluid and solid surface as a result of a temperature difference. There can be free or natural convection, such as the circulation pattern set up while boiling water in a kettle due

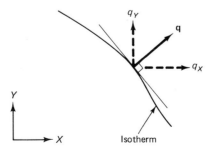

Figure 10.1   Heat flux in two dimensions.

to hot water rising and cooler water moving down, or there can be forced convection, such as when the fluid flow is caused by a fan. The governing equation is of the form

$$q = h(T_s - T_\infty) \tag{10.3}$$

where q is the convective heat flux (W/m^2), h is the convection heat transfer coefficient or film coefficient (W/m^2.°C), and T_s and T_∞ are the surface and fluid temperatures, respectively. The film coefficient h is a property of the flow, and depends on various factors, such as whether convection is natural or forced, whether the flow is laminar or turbulent, the type of fluid, and the geometry of the body.

In addition to conduction and convection, heat transfer can also occur in the form of thermal radiation. The radiation heat flux is proportional to the fourth power of the absolute temperature, which causes the problem to be nonlinear. This mode of heat transfer is not considered here.

One-Dimensional Heat Conduction

We now turn our attention to the steady-state heat conduction problem in one dimension. Our objective is to determine the temperature distribution. In one-dimensional steady-state problems, a temperature gradient exists along only one coordinate axis, and the temperature at each point is independent of time. Many engineering systems fall into this category.

Governing equation. Consider heat conduction in a plane wall with uniform heat generation (Fig. 10.2). Let A be the area normal to the direction of heat flow and let Q (W/m^3) be the internal heat generated per unit volume. A common example of heat generation is the heat produced in a wire carrying a current I and having a resistance R through a volume V, which results in $Q = I^2R/V$. A control vol-

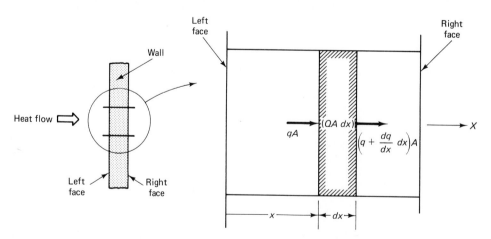

Figure 10.2 One-dimensional heat conduction.

ume is shown in Fig. 10.2. Since the heat rate (heat flux × area) that is entering the control volume plus the heat rate generated equals the heat rate leaving the control volume, we have

$$qA + QA\,dx = \left(q + \frac{dq}{dx}\,dx\right)A \qquad (10.4)$$

Canceling qA from both sides yields

$$Q = \frac{dq}{dx} \qquad (10.5)$$

Substituting Fourier's law

$$q = -k\frac{dT}{dx} \qquad (10.6)$$

into Eq. 10.5 results in

$$\frac{d}{dx}\left(k\frac{dT}{dx}\right) + Q = 0 \qquad (10.7)$$

Usually, Q is called a *source* when positive (heat is generated) and is called a *sink* when negative (heat is consumed). Here, Q will simply be referred to as a source. In general, k in Eq. 10.7 is a function of x. Equation 10.7 has to be solved with appropriate boundary conditions.

Boundary conditions.   The boundary conditions are mainly of three kinds: specified temperature, specified heat flux (or insulated), and convection. For example, consider the wall of a tank containing a hot liquid at a temperature T_0, with an airstream of temperature T_∞ passed on the outside, maintaining a wall temperature of T_L at the boundary (Fig. 10.3a). The boundary conditions for this problem are

$$T|_{x=0} = T_0 \qquad (10.8)$$

$$q|_{x=L} = h(T_L - T_\infty) \qquad (10.9)$$

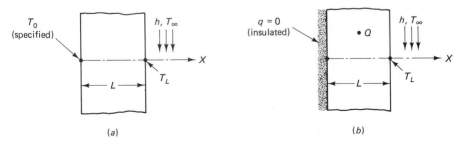

Figure 10.3   Examples of boundary conditions in 1-D heat conduction.

As another example, consider a wall, as shown in Fig. 10.3b, where the inside surface is insulated and the outside is a convection surface. Then, the boundary conditions are

$$q|_{x=0} = 0 \qquad q|_{x=L} = h(T_L - T_\infty) \tag{10.10}$$

We will consider the various types of boundary conditions in the following sections.

The one-dimensional element.   The two-node element with linear shape functions is considered below. Extension to three-node quadratic elements follows the same procedures as discussed in Chapter 3. Now, to apply the finite element method, the problem is discretized in the x dimension, as shown in Fig. 10.4a. The temperatures at the various nodal points, denoted by **T**, are the unknowns (except at node 1, where $T_1 = T_0$). Within a typical element e (Fig. 10.4b), whose local node numbers are 1 and 2, the temperature field is approximated using shape functions N_1 and N_2 as

$$
\begin{aligned}
T(\xi) &= N_1 T_1 + N_2 T_2 \\
&= \mathbf{N T}^e
\end{aligned}
\tag{10.11}
$$

where $N_1 = (1 - \xi)/2$, $N_2 = (1 + \xi)/2$, ξ varies from -1 to $+1$, $\mathbf{N} = [N_1, N_2]$, and $\mathbf{T}^e = [T_1, T_2]^{\mathrm{T}}$. Noting the relations

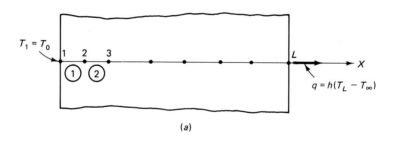

(a)

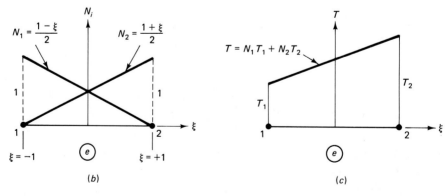

(b)   (c)

Figure 10.4   Finite element modeling and shape functions for linear interpolation of the temperature field.

$$\xi = \frac{2}{x_2 - x_1}(x - x_1) - 1$$

$$d\xi = \frac{2}{x_2 - x_1}dx \tag{10.12}$$

we have

$$\frac{dT}{dx} = \frac{dT}{d\xi}\frac{d\xi}{dx}$$

$$= \frac{2}{x_2 - x_1}\frac{d\mathbf{N}}{d\xi}\cdot\mathbf{T}^e \tag{10.13a}$$

$$= \frac{1}{x_2 - x_1}[-1,\ 1]\mathbf{T}^e$$

or

$$\frac{dT}{dx} = \mathbf{B}_T\mathbf{T}^e \tag{10.13b}$$

where

$$\mathbf{B}_T = \frac{1}{x_2 - x_1}[-1,\ 1] \tag{10.14}$$

Functional approach for heat conduction.   We now solve the heat conduction equation with convection boundary conditions given by

$$\frac{d}{dx}\left(k\frac{dT}{dx}\right) + Q = 0$$

$$T|_{x=0} = T_0 \qquad q|_{x=L} = h(T_L - T_\infty) \tag{10.15}$$

Similar to the potential energy for stress analysis problems, the solution of the above problem is equivalent to minimizing a functional, which is the energy form,

$$\Pi_T = \int_0^L \tfrac{1}{2}k\left(\frac{dT}{dx}\right)^2 dx - \int_0^L QT\,dx + \tfrac{1}{2}h(T_L - T_\infty)^2 \tag{10.16}$$

subject to $T|_{x=0} = T_0$. We now assume that heat source $Q = Q_e$ and thermal conductivity $k = k_e$ are constant within the element. Using $dx = \ell_e/2\,d\xi$, $T = \mathbf{N}\mathbf{T}^e$, and $dT/dx = \mathbf{B}_T\mathbf{T}^e$, the functional Π_T reduces to

$$\Pi_T = \sum_e \tfrac{1}{2}\mathbf{T}^{eT}\left[\frac{k_e\ell_e}{2}\int_{-1}^1 \mathbf{B}_T^T\mathbf{B}_T\,d\xi\right]\mathbf{T}^e - \sum_e \left[\frac{Q_e\ell_e}{2}\int_{-1}^1 \mathbf{N}\,d\xi\right]\mathbf{T}^e + \tfrac{1}{2}h(T_L - T_\infty)^2 \tag{10.17}$$

From above, we identify the *element conductivity matrix* as

$$\mathbf{k}_T = \frac{k_e \ell_e}{2} \int_{-1}^{1} \mathbf{B}_T^T \mathbf{B}_T \, d\xi \tag{10.18a}$$

or

$$\mathbf{k}_T = \frac{k_e}{\ell_e} \begin{bmatrix} 1 & -1 \\ -1 & 1 \end{bmatrix} \tag{10.18b}$$

Also, the *element heat rate vector* due to the heat source is

$$\mathbf{r}_Q^T = \frac{Q_e \ell_e}{2} \int_{-1}^{1} \mathbf{N} \, d\xi \tag{10.19a}$$

or, since $\int_{-1}^{1} N_i \, d\xi = 1$, we have

$$\mathbf{r}_Q = \frac{Q_e \ell_e}{2} \begin{Bmatrix} 1 \\ 1 \end{Bmatrix} \tag{10.19b}$$

Note that global matrices $\mathbf{K}_T$ and $\mathbf{R}$ are assembled from element matrices $\mathbf{k}_T$ and $\mathbf{r}_Q$ in the usual manner using element connectivity information.

The last term in Π_T can be written as

$$\tfrac{1}{2} T_L h T_L - (h T_\infty) T_L + \tfrac{1}{2} h T_\infty^2 \tag{10.20}$$

The term $\tfrac{1}{2} h T_\infty^2$ is a constant which drops out while minimizing Π_T. From above, we see that h gets added to the (L, L)th location of $\mathbf{K}_T$ and $(h T_\infty)$ gets added to the Lth row of $\mathbf{R}$.

Lastly, the handling of the specified temperature boundary condition $T_1 = T_0$ is analogous to the handling of a nonzero specified displacement as discussed in Chapter 3; either the elimination or the penalty approach can be adopted. For example, if the penalty approach is used, then a large conductivity C is added to $\mathbf{K}$ at $(1, 1)$ and a heat source $C T_0$ is added to the first row of $\mathbf{R}$. The finite element equations are

$$\begin{bmatrix} (K_{11} + C) & K_{12} & \cdots & K_{1L} \\ K_{21} & K_{22} & \cdots & K_{2L} \\ \vdots & & & \vdots \\ \vdots & & & \vdots \\ K_{L1} & K_{L2} & \cdots & (K_{LL} + h) \end{bmatrix} \begin{Bmatrix} T_1 \\ T_2 \\ \vdots \\ \vdots \\ T_L \end{Bmatrix} = \begin{Bmatrix} (R_1 + C T_0) \\ R_2 \\ \vdots \\ \vdots \\ (R_L + h T_\infty) \end{Bmatrix} \tag{10.21}$$

Galerkin's approach for heat conduction.   The element matrices $\mathbf{k}_T$, $\mathbf{r}_Q$, etc., that were derived earlier using the functional approach will now be derived using Galerkin's approach. The problem is

$$\frac{d}{dx}\left(k \frac{dT}{dx} \right) + Q = 0 \tag{10.22}$$

$$T|_{x=0} = T_0 \qquad q|_{x=L} = h(T_L - T_\infty)$$

If an approximate solution T is desired, Galerkin's approach is to solve

$$\int_0^L \phi \left[\frac{d}{dx}\left(k\frac{dT}{dx} \right) + Q \right] dx = 0 \qquad (10.23)$$

for every ϕ constructed from the same basis functions as those of T, with $\phi(0) = 0$. Integrating the first term by parts, we have

$$\phi k \frac{dT}{dx}\bigg|_0^L - \int_0^L k\frac{d\phi}{dx}\frac{dT}{dx}\,dx + \int_0^L \phi Q\,dx = 0 \qquad (10.24)$$

Now

$$\phi k \frac{dT}{dx}\bigg|_0^L = \phi(L)k(L)\frac{dT}{dx}(L) - \phi(0)k(0)\frac{dT}{dx}(0) \qquad (10.25a)$$

Since $\phi(0) = 0$ and $q = -k(L)(dT(L)/dx) = h(T_L - T_\infty)$, we get

$$\phi k \frac{dT}{dx}\bigg|_0^L = -\phi(L)h(T_L - T_\infty) \qquad (10.25b)$$

Thus, Eq. 10.24 becomes

$$-\phi(L)h(T_L - T_\infty) - \int_0^L k\frac{d\phi}{dx}\frac{dT}{dx}\,dx + \int_0^L \phi Q\,dx = 0 \qquad (10.26)$$

We now use the isoparametric relations $T = \mathbf{N}\mathbf{T}^e$, etc., defined in Eqs. 10.11–10.14. Further, a global virtual temperature vector is denoted as $\mathbf{\Psi} = [\Psi_1, \Psi_2, \ldots, \Psi_L]^T$, and the test function within each element is interpolated as

$$\phi = \mathbf{N}\boldsymbol{\psi} \qquad (10.27)$$

Analogous to $dT/dx = \mathbf{B}_T\mathbf{T}^e$ in Eq. 10.13b, we have

$$\frac{d\phi}{dx} = \mathbf{B}_T\boldsymbol{\psi} \qquad (10.28)$$

Thus, Eq. 10.26 becomes

$$-\Psi_L h(T_L - T_\infty) - \sum_e \boldsymbol{\psi}^T \left(\frac{k_e\ell_e}{2} \int_{-1}^1 \mathbf{B}_T^T\mathbf{B}_T\,d\xi \right)\mathbf{T}^e + \sum_e \boldsymbol{\psi}^T \frac{Q_e\ell_e}{2}\int_{-1}^1 \mathbf{N}^T\,d\xi = 0 \qquad (10.29)$$

$$-\Psi_L hT_L + \Psi_L hT_\infty - \mathbf{\Psi}^T\mathbf{K}_T\mathbf{T} + \mathbf{\Psi}^T\mathbf{R} = 0 \qquad (10.30)$$

which should be satisfied for all $\mathbf{\Psi}$ with $\Psi_1 = 0$. The global matrices $\mathbf{K}_T$ and $\mathbf{R}$ are assembled from element matrices $\mathbf{k}_T$ and $\mathbf{r}_Q$, as given in Eqs. 10.18b and 10.19b, respectively. When each $\mathbf{\Psi}$ is chosen in turn as $[0, 1, 0, \ldots, 0]^T$, $[0, 0, 1, 0, \ldots, 0]^T, \ldots, [0, 0, , \ldots, 0, 1]^T$ and since $T_1 = T_0$, then Eq. 10.30 yields

$$
\begin{bmatrix}
K_{22} & K_{23} & \cdots & K_{2L} \\
K_{32} & K_{33} & \cdots & K_{3L} \\
\vdots & & & \vdots \\
K_{L2} & K_{L3} & \cdots & (K_{LL} + h)
\end{bmatrix}
\begin{Bmatrix}
T_2 \\ T_3 \\ \vdots \\ \vdots \\ T_L
\end{Bmatrix}
=
\begin{Bmatrix}
R_2 \\ R_3 \\ \vdots \\ \vdots \\ (R_L + hT_\infty)
\end{Bmatrix}
-
\begin{Bmatrix}
K_{21}T_0 \\ K_{31}T_0 \\ \vdots \\ \vdots \\ K_{L1}T_0
\end{Bmatrix}
\tag{10.31}
$$

We observe that Eq. 10.31 can be solved for $T_2, T_3, \ldots, T_L$. We thus note that the Galerkin approach naturally leads to the elimination approach for handling nonzero specified temperature $T = T_0$ at node 1. However, it is also possible to develop Galerkin's method with a penalty approach to handle $T_1 = T_0$. In this case, the equations are as given by Eq. 10.21.

Example 10.1

A composite wall consists of three materials, as shown in Fig. E10.1a. The outer temperature is $T_0 = 20°C$. Convection heat transfer takes place on the inner surface of the wall with $T_\infty = 800°C$ and $h = 25$ W/m².°C. Determine the temperature distribution in the wall.

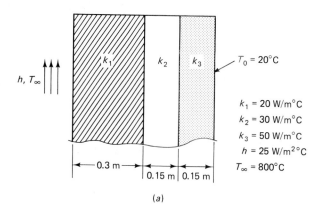

$k_1 = 20$ W/m°C
$k_2 = 30$ W/m°C
$k_3 = 50$ W/m°C
$h = 25$ W/m²°C
$T_\infty = 800°C$

(a)

(b)   **Figure E10.1**

Solution A three-element finite element model of the wall is shown in Fig. E10.1b. The element conductivity matrices are

$$
\mathbf{k}_T^{(1)} = \frac{20}{0.3}\begin{bmatrix} 1 & -1 \\ -1 & 1 \end{bmatrix} \qquad
\mathbf{k}_T^{(2)} = \frac{30}{0.15}\begin{bmatrix} 1 & -1 \\ -1 & 1 \end{bmatrix}
$$

$$
\mathbf{k}_T^{(3)} = \frac{50}{0.15}\begin{bmatrix} 1 & -1 \\ -1 & 1 \end{bmatrix}
$$

The global $\mathbf{K} = \Sigma\, \mathbf{k}_T$ is obtained from above as

$$\mathbf{K} = 66.7 \begin{bmatrix} 1 & -1 & 0 & 0 \\ -1 & 4 & -3 & 0 \\ 0 & -3 & 8 & -5 \\ 0 & 0 & -5 & 5 \end{bmatrix}$$

Now, since convection occurs at node 1, the constant $h = 25$ is added to the $(1, 1)$ location of $\mathbf{K}$. This results in

$$\mathbf{K} = 66.7 \begin{bmatrix} 1.375 & -1 & 0 & 0 \\ -1 & 4 & -3 & 0 \\ 0 & -3 & 8 & -5 \\ 0 & 0 & -5 & 5 \end{bmatrix}$$

Since no heat generation Q occurs in this problem, the heat rate vector $\mathbf{R}$ consists only of hT_∞ in the first row. That is,

$$\mathbf{R} = [25 \times 800, 0, 0, 0]^\mathsf{T}$$

The specified temperature boundary condition $T_4 = 20°C$, will now be handled by the penalty approach. We choose C based on

$$C = \max|\mathbf{K}_{ij}| \times 10^4$$
$$= 66.7 \times 8 \times 10^4$$

Now, C gets added to $(4, 4)$ location of $\mathbf{K}$, while CT_4 is added to the fourth row of $\mathbf{R}$. The resulting equations are

$$66.7 \begin{bmatrix} 1.375 & -1 & 0 & 0 \\ -1 & 4 & -3 & 0 \\ 0 & -3 & 8 & -5 \\ 0 & 0 & -5 & 80\,005 \end{bmatrix} \begin{Bmatrix} T_1 \\ T_2 \\ T_3 \\ T_4 \end{Bmatrix} = \begin{Bmatrix} 25 \times 800 \\ 0 \\ 0 \\ 10\,672 \times 10^4 \end{Bmatrix}$$

The solution is

$$\mathbf{T} = [304.6, 119.0, 57.1, 20.0]^\mathsf{T}°C$$

Comment. The boundary condition $T_4 = 20°C$ can also be handled by the elimination approach. The fourth row and column of $\mathbf{K}$ is deleted, and $\mathbf{R}$ is modified according to Eq. 3.70. The resulting equations are

$$66.7 \begin{bmatrix} 1.375 & -1 & 0 \\ -1 & 4 & -3 \\ 0 & -3 & 8 \end{bmatrix} \begin{bmatrix} T_1 \\ T_2 \\ T_3 \end{bmatrix} = \begin{bmatrix} 25 \times 800 \\ 0 \\ 0 + 6670 \end{bmatrix}$$

which yields

$$[T_1, T_2, T_3] = [304.6, 119.0, 57.1]°C \qquad \blacksquare$$

Heat flux boundary condition.   Certain physical situations are modeled using the boundary condition

$$q = q_0 \qquad \text{at } x = 0 \tag{10.32}$$

where q_0 is a specified heat flux on the boundary. If $q = 0$, then the surface is perfectly insulated. A nonzero value of q_0 occurs, for example, due to an electrical heater or pad where one face is in contact with the wall and the other face is insulated. It is important to note that the input heat flux q_0 has a sign convention associated with it: q_0 is input as a positive value if heat is flowing out of the body and as a negative value if heat is flowing into the body. The boundary condition in Eq. 10.32 is handled by adding $(-q_0)$ to the heat rate vector. The resulting equations are

$$\mathbf{KT} = \mathbf{R} + \begin{Bmatrix} -q_0 \\ 0 \\ \cdot \\ \cdot \\ \cdot \\ 0 \end{Bmatrix} \tag{10.33}$$

The sign convention for specified heat flux given above is clear if we consider the heat transfer occurring at a boundary. Let n be the outward normal (in 1-D problems, $n = +x$ or $-x$). The heat flow in the body towards the $+n$ direction is $q = -k\, \partial T/\partial n$, where $\partial T/\partial n < 0$. Thus, q is > 0 and since this heat flows out of the body, we have the boundary condition $q = q_0$ with the stated sign convention.

Comment on forced and natural boundary conditions.   In the above problem, boundary conditions of the type $T = T_0$, which is on the field variable itself, are called *forced* boundary conditions. On the other hand, the boundary condition $q|_{x=0} = q_0$, or equivalently, $-k\, dT/dx|_{x=0} = q_0$, is called a *natural* boundary condition involving the derivative of the field variable. Further, it is evident from Eq. 10.33 that the homogeneous natural boundary condition $q = q_0 = 0$ does not require any modifications in the element matrices. These are automatically satisfied at the boundary, in an average sense.

Example 10.2

Heat is generated in a large plate ($k = 0.8$ W/m.°C) at the rate of 4000 W/m³. The plate is 25 cm thick. The outside surfaces of the plate are exposed to ambient air at 30°C with a convective heat transfer coefficient of 20 W/m².°C. Determine the temperature distribution in the wall.

Solution   The problem is symmetric about the centerline of the plate. A two-element finite element model is shown in Fig. E10.2. The left end is insulated ($q = 0$) because no heat can flow across a line of symmetry. Noting that $k/\ell = 0.8/.0625 = 12.8$, we have

$$\mathbf{K} = \begin{bmatrix} 12.8 & -12.8 & 0 \\ -12.8 & 25.6 & -12.8 \\ 0 & -12.8 & (12.8 + 20) \end{bmatrix}$$

The heat rate vector is assembled from the heat source (Eq. 10.19b) as well as due to convection as

$$\mathbf{R} = [125 \quad 250 \quad (125 + 20 \times 30)]^{\mathrm{T}}$$

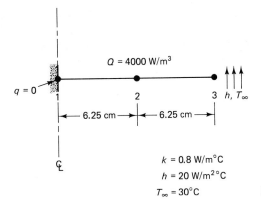

$Q = 4000 \text{ W/m}^3$

$q = 0$

6.25 cm — 6.25 cm

$k = 0.8 \text{ W/m}°\text{C}$

$h = 20 \text{ W/m}^2°\text{C}$

$T_\infty = 30°\text{C}$ **Figure E10.2**

Solution of $\mathbf{KT} = \mathbf{R}$ yields

$$[T_1, T_2, T_3] = [94.0, 84.3, 55.0]°\text{C} \qquad \blacksquare$$

Heat Transfer in Thin Fins

A fin is an *extended surface* which is added onto a structure to increase the rate of heat removal. A familiar example is in the motorcycle where fins extend from the cylinder head to quickly dissipate heat through convection. We present here the finite element method for analyzing heat transfer in thin rectangular fins (Fig. 10.5). This problem differs from the conduction problem discussed previously in that both conduction and convection occur within the body.

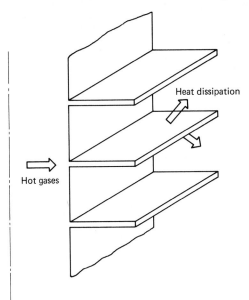

Heat dissipation

Hot gases

Figure 10.5 An array of thin rectangular fins.

Consider a thin rectangular fin as shown in Fig. 10.6. The problem can be treated as one-dimensional because the temperature gradients along the width and across the thickness are negligible. The governing equation may be derived from the conduction equation with heat source, given by

$$\frac{d}{dx}\left(k\frac{dT}{dx}\right) + Q = 0$$

The convection heat loss in the fin can be considered as a negative heat source

$$Q = -\frac{(P\,dx)h(T - T_\infty)}{A_c\,dx}$$

$$= -\frac{Ph}{A_c}(T - T_\infty) \tag{10.34}$$

where P = perimeter of fin, A_c = area of cross section. Thus, the governing equation is

$$\frac{d}{dx}\left(k\frac{dT}{dx}\right) - \frac{Ph}{A_c}(T - T_\infty) = 0 \tag{10.35}$$

We present our analysis for the case when the base of the fin is held at T_0 and the tip of the fin is insulated (heat going out of the tip is negligible). The boundary conditions are then given by

$$T = T_0 \qquad \text{at } x = 0 \tag{10.36a}$$

$$q = 0 \qquad \text{at } x = L \tag{10.36b}$$

The finite element method: Galerkin approach.   The element matrices and heat rate vectors for solving Eq. 10.35 with the boundary conditions in Eqs.

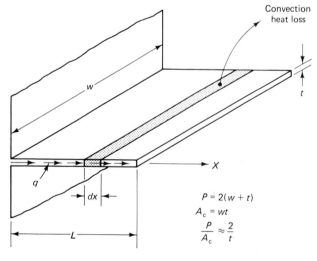

$$P = 2(w + t)$$
$$A_c = wt$$
$$\frac{P}{A_c} \approx \frac{2}{t}$$

Figure 10.6 Heat flow in a thin rectangular fin.

10.36 will now be developed. Galerkin's approach is attractive since we do not have to set up the functional that is to be minimized. Element matrices can be derived directly from the differential equation. Let $\phi(x)$ be any function satisfying $\phi(0) = 0$ using same basis as T. We require

$$\int_0^L \phi\left[\frac{d}{dx}\left(k\frac{dT}{dx}\right) - \frac{Ph}{A_c}(T - T_\infty)\right] dx = 0 \tag{10.37}$$

Integrating the first term by parts, we have

$$\phi k\frac{dT}{dx}\bigg|_0^L - \int_0^L k\frac{d\phi}{dx}\frac{dT}{dx}\, dx - \frac{Ph}{A_c}\int_0^L \phi T\, dx + \frac{Ph}{A_c}T_\infty\int_0^L \phi\, dx = 0 \tag{10.38}$$

Using $\phi(0) = 0$, $k(L)[dT(L)/dx] = 0$, and the isoparametric relations

$$dx = \frac{\ell_e}{2}\, d\xi \qquad T = \mathbf{N}\mathbf{T}^e \qquad \phi = \mathbf{N}\boldsymbol{\psi} \qquad \frac{dT}{dx} = \mathbf{B}_T\mathbf{T}^e \qquad \frac{d\phi}{dx} = \mathbf{B}_T\boldsymbol{\psi}$$

we get

$$-\sum_e \boldsymbol{\psi}^T\left[\frac{k_e\ell_e}{2}\int_{-1}^1 \mathbf{B}_T^T\mathbf{B}_T\, d\xi\right]\mathbf{T}^e - \frac{Ph}{A_c}\sum_e \boldsymbol{\psi}^T\int_{-1}^1 \mathbf{N}^T\mathbf{N}\, d\xi\, \mathbf{T}^e$$

$$+ \frac{PhT_\infty}{A_c}\sum_e \boldsymbol{\psi}^T\frac{\ell_e}{2}\int_{-1}^1 \mathbf{N}^T\, d\xi = 0 \tag{10.39}$$

We define

$$\mathbf{h}_T = \frac{Ph}{A_c}\frac{\ell_e}{2}\int_{-1}^1 \mathbf{N}^T\mathbf{N}\, d\xi = \frac{Ph}{A_c}\frac{\ell_e}{6}\begin{bmatrix} 2 & 1 \\ 1 & 2 \end{bmatrix} \tag{10.40a}$$

or, since $P/A_c \approx 2/t$ (Fig. 10.6),

$$\mathbf{h}_T \approx \frac{h\ell_e}{3t}\begin{bmatrix} 2 & 1 \\ 1 & 2 \end{bmatrix} \tag{10.40b}$$

and

$$\mathbf{r}_\infty = \frac{Ph}{A_c}T_\infty\frac{\ell_e}{2}\int_{-1}^1 \mathbf{N}^T\, d\xi = \frac{PhT_\infty}{A_c}\frac{\ell_e}{2}\begin{Bmatrix} 1 \\ 1 \end{Bmatrix} \tag{10.41a}$$

or

$$\mathbf{r}_\infty \approx \frac{hT_\infty\ell_e}{t}\begin{Bmatrix} 1 \\ 1 \end{Bmatrix} \tag{10.41b}$$

Equation 10.39 reduces to

$$-\sum_e \boldsymbol{\psi}^T(\mathbf{k}_T + \mathbf{h}_T)\mathbf{T}^e + \sum_e \boldsymbol{\psi}^T\mathbf{r}_\infty = 0$$

or   (10.42)

$$-\boldsymbol{\Psi}^{\mathrm{T}}(\mathbf{K}_T + \mathbf{H}_T) + \boldsymbol{\Psi}^{\mathrm{T}}\mathbf{R}_\infty = 0$$

which should hold for all $\boldsymbol{\Psi}$ satisfying $\Psi_1 = 0$.

Denoting $K_{ij} = (K_T + H_T)_{ij}$, we obtain

$$\begin{bmatrix} K_{22} & K_{23} & \cdots & K_{2L} \\ K_{32} & K_{33} & \cdots & K_{3L} \\ \cdot & \cdot & & \cdot \\ \cdot & \cdot & & \cdot \\ \cdot & \cdot & & \cdot \\ K_{L2} & L_{L3} & \cdots & K_{LL} \end{bmatrix} \begin{Bmatrix} T_2 \\ T_3 \\ \cdot \\ \cdot \\ \cdot \\ T_L \end{Bmatrix} = \begin{Bmatrix} \\ \\ \mathbf{R}_\infty \\ \\ \\ \end{Bmatrix} - \begin{Bmatrix} K_{21}T_0 \\ K_{31}T_0 \\ \cdot \\ \cdot \\ \cdot \\ K_{L1}T_0 \end{Bmatrix} \qquad (10.43)$$

which can be solved for $\mathbf{T}$. The above equations incorporate the elimination approach for handling the boundary condition $T = T_0$. Other types of boundary conditions as discussed for heat conduction can also be considered for fin problems.

Example 10.3

A metallic fin, with thermal conductivity $k = 360$ W/m.°C, 0.1 cm thick, and 10 cm long, extends from a plane wall whose temperature is 235°C. Determine the temperature distribution and amount of heat transferred from the fin to the air at 20°C with $h = 9$ W/m²°C. Take the width of fin to be 1 m.

Solution   Assume that the tip of the fin is insulated. Using a three-element finite element model (Fig. E10.3) and assembling $\mathbf{K}_T$, $\mathbf{H}_T$, $\mathbf{R}_\infty$ as given above, we find that Eq. 10.43 yields

$$\left[\frac{360}{3.33 \times 10^{-2}} \begin{bmatrix} 2 & -1 & 0 \\ -1 & 2 & -1 \\ 0 & -1 & 1 \end{bmatrix} + \frac{9 \times 3.33 \times 10^{-2}}{3 \times 10^{-3}} \begin{bmatrix} 4 & 1 & 0 \\ 1 & 4 & 1 \\ 0 & 1 & 2 \end{bmatrix} \right] \begin{Bmatrix} T_2 \\ T_3 \\ T_4 \end{Bmatrix}$$

$$= \frac{9 \times 20 \times 3.33 \times 10^{-2}}{10^{-3}} \begin{Bmatrix} 2 \\ 2 \\ 1 \end{Bmatrix} - \frac{360 \times 235}{3.33 \times 10^{-2}} \begin{Bmatrix} -1 \\ 0 \\ 0 \end{Bmatrix}$$

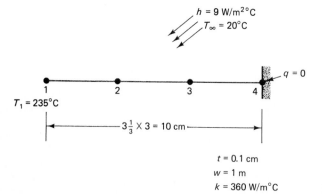

$h = 9$ W/m²°C

$T_\infty = 20$°C

$q = 0$

$T_1 = 235$°C

1   2   3   4

$3\frac{1}{3} \times 3 = 10$ cm

$t = 0.1$ cm

$w = 1$ m

$k = 360$ W/m°C   **Figure E10.3**

The solution is

$$[T_2, T_3, T_4] = [211.7, 197.0, 192.2]\,°C$$

The total heat loss in the fin can now be computed as

$$H = \sum_e H_e$$

The loss H_e in each element is

$$H_e = h(T_{av} - T_\infty)A_s$$

where $A_s = 2 \times (1 \times 0.0333)$ m^2, and T_{av} is the average temperature within the element. We obtain

$$H_{loss} = 333 \text{ W/m} \qquad \blacksquare$$

Two-Dimensional Steady-State Heat Conduction

Our objective here is to determine the temperature distribution $T(x, y)$ in a long, prismatic solid in which two-dimensional conduction effects are important. An example is a chimney of rectangular cross section, as shown in Fig. 10.7. Once the temperature distribution is known, the heat flux can be determined from Fourier's law.

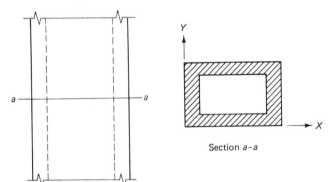

Section a–a

Figure 10.7   Two-dimensional model for heat conduction in a chimney.

Differential equation.   Consider a differential control volume in the body, as shown in Fig. 10.8. The control volume has a constant thickness τ in the z direction. The heat generation Q is denoted by Q (W/m^3). Since the heat rate (= heat flux × area) entering the control volume plus the heat rate generated equals the heat rate coming out, we have (Fig.10.8)

$$q_x \, dy\tau + q_y \, dx\tau + Q \, dx \, dy\tau = \left(q_x + \frac{\partial q_x}{\partial x} \, dx\right) dy\tau + \left(q_y + \frac{\partial q_y}{\partial y} \, dy\right) dx\tau$$

$$(10.44)$$

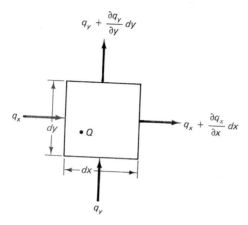

Figure 10.8 A differential control volume for heat transfer.

or, upon canceling terms,

$$\frac{\partial q_x}{\partial x} + \frac{\partial q_y}{\partial y} - Q = 0 \tag{10.45}$$

Substituting for $q_x = -k\,\partial T/\partial x$ and $q_y = -k\,\partial T/\partial y$ into the above, we get the **heat diffusion equation**

$$\frac{\partial}{\partial x}\left(k\frac{\partial T}{\partial x}\right) + \frac{\partial}{\partial y}\left(k\frac{\partial T}{\partial y}\right) + Q = 0 \tag{10.46}$$

We note that the partial differential equation above is a special case of the Helmholtz equation given in Eq. 10.1.

Boundary conditions.   The governing equation above has to be solved together with certain boundary conditions. These boundary conditions are of three types, as shown in Fig. 10.9: (1) specified temperature $T = T_0$ on S_T, (2) specified

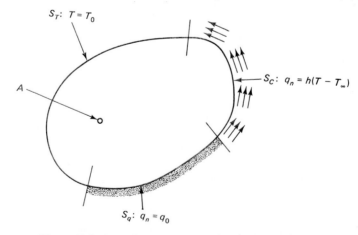

Figure 10.9 Boundary conditions for 2-D heat conduction.

heat flux $q_n = q_0$ on S_q, and (3) convection $q_n = h(T - T_\infty)$ on S_c. The interior of the body is denoted by A, and the boundary is denoted as $S = (S_T + S_q + S_c)$. Further, q_n is the heat flux normal to the boundary. The sign convention adopted here for specifying q_0 is that $q_0 > 0$ if heat is flowing out of the body, while $q_0 < 0$ if heat is flowing into the body.

The triangular element.   The triangular element (Fig. 10.10) will be used to solve the heat conduction problem. Extension to quadrilateral or other isoparametric elements follows in a similar manner as discussed earlier for stress analysis.

Consider a constant length of the body perpendicular to the x, y plane. The temperature field within an element is given by

$$T = N_1 T_1 + N_2 T_2 + N_3 T_3$$

or   (10.47)

$$T = \mathbf{N} \mathbf{T}^e$$

where $\mathbf{N} = [\xi, \ \eta, \ 1 - \xi - \eta]$ are the element shape functions and $\mathbf{T}^e = [T_1, T_2, T_3]^T$. Referring to Chapter 5, we also have

$$x = N_1 x_1 + N_2 x_2 + N_3 x_3$$
$$y = N_1 y_1 + N_2 y_2 + N_3 y_3$$ (10.48)

Further, the chain rule of differentiation yields

$$\frac{\partial T}{\partial \xi} = \frac{\partial T}{\partial x}\frac{\partial x}{\partial \xi} + \frac{\partial T}{\partial y}\frac{\partial y}{\partial \xi}$$

$$\frac{\partial T}{\partial \eta} = \frac{\partial T}{\partial x}\frac{\partial x}{\partial \eta} + \frac{\partial T}{\partial y}\frac{\partial y}{\partial \eta}$$ (10.49)

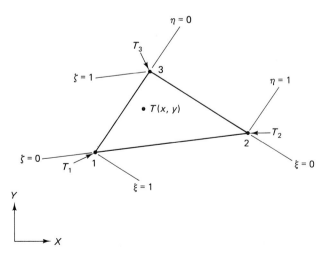

Figure 10.10   The linear triangular element for scalar field problems.

or

$$\left\{\begin{array}{c}\dfrac{\partial T}{\partial \xi}\\[2mm]\dfrac{\partial T}{\partial \eta}\end{array}\right\} = \mathbf{J}\left\{\begin{array}{c}\dfrac{\partial T}{\partial x}\\[2mm]\dfrac{\partial T}{\partial y}\end{array}\right\} \qquad (10.50)$$

Above, $\mathbf{J}$ is the Jacobian matrix given by

$$\mathbf{J} = \begin{bmatrix} x_{13} & y_{13} \\ x_{23} & y_{23} \end{bmatrix} \qquad (10.51)$$

where $x_{ij} = x_i - x_j$, $y_{ij} = y_i - y_j$, and $|\det \mathbf{J}| = 2A_e$, where A_e is the area of the triangle. Equation 10.50 yields

$$\left\{\begin{array}{c}\dfrac{\partial T}{\partial x}\\[2mm]\dfrac{\partial T}{\partial y}\end{array}\right\} = \mathbf{J}^{-1}\left\{\begin{array}{c}\dfrac{\partial T}{\partial \xi}\\[2mm]\dfrac{\partial T}{\partial \eta}\end{array}\right\} \qquad (10.52a)$$

$$= \frac{1}{\det \mathbf{J}}\begin{bmatrix} y_{23} & -y_{13} \\ -x_{23} & x_{13} \end{bmatrix}\begin{bmatrix} 1 & 0 & -1 \\ 0 & 1 & -1 \end{bmatrix}\mathbf{T}^e \qquad (10.52b)$$

which can be written as

$$\left\{\begin{array}{c}\dfrac{\partial T}{\partial x}\\[2mm]\dfrac{\partial T}{\partial y}\end{array}\right\} = \mathbf{B}_T\mathbf{T}^e \qquad (10.53)$$

where

$$\mathbf{B}_T = \frac{1}{\det \mathbf{J}}\begin{bmatrix} y_{23} & -y_{13} & (y_{13} - y_{23}) \\ -x_{23} & x_{13} & (x_{23} - x_{13}) \end{bmatrix} \qquad (10.54a)$$

$$= \frac{1}{\det \mathbf{J}}\begin{bmatrix} y_{23} & y_{31} & y_{12} \\ x_{32} & x_{13} & x_{21} \end{bmatrix} \qquad (10.54b)$$

Consequently,

$$\left(\frac{\partial T}{\partial x}\right)^2 + \left(\frac{\partial T}{\partial y}\right)^2 = \begin{bmatrix}\dfrac{\partial T}{\partial x} & \dfrac{\partial T}{\partial y}\end{bmatrix}\left\{\begin{array}{c}\dfrac{\partial T}{\partial x}\\[2mm]\dfrac{\partial T}{\partial y}\end{array}\right\} \qquad (10.55a)$$

$$= \mathbf{T}^{eT}\mathbf{B}_T^T\mathbf{B}_T\mathbf{T}^e \qquad (10.55b)$$

Functional approach for two-dimensional heat conduction.   The problem is to solve

$$\frac{\partial}{\partial x}\left(k\frac{\partial T}{\partial x}\right) + \frac{\partial}{\partial y}\left(k\frac{\partial T}{\partial y}\right) + Q = 0 \tag{10.56}$$

with the boundary conditions

$$T = T_0 \quad \text{on } S_T \qquad q_n = q_0 \quad \text{on } S_q \qquad q_n = h(T - T_\infty) \quad \text{on } S_c \tag{10.57}$$

The solution of the above problem is equivalent to minimizing the functional

$$\Pi_T = \frac{1}{2}\iint_A \left[k\left(\frac{\partial T}{\partial x}\right)^2 + k\left(\frac{\partial T}{\partial y}\right)^2 - 2QT\right] dA$$

$$+ \int_{S_q} q_0 T \, dS + \int_{S_c} \frac{1}{2}h(T - T_\infty)^2 \, dS \tag{10.58}$$

while satisfying $T = T_0$ on S_T.

In view of Eq. 10.55b, the first term in Π_T above is

$$\frac{1}{2}\iint_A \left[k\left(\frac{\partial T}{\partial x}\right)^2 + k\left(\frac{\partial T}{\partial y}\right)^2\right] dA = \sum_e \frac{1}{2}\int_e k\mathbf{T}^{e^T}\mathbf{B}_T^T\mathbf{B}_T\mathbf{T}^e \, dA$$

$$= \sum_e \frac{1}{2}\mathbf{T}^{e^T}k_e A_e \mathbf{B}_T^T \mathbf{B}_T \mathbf{T}^e \tag{10.59}$$

$$= \frac{1}{2}\mathbf{T}^T\mathbf{K}_T\mathbf{T}$$

where $\mathbf{K}_T = \sum_e \mathbf{k}_T$ is assembled in the usual manner using element connectivity matrix. The element conductivity matrix $\mathbf{k}_T$ is given by

$$\mathbf{k}_T = k_e A_e \mathbf{B}_T^T\mathbf{B}_T \tag{10.60}$$

where k_e is the thermal conductivity of element e.

The second (heat source) term in Π_T in Eq. 10.58 is now considered in detail. Three different distributions of the heat source Q within an element are now considered. First, assume that $Q = Q_e$ is **constant** within the element. Then,

$$-\int_e QT \, dA = -\sum_e \left(Q_e \int_e \mathbf{N} \, dA\right)\mathbf{T}^e$$

$$= -\sum_e \mathbf{r}_Q^T\mathbf{T}^e \tag{10.61}$$

Since $\int_e N_i \, dA = A_e/3$ (Fig. 5.6), the element heat rate vector is

$$\mathbf{r}_Q = \frac{Q_e A_e}{3}[1, 1, 1]^T \tag{10.62}$$

which simply states that the heat rate is distributed equally at the three nodes.

Second, consider the case when Q is **linearly distributed**, with $\mathbf{Q}^e = [Q_1, Q_2, Q_3]^T$ being the nodal values. Then we can write

$$Q(\xi, \eta) = \mathbf{N}\mathbf{Q}^e \tag{10.63}$$

which when substituted into $-\Sigma_e \int_e QT \, dA$ yields

$$\mathbf{r}_Q = \frac{A_e}{12}[(2Q_1 + Q_2 + Q_3), (Q_1 + 2Q_2 + Q_3), (Q_1 + Q_2 + 2Q_3)]^T \tag{10.64}$$

Third, consider the case when the heat source is the result of heat generated by electrical wiring or hot water pipes running along the length of the body. In the x, y plane these have to be treated as **point sources**. Let Q_0 be the magnitude of a point source located at (ξ_0, η_0) within the element having units of watts per unit length. The easiest way of incorporating such point sources is to have a node of the finite element mesh at the location; Q_0 is then simply added to the node in forming the right-hand side load vector. However, if the point source is located inside an element e, at the location (ξ_0, η_0), then the nodal contributions are obtained as

$$\begin{aligned} \int_e QT \, dA &= Q_0 T_0 \\ &= Q_0 \mathbf{N}(\xi_0, \eta_0)\mathbf{T}^e \\ &= \mathbf{r}_Q^T \mathbf{T}^e \end{aligned} \tag{10.65}$$

where

$$\mathbf{r}_Q = Q_0 \mathbf{N}^T(\xi_0, \eta_0) \tag{10.66a}$$

or

$$\mathbf{r}_Q = Q_0[\xi_0, \eta_0, 1 - \xi_0, -\eta_0]^T \tag{10.66b}$$

Next, the flux boundary term in Π_T is

$$\int_{S_q} q_0 T \, ds = \sum_e \left(\int_e q_0 \mathbf{N} \, dS_q\right)\mathbf{T}^e \tag{10.67}$$

Assume that the flux boundary is specified normal to edge 2–3 of the element, as shown in Fig. 10.11. Along this edge $\mathbf{N} = [0, \eta, 1 - \eta)$ and $dS = \ell_{2-3} \, d\eta$, where ℓ_{2-3} is the length of edge 2–3. Thus,

$$\begin{aligned} \sum_e \left(\int_e q_0 \mathbf{N} \, dS_q\right)\mathbf{T}^e &= \sum_e \left[q_0 \ell_{2-3} \int_0^1 \left[0, \eta, 1 - \eta\right] d\eta\right]\mathbf{T}^e \\ &= \sum_e \mathbf{r}_q^T \mathbf{T}^e \end{aligned} \tag{10.68}$$

where

$$\mathbf{r}_q = \frac{q_0 \ell_{2-3}}{2}[0, 1, 1]^T \tag{10.69}$$

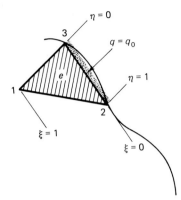

Figure 10.11 Specified heat flux boundary conduction on edge 2–3 of a triangular element.

Similar expressions can be given if heat flux is specified on edges 1–2 or 3–1.
The convection boundary terms in Π_T are now discussed. We have

$$
\frac{1}{2} \int_{S_c} hT^2 \, dS = \sum_e \frac{1}{2} \mathbf{T}^{e\mathrm{T}} \left(\int_e h \mathbf{N}^{\mathrm{T}} \mathbf{N} \, dS_c \right) \mathbf{T}^e
$$
$$
\equiv \sum_e \frac{1}{2} \mathbf{T}^{e\mathrm{T}} \mathbf{h}_T \mathbf{T}^e \tag{10.70}
$$

We see that the above expression involving $\mathbf{h}_T$ is quadratic in $\mathbf{T}^e$ and, hence, is added to $\mathbf{k}_T$. This situation is analogous to a foundation stiffness in structural problems. Again, if we assume that edge 2–3 is the convection surface of an element, then $\mathbf{N} = [0, \eta, 1 - \eta]$, and

$$
\mathbf{h}_T = \frac{h\ell_{2-3}}{6} \begin{bmatrix} 0 & 0 & 0 \\ 0 & 2 & 1 \\ 0 & 1 & 2 \end{bmatrix} \tag{10.71}
$$

The last convection term is

$$
-\int_{S_c} hTT_\infty \, dS = -\sum_e \left(\int_e hT_\infty \mathbf{N} \, dS_c \right) \mathbf{T}^e
$$
$$
= -\sum_e \mathbf{r}_\infty^{\mathrm{T}} \mathbf{T}^e \tag{10.72}
$$

where

$$
\mathbf{r}_\infty = \frac{hT_\infty \ell_{2-3}}{2} [0, 1, 1]^{\mathrm{T}} \tag{10.73}
$$

The functional Π_T in Eq. 10.58 can now be written in the form

$$
\Pi_T = \tfrac{1}{2} \mathbf{T}^{\mathrm{T}} \mathbf{K} \mathbf{T} - \mathbf{R}^{\mathrm{T}} \mathbf{T} \tag{10.74}
$$

where

$$\mathbf{K} = \sum_e \mathbf{k}_T + \sum_e \mathbf{h}_T \tag{10.75}$$

and

$$\mathbf{R} = \sum_e (\mathbf{r}_Q - \mathbf{r}_q + \mathbf{r}_\infty) \tag{10.76}$$

are assembled in the usual manner using element connectivity. The minimization of Π_T is to be carried out while satisfying $T = T_0$ at all nodes on S_T. Either the penalty or elimination approach can be used to account for these specified temperatures.

Galerkin approach.   Consider the heat conduction problem

$$\frac{\partial}{\partial x}\left(k\frac{\partial T}{\partial x}\right) + \frac{\partial}{\partial y}\left(k\frac{\partial T}{\partial y}\right) + Q = 0 \tag{10.77}$$

with the boundary conditions

$$T = T_0 \quad \text{on } S_T \qquad q_n = q_0 \quad \text{on } S_q \qquad q_n = h(T - T_\infty) \quad \text{on } S_c \tag{10.78}$$

In Galerkin's approach we seek an approximate solution T such that

$$\iint_A \phi\left[\frac{\partial}{\partial x}\left(k\frac{\partial T}{\partial x}\right) + \frac{\partial}{\partial y}\left(k\frac{\partial T}{\partial y}\right)\right] dA + \iint_A \phi Q \, dA = 0 \tag{10.79}$$

for every $\phi(x, y)$ constructed from the same basis functions as those used for T and satisfying $\phi = 0$ on S_T. Noting that

$$\phi\frac{\partial}{\partial x}\left(k\frac{\partial T}{\partial x}\right) = \frac{\partial}{\partial x}\left(\phi k\frac{\partial T}{\partial x}\right) - k\frac{\partial \phi}{\partial x}\frac{\partial T}{\partial x}$$

We find that Eq. 10.79 gives

$$\iint_A \left\{\left[\frac{\partial}{\partial x}\left(\phi k\frac{\partial T}{\partial x}\right) + \frac{\partial}{\partial y}\left(\phi k\frac{\partial T}{\partial y}\right)\right] - \left[k\frac{\partial \phi}{\partial x}\frac{\partial T}{\partial x} + k\frac{\partial \phi}{\partial y}\frac{\partial T}{\partial y}\right]\right\} dA$$

$$+ \iint_A \phi Q \, dA = 0 \tag{10.80}$$

From the notation $q_x = -k(\partial T/\partial x)$ and $q_y = -k(\partial T/\partial y)$, and the divergence theorem, the first term in Eq. 10.80 above is

$$-\iint_A \left[\frac{\partial}{\partial x}(\phi q_x) + \frac{\partial}{\partial y}(\phi q_y)\right] dA = -\int_S \phi[q_x n_x + q_y n_y] \, dS$$

$$= -\int_S \phi q_n \, dS \tag{10.81}$$

where n_x, n_y are the direction cosines of the unit normal $\mathbf{n}$ to the boundary and $q_n = q_x n_x + q_y n_y = \mathbf{q} \cdot \mathbf{n}$ is the normal heat flow along the unit outward normal, which is specified by boundary conditions. Since $S = S_T + S_q + S_c$, $\phi = 0$ on S_T, $q_n = q_0$ on S_q, and $q_n = h(T - T_\infty)$ on S_c, Eq. 10.80 reduces to

$$
-\int_{S_q} \phi q_0 \, dS - \int_{S_c} \phi h(T - T_\infty) \, dS - \iint_A \left(k\frac{\partial \phi}{\partial x}\frac{\partial T}{\partial x} + k\frac{\partial \phi}{\partial y}\frac{\partial T}{\partial y} \right) dA
$$

$$
+ \iint_A \phi Q \, dA = 0 \tag{10.82}
$$

Now, we introduce the isoparametric relations for the triangular element such as $T = \mathbf{N}\mathbf{T}^e$, given in Eqs. 10.47–10.55. Further, we denote the global virtual temperature vector as $\boldsymbol{\Psi}$ whose dimension equals number of nodes in the finite element model. The virtual temperature distribution within each element is interpolated as

$$
\phi = \mathbf{N}\boldsymbol{\psi} \tag{10.83a}
$$

Moreover, just as $[\partial T/\partial x \quad \partial T/\partial y]^{\mathrm{T}} = \mathbf{B}_T \mathbf{T}^e$, we have

$$
\left[\frac{\partial \phi}{\partial x} \quad \frac{\partial \phi}{\partial y} \right]^{\mathrm{T}} = \mathbf{B}_T \boldsymbol{\psi} \tag{10.83b}
$$

Now, consider the first term in Eq. 10.82:

$$
\int_{S_q} \phi q_0 \, dS = \sum_e \boldsymbol{\psi}^{\mathrm{T}} q_0 \mathbf{N}^{\mathrm{T}} \, dS \tag{10.84}
$$

If edge 2–3 is on the boundary (Fig. 10.11), we have $\mathbf{N} = [0, \eta, 1 - \eta]$, $dS = \ell_{2-3} \, d\eta$, and

$$
\int_{S_q} \phi q_0 \, dS = \sum_e \boldsymbol{\psi}^{\mathrm{T}} q_0 \ell_{2-3} \int_0^1 \mathbf{N}^{\mathrm{T}} \, d\eta \tag{10.85a}
$$

$$
= \sum_e \boldsymbol{\psi}^{\mathrm{T}} \mathbf{r}_q \tag{10.85b}
$$

where

$$
\mathbf{r}_q = \frac{q_0 \ell_{2-3}}{2} [0 \quad 1 \quad 1]^{\mathrm{T}} \tag{10.86}
$$

Next, consider

$$
\int_{S_c} \phi h(T - T_\infty) \, dS = \int_{S_c} \phi h T \, dS - \int_{S_c} \phi h T_\infty \, dS \tag{10.87a}
$$

If edge 2–3 is the convection edge of the element, then

$$
\int_{S_c} \phi h(T - T_\infty) \, dS = \sum_e \boldsymbol{\psi}^{\mathrm{T}} \left[h\ell_{2-3} \int_0^1 \mathbf{N}^{\mathrm{T}} \mathbf{N} \, d\eta \right] \mathbf{T}^e - \sum_e \boldsymbol{\psi}^{\mathrm{T}} h T_\infty \ell_{2-3} \int_0^1 \mathbf{N}^{\mathrm{T}} \, d\eta
$$

$$
= \sum_e \boldsymbol{\psi}^{\mathrm{T}} \mathbf{h}_T \mathbf{T}^e - \sum_e \boldsymbol{\psi}^{\mathrm{T}} \mathbf{r}_\infty \tag{10.87b}
$$

Substituting for $\mathbf{N} = [0, \eta, 1 - \eta]$, we get

$$\mathbf{h}_T = \frac{h\ell_{2-3}}{6} \begin{bmatrix} 0 & 0 & 0 \\ 0 & 2 & 1 \\ 0 & 1 & 2 \end{bmatrix} \tag{10.88}$$

$$\mathbf{r}_\infty = \frac{hT_\infty \ell_{2-3}}{2} [0 \quad 1 \quad 1]^T \tag{10.89}$$

Next,

$$\iint_A k\left(\frac{\partial\phi}{\partial x}\frac{\partial T}{\partial x} + \frac{\partial\phi}{\partial y}\frac{\partial T}{\partial y}\right) dA = \iint_A k\left[\frac{\partial\phi}{\partial x}\frac{\partial\phi}{\partial y}\right]\begin{Bmatrix} \dfrac{\partial T}{\partial x} \\[2mm] \dfrac{\partial T}{\partial y} \end{Bmatrix} dA \tag{10.90a}$$

$$= \sum_e \boldsymbol{\psi}^T \left[k_e \int_e \mathbf{B}_T^T \mathbf{B}_T \, dA \right] \mathbf{T}^e \tag{10.90b}$$

$$= \sum_e \boldsymbol{\psi}^T \mathbf{k}_T \mathbf{T}^e \tag{10.90c}$$

where

$$\mathbf{k}_T = k_e A_e \mathbf{B}_T^T \mathbf{B}_T \tag{10.91}$$

Finally, if $Q = Q_e$ is constant within the element,

$$\iint_A \phi Q \, dA = \sum_e \boldsymbol{\psi}^T Q_e \int_e \mathbf{N} \, dA = \sum_e \boldsymbol{\psi}^T \mathbf{r}_Q$$

where

$$\mathbf{r}_Q = \frac{Q_e A_e}{3}[1 \quad 1 \quad 1]^T \tag{10.92}$$

Other distributions of Q within the element are considered in Eqs. 10.64 and 10.66. Thus, Eq. 10.82 is of the form

$$-\sum_e \boldsymbol{\psi}^T \mathbf{r}_q - \sum_e \boldsymbol{\psi}^T \mathbf{h}_T \mathbf{T}^e + \sum_e \boldsymbol{\psi}^T \mathbf{r}_\infty - \sum_e \boldsymbol{\psi}^T \mathbf{k}_T \mathbf{T}^e + \sum_e \boldsymbol{\psi}^T \mathbf{r}_Q = 0 \tag{10.93}$$

or

$$\boldsymbol{\Psi}^T(\mathbf{R}_\infty - \mathbf{R}_q + \mathbf{R}_Q) - \boldsymbol{\Psi}^T(\mathbf{H}_T + \mathbf{K}_T)\mathbf{T} = 0 \tag{10.94}$$

which is to hold for all $\boldsymbol{\Psi}$ satisfying $\boldsymbol{\Psi} = \mathbf{0}$ at nodes on S_T. We thus obtain

$$\mathbf{K}^E \mathbf{T}^E = \mathbf{R}^E \tag{10.95}$$

where $\mathbf{K} = \Sigma_e\,(\mathbf{k}_T + \mathbf{h}_T)$, $\mathbf{R} = \Sigma_e\,(\mathbf{r}_\infty - \mathbf{r}_q + \mathbf{r}_Q)$, and superscript E represents the familiar modifications made to $\mathbf{K}$ and $\mathbf{R}$ to handle $T = T_0$ on S_T by the elimination approach.

Example 10.4

A long bar of rectangular cross section, having thermal conductivity of 1.5 W/m °C is subjected to the boundary conditions shown in Fig. E10.4a. Two opposite sides are maintained at a uniform temperature of 180°C; one side is insulated, and the remaining side is subjected to a convection process with $T_\infty = 25$°C and $h = 50$ W/m² °C. Determine the temperature distribution in the bar.

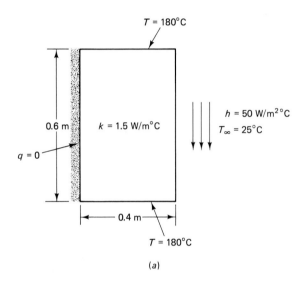

(a)

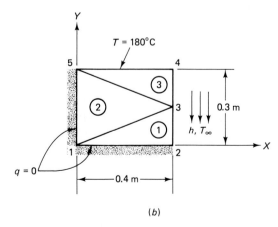

(b)   **Figure E10.4**

Solution A five-node, three-element finite element model of the problem is shown in Fig. E10.4*b*, where symmetry about the horizontal axis is used. Note that the line of symmetry is shown as insulated, since no heat can flow across it.

The element matrices are developed as follows. The element connectivity is defined as

Element	1	2	3	← local
1	1	2	3	↑
2	5	1	3	global
3	5	4	3	↓

We have

$$\mathbf{B}_T = \frac{1}{\det \mathbf{J}}\begin{bmatrix} y_{23} & y_{31} & y_{12} \\ x_{32} & x_{13} & x_{21} \end{bmatrix}$$

For each element,

$$\mathbf{B}_T^{(1)} = \frac{1}{0.06}\begin{bmatrix} -0.15 & 0.15 & 0 \\ 0 & -0.4 & 0.4 \end{bmatrix}$$

$$\mathbf{B}_T^{(2)} = \frac{1}{0.12}\begin{bmatrix} -0.15 & -0.15 & 0.3 \\ 0.4 & -0.4 & 0 \end{bmatrix}$$

$$\mathbf{B}_T^{(3)} = \frac{-1}{0.06}\begin{bmatrix} 0.15 & -0.15 & 0 \\ 0 & -0.4 & 0.4 \end{bmatrix}$$

Then, $\mathbf{k}_T = kA_e\mathbf{B}_T^\mathsf{T}\mathbf{B}_T$ yields

$$\mathbf{k}_T^{(1)} = (1.5)(0.03)\mathbf{B}_T^{(1)^\mathsf{T}}\mathbf{B}_T^{(1)}$$

$$\begin{array}{ccc} \quad 1 & 2 & 3 \end{array}$$

$$= \begin{bmatrix} 0.28125 & -0.28125 & 0 \\ -0.28125 & 2.28125 & -2.0 \\ 0 & -2.0 & 2.0 \end{bmatrix}$$

$$\begin{array}{ccc} \quad 5 & 1 & 3 \end{array}$$

$$\mathbf{k}_T^{(2)} = \begin{bmatrix} 1.14 & -0.86 & -0.28125 \\ -0.86 & 1.14 & -0.28125 \\ -0.28125 & -0.28125 & 0.5625 \end{bmatrix}$$

$$\begin{array}{ccc} \quad 5 & 4 & 3 \end{array}$$

$$\mathbf{k}_T^{(3)} = \begin{bmatrix} 0.28125 & -0.28125 & 0 \\ -0.28125 & 2.28125 & -2.0 \\ 0 & -2.0 & 2.0 \end{bmatrix}$$

Now the matrices $\mathbf{h}_T$ for elements with convection edges are developed. Since both elements 1 and 3 have edges 2–3 (in local node numbers) as convection edges, the formula

$$\mathbf{h}_T = \frac{h\ell_{2-3}}{6} \begin{bmatrix} 0 & 0 & 0 \\ 0 & 2 & 1 \\ 0 & 1 & 2 \end{bmatrix}$$

can be used, resulting in

$$\mathbf{h}_T^{(1)} = \begin{array}{ccc} 1 & 2 & 3 \end{array} \begin{bmatrix} 0 & 0 & 0 \\ 0 & 2.5 & 1.25 \\ 0 & 1.25 & 2.5 \end{bmatrix} \qquad \mathbf{h}_T^{(2)} = \begin{array}{ccc} 5 & 4 & 3 \end{array} \begin{bmatrix} 0 & 0 & 0 \\ 0 & 2.5 & 1.25 \\ 0 & 1.25 & 2.5 \end{bmatrix}$$

The matrix $\mathbf{K} = \Sigma \, (\mathbf{k}_T + \mathbf{h}_T)$ is now assembled. The elimination approach for handling the boundary conditions $T = 180°C$ at nodes 4 and 5 results in striking out these rows and columns. However, these fourth and fifth rows are used subsequently for modifying the $\mathbf{R}$ vector. The result is

$$\mathbf{K} = \begin{array}{ccc} 1 & 2 & 3 \end{array} \begin{bmatrix} 1.42125 & -0.28125 & -0.28125 \\ -0.28125 & 4.78125 & -0.75 \\ -0.28125 & -0.75 & 9.5625 \end{bmatrix}$$

Now the heat rate vector $\mathbf{R}$ is assembled from element convection contributions. The formula

$$\mathbf{r}_\infty = \frac{hT_\infty\ell_{2-3}}{2} \begin{bmatrix} 0 & 1 & 1 \end{bmatrix}$$

results in

$$\mathbf{r}_\infty^{(1)} = \frac{(50)(25)(0.15)}{2} \begin{array}{ccc} 1 & 2 & 3 \end{array} \begin{bmatrix} 0 & 1 & 1 \end{bmatrix}$$

and

$$\mathbf{r}_\infty^{(3)} = \frac{(50)(25)(0.15)}{2} \begin{array}{ccc} 5 & 4 & 3 \end{array} \begin{bmatrix} 0 & 1 & 1 \end{bmatrix}$$

Thus,

$$\begin{array}{ccc} 1 & 2 & 3 \end{array}$$
$$\mathbf{R} = 93.75 \begin{bmatrix} 0 & 1 & 2 \end{bmatrix}^T$$

In the elimination approach, $\mathbf{R}$ gets modified according to Eq. 3.70. Solution of $\mathbf{KT} = \mathbf{R}$ then yields

$$[T_1, T_2, T_3] = [124.5, 34.0, 45.4]°C$$

Note. A large temperature gradient exists along the line connecting nodes 2 and 4. This is because node 4 is maintained at 180°C while node 2 has a temperature close to the ambient temperature of $T_\infty = 25°C$ because of the relatively large value of h. This fact implies that our finite element model should capture this large temperature gradient by having sufficient number of nodes along line 2–4. In fact, a model with only two nodes (as opposed to three as used here) will lead to an incorrect solution for the temperatures.   ■

It is also noted that a thermal stress analysis can now be performed once the temperature distribution is known, as discussed in Chapter 5.

10.3 TORSION

Consider a prismatic rod of arbitrary cross-sectional shape which is subjected to a twisting moment M as shown in Fig. 10.12. The problem is to determine shearing stresses τ_{xz}, τ_{yz} (Fig. 10.13) and the angle of twist per unit length, α. It can be shown that the solution of such problems, with simply connected cross sections, reduces to solving the two-dimensional equation

$$\frac{\partial^2 \theta}{\partial x^2} + \frac{\partial^2 \theta}{\partial y^2} + 2 = 0 \qquad \text{in } A \tag{10.96}$$

$$\theta = 0 \qquad \text{on } S \tag{10.97}$$

where A is interior and S is the boundary of the cross section. Again, we note that Eq. 10.96 is a special case of Helmholtz's equations given in Eq. 10.1. In Eq. 10.97, θ is called the **stress function**, since once θ is known, then shearing stresses are obtained as

$$\tau_{xz} = G\alpha \frac{\partial \theta}{\partial y} \qquad \tau_{yz} = -G\alpha \frac{\partial \theta}{\partial x} \tag{10.98}$$

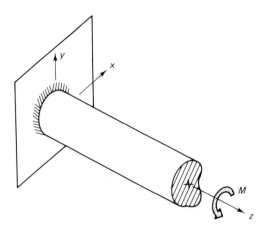

Figure 10.12   A rod of arbitrary cross section subjected to a torque.

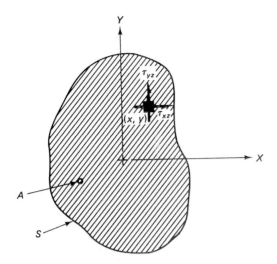

Figure 10.13   Shearing stresses in torsion.

with α determined from

$$M = 2G\alpha \iint_A \theta \, dA \qquad (10.99)$$

where G is the shear modulus of the material. The finite element method for solving Eqs. 10.96 and 10.97 will now be given.

Triangular Element

The stress function θ within a triangular element is interpolated as

$$\theta = \mathbf{N}\boldsymbol{\theta}^e \qquad (10.100)$$

where $\mathbf{N} = [\xi, \eta, 1 - \xi - \eta]$ are the usual shape functions, and $\boldsymbol{\theta}^e = [\theta_1, \theta_2, \theta_3]^T$ are the nodal values of θ. Furthermore, we have the isoparametric relations (Chapter 5):

$$x = N_1 x_1 + N_2 x_2 + N_3 x_3$$
$$y = N_1 y_1 + N_2 y_2 + N_3 y_3 \qquad (10.101)$$

$$\begin{Bmatrix} \dfrac{\partial \theta}{\partial \xi} \\[2mm] \dfrac{\partial \theta}{\partial \eta} \end{Bmatrix} = \begin{bmatrix} \dfrac{\partial x}{\partial \xi} & \dfrac{\partial y}{\partial \xi} \\[2mm] \dfrac{\partial x}{\partial \eta} & \dfrac{\partial y}{\partial \eta} \end{bmatrix} \begin{Bmatrix} \dfrac{\partial \theta}{\partial x} \\[2mm] \dfrac{\partial \theta}{\partial y} \end{Bmatrix} \qquad (10.102)$$

or

$$\begin{bmatrix} \dfrac{\partial \theta}{\partial \xi} & \dfrac{\partial \theta}{\partial \eta} \end{bmatrix}^T = \mathbf{J} \begin{bmatrix} \dfrac{\partial \theta}{\partial x} & \dfrac{\partial \theta}{\partial y} \end{bmatrix}^T \qquad (10.103)$$

where the Jacobean matrix $\mathbf{J}$ is given by

$$\mathbf{J} = \begin{bmatrix} x_{13} & y_{13} \\ x_{23} & y_{23} \end{bmatrix} \tag{10.104}$$

with $x_{ij} = x_i - x_j$, $y_{ij} = y_i - y_j$, and $|\det \mathbf{J}| = 2A_e$. Equations 10.100, 10.103, and 10.104 yield

$$\begin{bmatrix} \dfrac{\partial \theta}{\partial x} & \dfrac{\partial \theta}{\partial y} \end{bmatrix}^{\mathrm{T}} = \mathbf{B}\boldsymbol{\theta}^e \tag{10.105a}$$

or

$$\begin{bmatrix} -\tau_{yz} & \tau_{xz} \end{bmatrix}^{\mathrm{T}} = G\alpha\, \mathbf{B}\boldsymbol{\theta}^e \tag{10.105b}$$

where

$$\mathbf{B} = \frac{1}{\det \mathbf{J}} \begin{bmatrix} y_{23} & y_{31} & y_{12} \\ x_{32} & x_{13} & x_{21} \end{bmatrix} \tag{10.106}$$

The fact that identical relations also apply to the heat conduction problem in the previous section show the similarity of treating all field problems by the finite element method.

Complementary Potential Energy Approach

The solution of Eq. 10.96 is equivalent to minimizing the complementary potential energy Π in the rod. For unit length of the rod,

$$\Pi = G\alpha^2 \iint_A \left\{ \frac{1}{2} \left[\left(\frac{\partial \theta}{\partial x} \right)^2 + \left(\frac{\partial \theta}{\partial y} \right)^2 \right] - 2\theta \right\} dA \tag{10.107}$$

The element stiffness matrix and load vectors can now be derived from Π. The constant $G\alpha^2$ multiplying Π may be disregarded.

Using Eq. 10.105a, we have

$$\left(\frac{\partial \theta}{\partial x} \right)^2 + \left(\frac{\partial \theta}{\partial y} \right)^2 = \begin{bmatrix} \dfrac{\partial \theta}{\partial x} & \dfrac{\partial \theta}{\partial y} \end{bmatrix} \begin{Bmatrix} \dfrac{\partial \theta}{\partial x} \\ \dfrac{\partial \theta}{\partial y} \end{Bmatrix} \tag{10.108}$$

$$= \boldsymbol{\theta}^{e\mathrm{T}} \mathbf{B}^{\mathrm{T}} \mathbf{B} \boldsymbol{\theta}^e$$

Thus, the strain energy term in Π becomes

$$\iint_A \frac{1}{2} \left[\left(\frac{\partial \theta}{\partial x} \right)^2 + \left(\frac{\partial \theta}{\partial y} \right)^2 \right] dA = \sum_e \frac{1}{2} \boldsymbol{\theta}^{e\mathrm{T}} \mathbf{k} \boldsymbol{\theta}^e \tag{10.109}$$

where

$$\mathbf{k} = A_e \mathbf{B}^{\mathrm{T}} \mathbf{B} \tag{10.110}$$

The load term becomes

$$\iint_A 2\theta \, dA = \sum_e \left(2 \int_e \mathbf{N} \, dA \right) \mathbf{\theta}^e \tag{10.111}$$

Since $\int_e N_i \, dA = A_e/3$ (Fig. 5.6), we have

$$\iint_A 2\theta \, dA = \sum_e \mathbf{\theta}^{e\mathrm{T}} \mathbf{f} \tag{10.112}$$

where

$$\mathbf{f} = \frac{2A_e}{3} [1 \quad 1 \quad 1]^{\mathrm{T}} \tag{10.113}$$

The expression for Π is now given by

$$\Pi = \frac{1}{2} \mathbf{\Theta}^{\mathrm{T}} \mathbf{K} \mathbf{\Theta} - \mathbf{\Theta}^{\mathrm{T}} \mathbf{F} \tag{10.114}$$

where $\mathbf{K} = \sum_e \mathbf{k}$, $\mathbf{F} = \sum_e \mathbf{f}$ are assembled in the usual manner using element connectivity. The minimization of Π is subject to the boundary conditions $\Psi_i = 0$ for every node i on the boundary. The resulting equations are

$$\mathbf{K} \mathbf{\Theta} = \mathbf{F} \tag{10.115}$$

where $\mathbf{K}, \mathbf{F}$ are then modified to account for boundary conditions, either by the penalty or elimination approaches. Once $\mathbf{\Theta}$ is known, we can determine the shearing stresses from Eq. 10.105b.

Galerkin Approach

The problem in Eqs. 10.96–10.97 will now be solved using Galerkin's approach. The problem is to find the approximate solution θ such that

$$\iint_A \phi \left(\frac{\partial^2 \theta}{\partial x^2} + \frac{\partial^2 \theta}{\partial y^2} + 2 \right) dA = 0 \tag{10.116}$$

for every $\phi(x, y)$ constructed from the same basis as θ and satisfying $\phi = 0$ on S. Since

$$\phi \frac{\partial^2 \theta}{\partial x^2} = \frac{\partial}{\partial x}\left(\phi \frac{\partial \theta}{\partial x} \right) - \frac{\partial \phi}{\partial x}\frac{\partial \theta}{\partial x}$$

we have

$$\iint_A \left[\frac{\partial}{\partial x}\left(\phi \frac{\partial \theta}{\partial x} \right) + \frac{\partial}{\partial y}\left(\phi \frac{\partial \theta}{\partial y} \right) \right] dA - \iint_A \left(\frac{\partial \phi}{\partial x}\frac{\partial \theta}{\partial x} + \frac{\partial \phi}{\partial y}\frac{\partial \theta}{\partial y} \right) dA$$
$$+ \iint_A 2\phi \, dA = 0 \tag{10.117}$$

Using the divergence theorem, the first term in the above expression reduces to

$$\iint_A \left[\frac{\partial}{\partial x}\left(\phi \frac{\partial \theta}{\partial x} \right) + \frac{\partial}{\partial y}\left(\phi \frac{\partial \theta}{\partial y} \right) \right] dA = \int_S \phi \left(\frac{\partial \theta}{\partial x} n_x + \frac{\partial \theta}{\partial y} n_y \right) dS$$
$$= 0$$

(10.118)

where the right side is equated to zero owing to the boundary condition $\phi = 0$ on S. Equation 10.117 becomes

$$\iint_A \left[\frac{\partial \phi}{\partial x}\frac{\partial \theta}{\partial x} + \frac{\partial \phi}{\partial y}\frac{\partial \theta}{\partial y} \right] dA - \iint_A 2\phi \, dA = 0$$

(10.119)

Now, we introduce the isoparametric relations $\theta = \mathbf{N}\boldsymbol{\theta}^e$, etc., as given in Eqs. 10.100–10.106. Further, we denote the global virtual stress function vector as $\boldsymbol{\Psi}$ whose dimension equals number of nodes in the finite element model. The virtual stress function within each element is interpolated as

$$\phi = \mathbf{N}\boldsymbol{\psi}$$

(10.120)

Moreover, we have

$$\left[\frac{\partial \phi}{\partial x} \quad \frac{\partial \phi}{\partial y} \right]^T = \mathbf{B}\boldsymbol{\psi}$$

(10.121)

Substituting these into Eq. 10.119, noting that

$$\left(\frac{\partial \phi}{\partial x}\frac{\partial \theta}{\partial x} + \frac{\partial \phi}{\partial y}\frac{\partial \theta}{\partial y} \right) = \left(\frac{\partial \phi}{\partial x} \quad \frac{\partial \phi}{\partial y} \right) \begin{Bmatrix} \dfrac{\partial \theta}{\partial x} \\[2mm] \dfrac{\partial \theta}{\partial y} \end{Bmatrix}$$

we get

$$\sum_e \boldsymbol{\psi}^T \mathbf{k}\boldsymbol{\theta}^e - \sum_e \boldsymbol{\psi}^T \mathbf{f} = 0$$

(10.122)

where $\mathbf{k} = A_e \mathbf{B}^T \mathbf{B}$, $\mathbf{f} = (2A_e/3)[1 \quad 1 \quad 1]^T$. Equation 10.122 can be written as

$$\boldsymbol{\Psi}^T(\mathbf{K}\boldsymbol{\Theta} - \mathbf{F}) = 0$$

(10.123)

which should hold for all $\boldsymbol{\Psi}$ satisfying $\Psi_i = 0$ at nodes i on the boundary. We thus have

$$\mathbf{K}\boldsymbol{\Theta} = \mathbf{F}$$

(10.124)

where rows and columns of $\mathbf{K}$ and $\mathbf{F}$ that correspond to boundary nodes have been deleted.

Example 10.5

Consider the shaft with a rectangular cross section shown in Fig. E10.5a. Determine, in terms of M and G, the angle of twist per unit length.

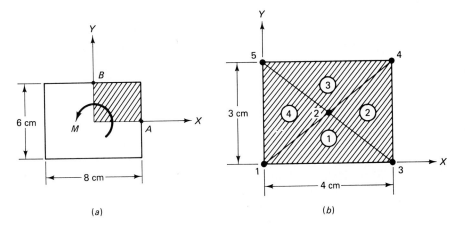

Figure E10.5

Solution   A finite element model of a quadrant of this cross section is shown in Fig. E10.5*b*. We define the element connectivity as

Element	1	2	3
1	1	3	2
2	3	4	2
3	4	5	2
4	5	1	2

Using the relations

$$\mathbf{B} = \frac{1}{\det \mathbf{J}} \begin{bmatrix} y_{23} & y_{31} & y_{12} \\ x_{32} & x_{13} & x_{21} \end{bmatrix}$$

and

$$\mathbf{k} = A_e \mathbf{B}^{\mathsf{T}} \mathbf{B}$$

we get

$$\mathbf{B}^{(1)} = \frac{1}{6} \begin{bmatrix} -1.5 & 1.5 & 0 \\ -2 & -2 & 4 \end{bmatrix} \qquad \mathbf{k}^{(1)} = \frac{1}{2} \begin{array}{c} \\ \end{array} \begin{matrix} & 1 & 2 & 3 \\ \\ \\ \end{matrix}$$

$$\mathbf{k}^{(1)} = \frac{1}{2} \begin{bmatrix} 1.042 & 0.292 & -1.333 \\ & 1.042 & -1.333 \\ \text{Symmetric} & & 2.667 \end{bmatrix} \begin{matrix} 1 \\ 2 \\ 3 \end{matrix}$$

Similarly,

$$\mathbf{k}^{(2)} = \frac{1}{2} \begin{bmatrix} 1.042 & -0.292 & -0.75 \\ & 1.042 & -0.75 \\ \text{Symmetric} & & 1.5 \end{bmatrix} \begin{matrix} 3 \\ 4 \\ 2 \end{matrix}$$

$$\mathbf{k}^{(3)} = \frac{1}{2} \begin{bmatrix} \overset{4}{1.042} & \overset{5}{0.292} & \overset{2}{-1.333} \\ & 1.042 & -1.333 \\ \text{Symmetric} & & 2.667 \end{bmatrix}$$

$$\mathbf{k}^{(4)} = \frac{1}{2} \begin{bmatrix} \overset{5}{1.042} & \overset{1}{-0.292} & \overset{2}{-0.75} \\ & 1.042 & -0.75 \\ \text{Symmetric} & & 1.5 \end{bmatrix}$$

Similarly, the element load vector $\mathbf{f} = (2A_e/3)[1, 1, 1]^T$ for each element is

$$\mathbf{f}^{(i)} = \begin{Bmatrix} 2 \\ 2 \\ 2 \end{Bmatrix} \qquad i = 1, 2, 3, 4$$

We can now assemble $\mathbf{K}$ and $\mathbf{F}$. Since the boundary conditions are

$$\Theta_3 = \Theta_4 = \Theta_5 = 0$$

we are interested only in degrees of freedom 1 and 2. Thus, the finite element equations are

$$\frac{1}{2} \begin{bmatrix} 2.084 & -2.083 \\ -2.083 & 8.334 \end{bmatrix} \begin{Bmatrix} \theta_1 \\ \theta_2 \end{Bmatrix} = \begin{Bmatrix} 4 \\ 8 \end{Bmatrix}$$

The solution is

$$[\Theta_1, \Theta_2] = [7.676, 3.838]$$

Consider the equation

$$M = 2G\alpha \iint_A \theta \, dA$$

Using $\theta = \mathbf{N}\boldsymbol{\theta}^e$, and noting that $\int_e \mathbf{N} \, dA = (A_e/3)[1, 1, 1]$, we get

$$M = 2G\alpha \left[\sum_e \frac{A_e}{3} (\theta_1^e + \theta_2^e + \theta_3^e) \right] \times 4$$

The multiplication by 4 above is because the finite element model represents only one-quarter of the rectangular cross section. Thus, we get the angle of twist per unit length to be

$$\alpha = 0.004 \frac{M}{G}$$

For given values of M and G, we can thus determine the value of α. Further, the shearing stresses in each element can be calculated from Eq. 10.105b. ∎

10.4 POTENTIAL FLOW, SEEPAGE, ELECTRIC AND MAGNETIC FIELDS, FLUID FLOW IN DUCTS

We have discussed steady-state heat conduction and torsion problems in some detail. Other examples of field problems occurring in engineering are briefly discussed below. Their solution follows the same procedure as for heat conduction and torsion problems, since the governing equations are special cases of the general Helmholtz equation, as discussed in the introduction to this chapter. In fact, the computer program HEAT2D can be used to solve the problems given below.

Potential Flow

Consider steady-state irrotational flow of an incompressible, nonviscous fluid around a cylinder, as shown in Fig. 10.14a. The velocity of the incoming flow is u_0. We want to determine the flow velocities near the cylinder. The solution of this problem is given by

$$\frac{\partial^2 \psi}{\partial x^2} + \frac{\partial^2 \psi}{\partial y^2} = 0 \qquad (10.125)$$

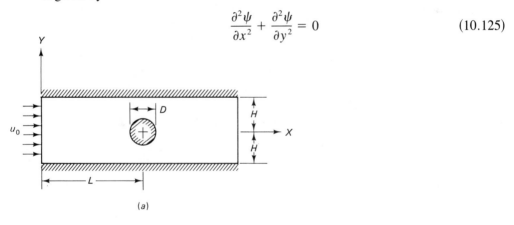

(a)

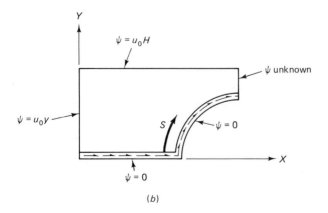

(b)

Figure 10.14 (a) Flow of an ideal fluid around a cylinder, and (b) boundary conditions for the finite element model.

where ψ is a stream function (m^3/s) per meter in the z direction. The value of ψ is constant along a stream line. A stream line is a line that is tangent to the velocity vector. By definition, there is no flow crossing a stream line. The flow between two adjacent stream lines can be thought of as the flow through a tube. Once the stream function $\psi = \psi(x, y)$ is known, the velocity components u and v along x and y, respectively, are obtained as

$$u = \frac{\partial \psi}{\partial y} \qquad v = \frac{-\partial \psi}{\partial x} \qquad (10.126)$$

Thus, the stream function ψ is analogous to the stress function in the torsion problem. Further, the rate of flow Q through a region bounded by two stream lines A and B is

$$Q = \psi_B - \psi_A \qquad (10.127)$$

To illustrate the boundary conditions and use of symmetry, we consider one quadrant of Fig. 10.14a as shown in Fig. 10.14b. First, note that velocities depend only on derivatives of ψ. Thus, we may choose the reference or base value of ψ; in Fig. 10.14b, we have chosen $\psi = 0$ at all nodes on the x axis. Then, along the y axis, we have $u = u_0$ or $\partial \psi / \partial y = u_0$. This is integrated to give the boundary condition $\psi = u_0 y$. That is, for each node i along the y axis, we have $\psi = u_0 y_i$. Along all nodes on $y = H$, we therefore have $\psi = u_0 H$. On the cylinder we now know that the velocity of the flow into the cylinder is zero. That is, $\partial \psi / \partial s = 0$ (Fig. 10.14b). Integrating this with the fact that $\psi = 0$ at the bottom of the cylinder results in $\psi = 0$ at all nodes along the cylinder. Thus, the fixed boundary is a stream line, as is to be expected.

Seepage

Flow of water that occurs in land drainage or seepage under dams can, under certain conditions, be described by Laplace's equation

$$\frac{\partial}{\partial x}\left(k_x \frac{\partial \phi}{\partial x}\right) + \frac{\partial}{\partial y}\left(k_y \frac{\partial \phi}{\partial y}\right) = 0 \qquad (10.128)$$

where $\phi = \phi(x, y)$ is the hydraulic potential (or hydraulic head) and k_x and k_y are the hydraulic conductivity in the x and y directions, respectively. The fluid velocity components are obtained from Darcy's law as $v_x = -k_x(\partial \phi / \partial x)$, $v_y = -k_y(\partial \phi / \partial y)$. Equation 10.128 is similar to the heat conduction equation. Lines of $\phi = $ constant are called equipotential surfaces, across which flow occurs. Equation 10.128 can include a source or sink Q (see Table 10.1), representing discharge per unit volume, to solve problems where pumps are removing water from an aquifer.

The appropriate boundary conditions associated with Eq. 10.128 are illustrated in the problem of water seepage through an earth dam (Fig. 10.15). The region to be modeled is shown shaded in the figure. Along the left and right surfaces, we have the boundary condition

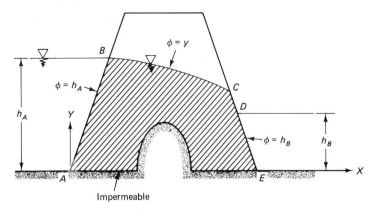

Figure 10.15 Seepage through an earth dam.

$$\phi = \text{constant} \tag{10.129}$$

The impermeable bottom surface corresponds to the natural boundary condition, $\partial\phi/\partial n = 0$ where n is the normal, and does not affect the element matrices; the values of ϕ there are unknowns. The top of the region is a *line of seepage* (free surface) where $\partial\phi/\partial n = 0$ and ϕ has its value equal to the y coordinate,

$$\phi = y \tag{10.130a}$$

This boundary condition requires iterative solution of the finite element analysis since the location of the boundary is unknown. We first assume a location for the line of seepage and impose the boundary condition $\phi = y_i$ at nodes i on the surface. Then, we solve for $\phi = \bar{\phi}$ and check the error $(\bar{\phi}_i - y_i)$. Based on this error, we update the locations of the nodes and obtain a new line of seepage. This process is repeated until the error is sufficiently small. Finally, portion CD in Fig. 10.15 is a *surface of seepage*. If no evaporation is taking place in this surface, then we have the boundary condition

$$\phi = \bar{y} \tag{10.130b}$$

where $\bar{y}$ is the coordinate of the surface.

Electrical and Magnetic Field Problems

In the area of electrical engineering, there are several interesting problems involving scalar and vector fields in two and three dimensions. We consider here some of the typical two-dimensional scalar field problems. In an isotropic dielectric medium with a permittivity of ϵ (F/m), and a volume charge density ρ (C/m^3) the electric potential u (V) must satisfy (Fig. 10.16)

$$\epsilon\left(\frac{\partial^2 u}{\partial x^2} + \frac{\partial^2 u}{\partial y^2}\right) = -\rho \tag{10.131}$$

$$u = a \quad \text{on } S_1 \qquad u = b \quad \text{on } S_2$$

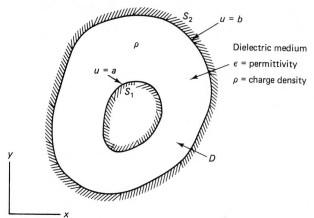

Figure 10.16   Electric potential problem.

Unit thickness may be assumed without loss of generality.

Finite element formulation may proceed from the minimization of the stored field energy

$$\Pi = \frac{1}{2} \iint_A \epsilon \left[\left(\frac{\partial u}{\partial x} \right)^2 + \left(\frac{\partial u}{\partial y} \right)^2 \right] dx \, dy - \int_A \rho u \, dA \qquad (10.132)$$

In Galerkin's formulation, we seek the approximate solution u such that

$$\iint_A \epsilon \left(\frac{\partial u}{\partial x} \frac{\partial \phi}{\partial x} + \frac{\partial u}{\partial y} \frac{\partial \phi}{\partial y} \right) dx \, dy - \int_A \rho \phi \, dA = 0 \qquad (10.133)$$

for every ϕ constructed from the basis functions of u, satisfying $\phi = 0$ on S_1 and S_2. In the above equation, integration by parts has been carried out.

Permittivity ϵ for various materials is defined in terms of relative permittivity ϵ_R and permittivity of free space ϵ_0 ($= 8.854 \times 10^{-12}$ F/m) as $\epsilon = \epsilon_R \epsilon_0$. Relative permittivity of rubber is in the range of 2.5–3. The coaxial cable problem is a typical example of Eq. 10.131 with $\rho = 0$. Figure 10.17 shows the section of a coaxial

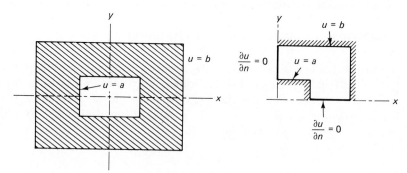

Figure 10.17   Rectangular coaxial cable.

cable of rectangular cross section. By symmetry, only a quarter of the section need be considered. On the separated boundary, $\partial u/\partial n = 0$ is a natural boundary condition which is satisfied automatically in the potential and Galerkin formulations. Another example is the determination of the electrical field distribution between two parallel plates (Fig. 10.18). Here, the field extends to infinity. Since the field drops as we move away from the plates, an arbitrary large domain D is defined, enclosing the plates symmetrically. The dimensions of this enclosure may be 5–10 times the plate dimensions. However, we may use larger elements away from the plates. On the boundary S, we may typically set $u = 0$.

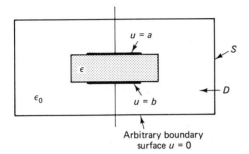

Arbitrary boundary
surface $u = 0$

Figure 10.18 Parallel strips separated by dielectric medium.

If u is the magnetic field potential, and μ is the permeability (H/m), the field equation is

$$\mu\left(\frac{\partial^2 u}{\partial x^2} + \frac{\partial^2 u}{\partial y^2}\right) = 0 \qquad (10.134)$$

where u is the scalar magnetic potential (A). Permeability μ is defined in terms of relative permeability μ_R and permeability of free space μ_0 ($=4\pi \times 10^{-7}$ H/m) as $\mu = \mu_R\mu_0$. μ_R for pure iron is about 4000, and for aluminum or copper it is about 1. Consider a typical application in an electric motor with no current flowing through the conductor, as shown in Fig. 10.19. We have $u = a$ and $u = b$ on the iron sur-

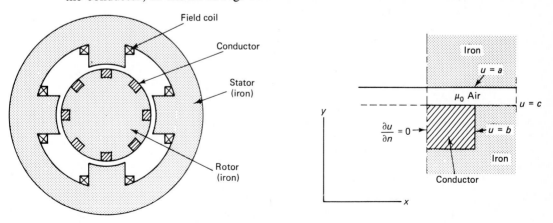

Figure 10.19 Model of a simple electric motor.

face; $u = c$ is used on an arbitrarily defined boundary ($u = 0$ may be used if the boundary is set at a large distance relative to the gap).

The ideas may be easily extended to axisymmetric coaxial cable problems. Problems in three dimensions can be considered using the steps developed in Chapter 9.

Fluid Flow in Ducts

The pressure drop occurring in the flow of a fluid in long, straight, uniform pipes and ducts is given by the equation

$$\Delta p = 2f\rho v_m^2 \frac{L}{D_h} \tag{10.135}$$

where f is the Fanning friction factor, ρ is the density, v_m is the mean velocity of fluid, L is the length of duct, and $D_h = (4 \times \text{area})/\text{perimeter}$ is the hydraulic diameter. The finite element method for determining the Fanning friction factor f for fully developed laminar flow in ducts of general cross-sectional shape will now be discussed.

Let fluid flow be in the z direction, with x, y being the plane of the cross section. A force balance (Fig. 10.20) yields

$$0 = pA - \left(p + \frac{dp}{dz}\Delta z\right)A - \tau_w P\Delta z \tag{10.136a}$$

or

$$-\frac{dp}{dz} = \frac{4\tau_w}{D_h} \tag{10.136b}$$

where τ_w is the shear stress at the wall. The friction factor f is defined as the ratio $f = \tau_w/(\rho v_m^2/2)$. The Reynolds number R_e is defined as $R_e = v_m D_h/\nu$, where $\nu = \mu/\rho$ is the kinematic viscosity, with μ representing absolute viscosity. Thus, from the above equations, we get

$$-\frac{dp}{dz} = \frac{2\mu v_m f R_e}{D_h^2} \tag{10.137}$$

The momentum equation is given by

$$\mu\left(\frac{\partial^2 w}{\partial x^2} + \frac{\partial^2 w}{\partial y^2}\right) - \frac{dp}{dz} = 0 \tag{10.138}$$

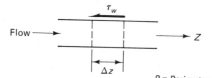

P = Perimeter

Figure 10.20   Force balance for fluid flow in a duct.

where $w = w(x, y)$ is the velocity of the fluid in the z direction. We introduce the nondimensional quantities

$$X = \frac{x}{D_h} \qquad Y = \frac{y}{D_h} \qquad W = \frac{w}{2v_m f R_e} \tag{10.139}$$

Equations 10.137–10.139 result in the equation

$$\frac{\partial^2 W}{\partial X^2} + \frac{\partial^2 W}{\partial Y^2} + 1 = 0 \tag{10.140}$$

Since the velocity of the fluid in contact with the wall of the duct is zero, we have $w = 0$ on the boundary and hence

$$W = 0 \quad \text{on boundary} \tag{10.141}$$

The solution of Eqs. 10.140 and 10.141 by the finite element method follows the same steps as for the heat conduction or torsion problems. Once W is known, then its average value may be determined from

$$W_m = \frac{\int_A W \, dA}{\int_A dA} \tag{10.142}$$

The integral $\int_A W \, dA$ may be readily evaluated using the element shape functions. For example, with CST elements, we get $\int_A W \, dA = \Sigma_e \, [A_e(w_1 + w_2 + w_3)/3]$. Once W_m is obtained, Eq. 10.139 is used

$$W_m = \frac{w_m}{2v_m f R_e} = \frac{v_m}{2v_m f R_e} \tag{10.143}$$

which yields

$$f = \frac{1/(2W_m)}{R_e} \tag{10.144}$$

Our aim is to determine the constant $1/(2W_m)$, which depends only on the cross-sectional shape. In preparing input data to solve Eqs. 10.140–10.141, we should remember that the nodal coordinates are in nondimensional form, as given in Eq. 10.139.

10.5 CONCLUSION

We have seen that all the field equations stem from the Helmholtz equations. Our presentation stressed the physical problems rather than considering one general equation with different variables and constants. This approach should help us identify the proper boundary conditions for modeling a variety of problems in engineering.

Example 10.6

The solution of Example 10.4 using program HEAT2D is shown below.

TWO-DIMENSIONAL HEAT CONDUCTION

```
GIVE NAME OF OUTPUT FILE=? eg10_4.dat
NUMBER OF ELEMENTS=? 3
NUMBER OF NODES=? 5
NUMBER OF MATERIALS=? 1
GIVE ELEMENT NO., 3 NODES, MATERIAL NO.
? 1,1,2,3,1
? 2,5,1,3,1
? 3,5,3,4,1
GIVE NODE NUMBER AND ITS TWO COORDINATES
? 1,0,0
? 2,.4,0
? 3,.4,.15
? 4,.4,.3
? 5,0,.3
GIVE MATERIAL NUMBER AND THERMAL CONDUCTIVITY
? 1,1.5
GIVE BOUNDARY CONDITION DATA BELOW
NO. OF NODES WITH SPECIFIED TEMPERATURES=? 2
GIVE NODE NO., SPECIFIED TEMPERATURE
? 4,180
? 5,180
NO. OF EDGES ON BOUNDARY WITH SPECIFIED HEAT FLUX=? 0
NO. OF EDGES ON BOUNDARY WITH SPECIFIED CONVECTION=? 2
GIVE TWO NODES OF EDGE, CONV. COEFF., AMBIENT TEMP.
? 2,3,50,25
? 3,4,50,25
NUMBER OF NODAL HEAT SOURCES =? 0
NO. OF ELEMENT HEAT SOURCES=? 0
Ok
```

```
NODE NO., TEMPERATURES
  1             124.4960058129545
  2             34.04508921672412
  3             45.3514441226464
  4             179.9998447258756
  5             179.9999805249125
** CONDUCTION HEAT FLOW PER UNIT AREA IN EACH ELEMENT **
ELEMENT NO., QX= -K*DT/DX, QY= -K*DT/DY
  1             339.1909372358638          -113.0635490592228
  2             400.8620589235765          -277.5198735597901
  3             5.092463882903076D-04      -1346.484006032292
```

Using HEAT2D with file input. The data file created by program MESHGEN together with the specification of the thermal conductivity for each material (property #1 when using DATAFEM) can be input into program HEAT2D. Once PLOT2D is used to obtain node numbers, the boundary conditions can then be specified interactively in HEAT2D.

Use of HEAT2D in solving other scalar field problems. It is possible to use HEAT2D to solve other problems discussed in this chapter. This is done by suitably identifying element heat sources, nodal heat sources, thermal conductivity, film coefficient, and ambient temperature. The TORSION program has been provided to solve torsion problems. ∎

```
1000 REM ******        HEAT1D        ******
1010 REM *** One dimensional Heat Conduction ***
1020 PRINT "======================================="
1030 PRINT "       Program written by           "
1040 PRINT "  T.R.Chandrupatla and A.D.Belegundu   "
1050 PRINT "======================================="
1060 DEFINT I-N
1070 INPUT "NUMBER OF ELEMENTS="; NE
1080 INPUT "NUMBER OF BOUNDARY CONDITIONS"; NBC
1090 INPUT "NUMBER OF NODES WITH HEAT SOURCE"; NQ
1100 NN = NE + 1
1110 NBW = 2   'NBW IS THE HALF-BAND-WIDTH
1120 DIM X(NN), S(NN, NBW), BC$(NBC), TC(NE), F(NN), V(NBC), H(NBC)
1130 DIM NB(NBC)
1140 PRINT "GIVE ELEMENT NO. AND ITS THERMAL CONDUCTIVITY"
1150 FOR I = 1 TO NE
1160 INPUT N, TC(N): NEXT I
1170 PRINT "GIVE NODE NO. AND ITS COORDINATE"
1180 FOR I = 1 TO NN
1190 INPUT N, X(N): NEXT I
1200 PRINT "GIVE BOUNDARY CONDITION (B.C.) DATA --"
1210 FOR I = 1 TO NBC
1220 INPUT "NODE NUMBER="; NB(I)
1230 INPUT "B.C. TYPE: TYPE 'TEMP' OR 'HFLUX' OR 'CONV'"; BC$(I)
1240 IF BC$(I) = "TEMP" OR BC$(I) = "temp" THEN INPUT "T0="; V(I)
1250 IF BC$(I) = "HFLUX" OR BC$(I) = "hflux" THEN INPUT "Q0="; V(I)
1260 IF BC$(I) = "CONV" OR BC$(I) = "conv" THEN INPUT "H,TINF="; H(I), V(I)
1270 NEXT I
1280 REM ** CALCULATE AND INPUT NODAL HEAT SOURCE VECTOR **
1290 FOR I = 1 TO NN: F(I) = 0!: NEXT I
1300 IF NQ = 0 THEN 1330
1310 FOR I = 1 TO NQ
1320 INPUT "GIVE NODE NO. AND HEAT SOURCE="; N, F(N): NEXT I
1330 FOR J = 1 TO NBW
1340 FOR I = 1 TO NN: S(I, J) = 0!: NEXT I: NEXT J
1350 REM ** STIFFNESS MATRIX **
1360 FOR I = 1 TO NE
1370 I1 = I: I2 = I + 1
1380 ELL = ABS(X(I2) - X(I1))
1390 EKL = TC(I) / ELL
1400 S(I1, 1) = S(I1, 1) + EKL
1410 S(I2, 1) = S(I2, 1) + EKL
1420 S(I1, 2) = S(I1, 2) - EKL: NEXT I
```

```
1430 REM ** ACCOUNT FOR B.C.'S **
1440 AMAX = 0!
1450 FOR I = 1 TO NN
1460 IF S(I, 1) > AMAX THEN AMAX = S(I, 1): NEXT I
1470 CNST = AMAX * 10000!
1480 FOR I = 1 TO NBC
1490 N = NB(I)
1500 IF BC$(I) = "CONV" OR BC$(I) = "conv" THEN GOTO 1540
1510 IF BC$(I) = "HFLUX" OR BC$(I) = "hflux" THEN GOTO 1560
1520 S(N, 1) = S(N, 1) + CNST
1530 F(N) = F(N) + CNST * V(I): GOTO 1570
1540 S(N, 1) = S(N, 1) + H(I)
1550 F(N) = F(N) + H(I) * V(I): GOTO 1570
1560 F(N) = F(N) - V(I)
1570 NEXT I
1580 GOSUB 5000
1590 REM ** F CONTAINS THE SOLUTION. 'RHS' IS OVER-WRITTEN"
1600 PRINT "        THE NODAL TEMPERATURES ARE"
1610 FOR I = 1 TO N: PRINT ; F(I): NEXT I
1620 END
1630 REM ** EQUATION SOLVING ... F CONTAINS THE TEMPERATURES ON EXIT **
5000 N = NN: HBW% = NBW
5010 FOR K = 1 TO N - 1
5020 NBK = N - K + 1
5030 IF N - K + 1 > HBW% THEN NBK = HBW%
5040 FOR I = K + 1 TO NBK + K - 1
5050 I1 = I - K + 1
5060 C = S(K, I1) / S(K, 1)
5070 FOR J = I TO NBK + K - 1
5080 J1 = J - I + 1
5090 J2 = J - K + 1
5100 S(I, J1) = S(I, J1) - C * S(K, J2): NEXT J
5110 F(I) = F(I) - C * F(K)
5120 NEXT I: NEXT K
5130 REM ** BACK SUBSTITUTION **
5140 F(N) = F(N) / S(N, 1)
5150 FOR II = 1 TO N - 1
5160 I = N - II
5170 NBI = N - I + 1
5180 IF N - I + 1 > HBW% THEN NBI = HBW%
5190 SUM = 0!
5200 FOR J = 2 TO NBI
5210 SUM = SUM + S(I, J) * F(I + J - 1): NEXT J
5220 F(I) = (F(I) - SUM) / S(I, 1): NEXT II
5230 RETURN


1000 REM ********        HEAT2D          ********
1010 REM   Two Dimensional Conduction with Temperature
1020 REM        HeatFlux and Convection BCs
1030 CLS
1040 PRINT "===================================="
1050 PRINT "       Program written by          "
1060 PRINT "  T.R.Chandrupatla and A.D.Belegundu   "
1070 PRINT "===================================="
1080 DEFINT I-N
1090 PRINT "====================================="
1100 PRINT "      TWO-DIMENSIONAL HEAT CONDUCTION      "
1110 PRINT "====================================="
```

```
1120 LOCATE 8,28: PRINT "1. Interactive Data Input"
1130 LOCATE 9,28: PRINT "2. Data Input (Except B.C.'s) From File"
1140 LOCATE 12,28: INPUT " Your Choice <1 or 2> "; IFL1
1150 IF IFL1 = 1 GOTO 1220
1160 PRINT "THE INPUT FILE SHOULD CONTAIN CONTROL INFO., COORDINATES,"
1170 PRINT " CONNECTIVITY, MATERIAL PROPERTIES(THERMAL CONDUCTIVITY)"
1180 PRINT
1190 INPUT "INPUT FILE NAME"; FILE1$
1200 OPEN FILE1$ FOR INPUT AS #2
1210 INPUT #2, NP,NE, ND,NL, NMAT, I, I : GOTO 1320
1220 INPUT "NUMBER OF ELEMENTS="; NE
1230 INPUT "NUMBER OF NODES="; NP
1240 INPUT "NUMBER OF MATERIALS="; NMAT
1250 REM ** NMAT= NUMBER OF MATERIALS
1260 REM ** NP  = NUMBER OF NODES = NO. OF DEGREES OF FREEDOM
1270 REM ** NE  = NUMBER OF ELEMENTS
1280 REM ** X   = X,Y COORDINATES OF EACH NODE
1290 REM ** NOC = ELEMENT CONNECTIVITY
1300 REM ** MATNO=MATERIAL NUMBER OF EACH ELEMENT
1310 REM ** TC  = THERMAL CONDUCTIVITY OF EACH MATERIAL
1320 DIM X(NP, 2), NOC(NE, 3), MATNO(NE), TC(NMAT)
1330 INPUT "GIVE NAME OF OUTPUT FILE="; FILENAME$
1340 OPEN FILENAME$ FOR OUTPUT AS #1
1350 IF IFL1 = 2 GOTO 1470
1360 PRINT "GIVE ELEMENT NO., 3 NODES, MATERIAL NO."
1370 FOR I = 1 TO NE
1380 INPUT N, NOC(N, 1), NOC(N, 2), NOC(N, 3), MATNO(N)
1390 NEXT I
1400 PRINT "GIVE NODE NUMBER AND ITS TWO COORDINATES"
1410 FOR I = 1 TO NP
1420 INPUT N, X(N, 1), X(N, 2): NEXT I
1430 PRINT "GIVE MATERIAL NUMBER AND THERMAL CONDUCTIVITY"
1440 FOR I = 1 TO NMAT
1450 INPUT N, TC(N): NEXT I
1460 GOTO 1590
1470 ' ** FILE INPUT FOR CONNECTIVITY, COORDINATES AND MATERIAL# **
1480 FOR I=1 TO NP
1490 INPUT #2, X(I,1), X(I,2) : NEXT I
1500 FOR I=1 TO NE
1510 INPUT #2, NOC(I,1), NOC(I,2), NOC(I,3) : NEXT I
1520 FOR I=1 TO NE: INPUT #2, MATNO(I): NEXT I
1530 REM ** "ND, NL NOT RELEVANT"; b.c.'s are read in interactively.
1540 FOR I=1 TO ND: INPUT #2, I,UI :NEXT I
1550 FOR I=1 TO NL: INPUT #2, I, UI:NEXT I
1560 FOR I=1 TO NMAT
1570 INPUT #2, TC(I), C, C :NEXT I
1580 CLOSE #2
1590 ' ** PRINT DATA **
1600 PRINT #1, "NO. OF NODES, NO. OF ELEM, NO. OF MATERIALS ="
1610 PRINT #1, NP, NE, NMAT
1620 PRINT #1, "NODE#,   X-COORD.,   Y-COORD"
1630 FOR I=1 TO NP
1640 PRINT #1, I, X(I,1), X(I,2) : NEXT I
1650 PRINT #1, "ELEMENT#, 3 NODES, MATERIAL#"
1660 FOR I=1 TO NE
1670 PRINT #1,I, NOC(I,1), NOC(I,2), NOC(I,3), MATNO(I): NEXT I
1680 PRINT #1, "MATERIAL#, THERMAL CONDUCTIVITY"
1690 FOR I=1 TO NMAT: PRINT #1, I, TC(I) : NEXT I
```

```
1700 REM ** BAND WIDTH CALCULATION
1710 IDIFF = 0
1720 FOR K = 1 TO NE
1730 FOR I = 1 TO 2
1740 FOR J = I + 1 TO 3
1750 II = ABS(NOC(K, J) – NOC(K, I))
1760 IF II > IDIFF THEN IDIFF = II
1770 NEXT J: NEXT I
1780 NEXT K
1790 NBW = IDIFF + 1
1800 PRINT #1, "THE BAND WIDTH IS"; NBW
1810 REM ** INITIALIZATION OF CONDUCTIVITY MATRIX AND HEAT RATE VECTOR
1820 DIM F(NP), S(NP, NBW)
1830 FOR I = 1 TO NP
1840 F(I) = 0!
1850 FOR J = 1 TO NBW
1860 S(I, J) = 0!
1870 NEXT J: NEXT I
1880 PRINT "GIVE BOUNDARY CONDITION DATA BELOW "
1890 PRINT #1, " ** BOUNDARY CONDITION DATA ** "
1900 REM ** ND= NO. OF SPECIFIED NODAL TEMPERATURES
1910 REM ** NT(.) = CORRESP. NODE NO., AND U(.) = VALUE OF SPEC. TEMP.
1920 INPUT "NO. OF NODES WITH SPECIFIED TEMPERATURES="; ND
1930 PRINT #1, "NO. OF NODES WITH SPEC. TEMP. ="; ND
1940 IF ND = 0 THEN 2010
1950 DIM NT(ND), U(ND)
1960 PRINT "GIVE NODE NO., SPECIFIED TEMPERATURE"
1970 PRINT #1, "NODE NO. AND SPEC. TEMPERATURE"
1980 FOR I = 1 TO ND
1990 INPUT NT(I), U(I)
2000 PRINT #1, NT(I), U(I): NEXT I
2010 INPUT "NO. OF EDGES ON BOUNDARY WITH SPECIFIED HEAT FLUX="; NHF
2020 PRINT #1, "NO. OF EDGES ON BOUNDARY WITH SPEC. HEAT FLUX="; NHF
2030 IF NHF = 0 THEN 2130
2040 PRINT "GIVE TWO NODES OF EDGE AND SPEC. HEAT FLUX(>0 IF OUT)"
2050 PRINT #1, "TWO NODES OF EDGE, SPEC. HEAT FLUX"
2060 FOR I = 1 TO NHF
2070 INPUT N1, N2, V
2080 PRINT #1, N1, N2, V
2090 ELEN = SQR((X(N1, 1) – X(N2, 1)) ^ 2 + (X(N1, 2) – X(N2, 2)) ^ 2)
2100 F(N1) = F(N1) – ELEN * V / 2!
2110 F(N2) = F(N2) – ELEN * V / 2!
2120 NEXT I
2130 INPUT "NO. OF EDGES ON BOUNDARY WITH SPECIFIED CONVECTION="; NCONV
2140 PRINT #1, "NO. OF EDGES ON BOUNDARY WITH SPEC. CONVECTION="; NCONV
2150 IF NCONV = 0 THEN 2310
2160 PRINT "GIVE TWO NODES OF EDGE, CONV. COEFF., AMBIENT TEMP."
2170 PRINT #1, "TWO NODES OF EDGE, SPECIFIED CONVECTION"
2180 FOR I = 1 TO NCONV
2190 INPUT N1, N2, H, TINF
2200 PRINT #1, N1, N2, H, TINF
2210 ELEN = SQR((X(N1, 1) – X(N2, 1)) ^ 2 + (X(N1, 2) – X(N2, 2)) ^ 2)
2220 F(N1) = F(N1) + ELEN * H * TINF / 2!
2230 F(N2) = F(N2) + ELEN * H * TINF / 2!
2240 S(N1, 1) = S(N1, 1) + H * ELEN / 3!
2250 S(N2, 1) = S(N2, 1) + H * ELEN / 3!
2260 IF N1 < N2 GOTO 2280
2270 N3 = N1: N1 = N2: N2 = N3
```

```
2280 S(N1, N2 - N1 + 1) = S(N1, N2 - N1 + 1) + H * ELEN / 6!
2290 NEXT I
2300 REM ** READ HEAT SOURCE VECTOR
2310 INPUT "NUMBER OF NODAL HEAT SOURCES ="; NQ
2320 PRINT #1, "NUMBER OF NODAL HEAT SOURCES="; NQ
2330 IF NQ = 0 THEN 2400
2340 PRINT "GIVE NODE NO. AND HEAT SOURCE (>0 IF SOURCE)"
2350 PRINT #1, "NODE NO., HEAT SOURCE"
2360 FOR I = 1 TO NQ
2370 INPUT N, Q
2380 PRINT #1, N, Q
2390 F(N) = F(N) + Q: NEXT I
2400 INPUT "NO. OF ELEMENT HEAT SOURCES="; NEQ
2410 PRINT #1, "NO. OF ELEMENT HEAT SOURCES="; NEQ
2420 IF NEQ = 0 THEN 2560
2430 PRINT "GIVE ELEMENT NO., AVERAGE HEAT RATE PER UNIT VOL."
2440 PRINT #1, "ELEMENT NO. AND AVERAGE HEAT RATE PER UNIT VOL."
2450 FOR I = 1 TO NEQ
2460 INPUT N, EQ
2470 PRINT #1, N, EQ
2480 I1 = NOC(N, 1): I2 = NOC(N, 2): I3 = NOC(N, 3)
2490 X32 = X(I3, 1)-X(I2, 1):X13=X(I1, 1)-X(I3, 1): X21 = X(I2, 1)-X(I1, 1)
2500 Y23=X(I2, 2)-X(I3, 2): Y31 = X(I3, 2)-X(I1, 2): Y12 =X(I1, 2)-X(I2, 2)
2510 AREA = .5 * ABS(X13 * Y23 - X32 * Y31)
2520 C = EQ * AREA / 3!
2530 F(I1) = F(I1) + C: F(I2) = F(I2) + C: F(I3) = F(I3) + C
2540 NEXT I
2550 REM ** CONDUCTIVITY MATRIX
2560 DIM BT(2, 3)
2570 FOR I = 1 TO NE
2580 I1 = NOC(I, 1): I2 = NOC(I, 2): I3 = NOC(I, 3)
2590 X32 =X(I3, 1)-X(I2, 1):X13=X(I1, 1)-X(I3, 1):X21=X(I2, 1) - X(I1, 1)
2600 Y23 =X(I2,2)-X(I3, 2):Y31=X(I3, 2)-X(I1, 2): Y12 = X(I1, 2)-X(I2, 2)
2610 DETJ = X13 * Y23 - X32 * Y31
2620 AREA = .5 * ABS(DETJ)
2630 BT(1, 1) = Y23: BT(1, 2) = Y31: BT(1, 3) = Y12
2640 BT(2, 1) = X32: BT(2, 2) = X13: BT(2, 3) = X21
2650 FOR II = 1 TO 3
2660 FOR JJ = 1 TO 2
2670 BT(JJ, II) = BT(JJ, II) / DETJ
2680 NEXT JJ: NEXT II
2690 FOR II = 1 TO 3
2700 FOR JJ = 1 TO 3
2710 II1 = NOC(I, II): II2 = NOC(I, JJ)
2720 IF II1 > II2 THEN 2770
2730 SUM = 0!
2740 FOR J = 1 TO 2
2750 SUM = SUM + BT(J, II) * BT(J, JJ): NEXT J
2760 S(II1, II2 - II1 + 1) = S(II1, II2 - II1 + 1) + SUM * AREA * TC(MATNO(I))
2770 NEXT JJ: NEXT II
2780 NEXT I
2790 IF ND = 0 THEN 2910
2800 REM ** MODIFY FOR TEMP. BOUNDARY CONDITIONS
2810 SUM = 0!
2820 FOR I = 1 TO NP
2830 SUM = SUM + S(I, 1): NEXT I
2840 SUM = SUM / NP
2850 CNST = SUM * 1000000!
```

```
2860 FOR I = 1 TO ND
2870 N = NT(I)
2880 S(N, 1) = S(N, 1) + CNST
2890 F(N) = F(N) + CNST * U(I): NEXT I
2900 REM ** EQUATION SOLVING
2910 N = NP
2920 GOSUB 5000
2930 PRINT #1, "NODE NO., TEMPERATURES"
2940 FOR I = 1 TO NP
2950 PRINT #1, I, F(I)
2960 NEXT I
2970 INPUT "DO YOU WISH TO USE PROGRAM CONTOUR FOR ISOTHERMS<Y OR N>";ANS$
2980 IF ANS$ <> "Y" AND ANS$ <> "y" GOTO 3030
2990 INPUT "ENTER FILE NAME FOR NODAL TEMPERATURE DATA"; FILET$
3000 OPEN FILET$ FOR OUTPUT AS #3
3010 FOR I=1 TO NP: PRINT #3, F(I): NEXT I
3020 CLOSE #3
3030 PRINT #1, " ** CONDUCTION HEAT FLOW PER UNIT AREA IN EACH ELEMENT ** "
3040 PRINT #1, "ELEMENT NO., QX= -K*DT/DX, QY= -K*DT/DY "
3050 FOR I = 1 TO NE
3060 I1 = NOC(I, 1): I2 = NOC(I, 2): I3 = NOC(I, 3)
3070 X32=X(I3, 1)–X(I2, 1):X13=X(I1, 1)–X(I3, 1):X21 = X(I2, 1)–X(I1, 1)
3080 Y23 =X(I2, 2)– X(I3, 2):Y31=X(I3, 2)–X(I1, 2):Y12=X(I1, 2)–X(I2, 2)
3090 DETJ = X13 * Y23 – X32 * Y31
3100 BT(1, 1) = Y23: BT(1, 2) = Y31: BT(1, 3) = Y12
3110 BT(2, 1) = X32: BT(2, 2) = X13: BT(2, 3) = X21
3120 FOR II = 1 TO 3
3130 FOR JJ = 1 TO 2
3140 BT(JJ, II) = BT(JJ, II) / DETJ
3150 NEXT JJ: NEXT II
3160 QX = BT(1, 1) * F(I1) + BT(1, 2) * F(I2) + BT(1, 3) * F(I3)
3170 QX = –QX * TC(MATNO(I))
3180 QY = BT(2, 1) * F(I1) + BT(2, 2) * F(I2) + BT(2, 3) * F(I3)
3190 QY = –QY * TC(MATNO(I))
3200 PRINT #1, I, QX, QY
3210 NEXT I
3220 END
5000 REM ** FORWARD ELIMINATION
5010 FOR K = 1 TO N – 1
5020 NBK = N – K + 1
5030 IF N – K + 1 > NBW THEN NBK = NBW
5040 FOR I = K + 1 TO NBK + K – 1
5050 I1 = I – K + 1
5060 C = S(K, I1) / S(K, 1)
5070 FOR J = I TO NBK + K – 1
5080 J1 = J – I + 1
5090 J2 = J – K + 1
5100 S(I, J1) = S(I, J1) – C * S(K, J2): NEXT J
5110 F(I) = F(I) – C * F(K)
5120 NEXT I: NEXT K
5130 REM ** BACK SUBSTITUTION **
5140 F(N) = F(N) / S(N, 1)
5150 FOR II = 1 TO N – 1
5160 I = N – II
5170 NBI = N – I + 1
5180 IF N – I + 1 > NBW THEN NBI = NBW
5190 SUM = 0!
5200 FOR J = 2 TO NBI
```

```
5210 SUM = SUM + S(I, J) * F(I + J – 1): NEXT J
5220 F(I) = (F(I) – SUM) / S(I, 1): NEXT II
5230 REM ** F CONTAINS THE SOLUTION. 'S' IS OVER-WRITTEN"
5240 RETURN

1000 REM *****          TORSION          *****
1010 CLS
1020 PRINT "======================================="
1030 PRINT "   TWO-DIMENSIONAL TORSION OF RODS    "
1040 PRINT "======================================="
1050 PRINT "       Program written by            "
1060 PRINT "  T.R.Chandrupatla and A.D.Belegundu   "
1070 PRINT "======================================="
1080 DEFINT I–N
1090 PRINT "1. Interactive Data Input"
1100 PRINT "2. Data Input (Except B.C. 's) From File"
1110 INPUT " Your Choice <1 or 2> "; IFL1
1120 IF IFL1 = 1 GOTO 1160
1130 INPUT "INPUT FILE NAME "; FILE2$
1140 OPEN FILE2$ FOR INPUT AS #2
1150 INPUT #2, NP, NE, ND, NL, NMAT, I, I: GOTO 1220
1160 INPUT "NUMBER OF ELEMENTS="; NE
1170 INPUT "NUMBER OF NODES="; NP
1180 REM ** NP  = NUMBER OF NODES = NO. OF DEGREES OF FREEDOM
1190 REM ** NE  = NUMBER OF ELEMENTS
1200 REM ** X   = X,Y COORDINATES OF EACH NODE
1210 REM ** NOC = ELEMENT CONNECTIVITY
1220 DIM X(NP, 2), NOC(NE, 3)
1230 INPUT "GIVE NAME OF OUTPUT FILE="; FILE1$
1240 OPEN FILE1$ FOR OUTPUT AS #1
1250 IF IFL1 = 2 GOTO 1390
1260 PRINT #1, "ELEM. NO., 3 NODES "
1270 PRINT "GIVE ELEMENT NO., 3 NODES "
1280 FOR I = 1 TO NE
1290 INPUT N, NOC(N, 1), NOC(N, 2), NOC(N, 3)
1300 PRINT #1, N, NOC(N, 1), NOC(N, 2), NOC(N, 3)
1310 NEXT I
1320 PRINT "GIVE NODE NUMBER AND ITS TWO COORDINATES"
1330 PRINT #1, "NODE NO. AND TWO COORDINATES"
1340 FOR I = 1 TO NP
1350 INPUT N, X(N, 1), X(N, 2)
1360 PRINT #1, N, X(N, 1), X(N, 2): NEXT I
1370 GOTO 1450
1380 ' ** FILE INPUT **
1390 FOR I = 1 TO NP
1400 INPUT #2, X(I, 1), X(I, 2): NEXT I
1410 FOR I = 1 TO NE
1420 INPUT #2, NOC(I, 1), NOC(I, 2), NOC(I, 3): NEXT I
1430 CLOSE #2
1440 REM ** BAND WIDTH CALCULATION
1450 IDIFF = 0
1460 FOR K = 1 TO NE
1470 FOR I = 1 TO 2
1480 FOR J = I + 1 TO 3
1490 II = ABS(NOC(K, J) – NOC(K, I))
1500 IF II > IDIFF THEN IDIFF = II
1510 NEXT J: NEXT I
1520 NEXT K
```

```
1530 NBW = IDIFF + 1
1540 PRINT #1, "THE BAND WIDTH IS"; NBW
1550 REM ** INITIALIZATION OF STIFFNESS AND LOAD MATRICES
1560 DIM F(NP), S(NP, NBW)
1570 FOR I = 1 TO NP
1580 F(I) = 0!
1590 FOR J = 1 TO NBW
1600 S(I, J) = 0!
1610 NEXT J: NEXT I
1620 PRINT "GIVE BOUNDARY CONDITION DATA BELOW "
1630 INPUT "NO. OF BOUNDARY NODES "; ND
1640 PRINT #1, "NO. OF BOUNDARY NODES ="; ND
1650 DIM NT(ND), U(ND)
1660 PRINT "GIVE BOUNDARY NODE NO., SPECIFIED VALUE(=0 FOR TORSION)"
1670 PRINT #1, "NODE NO. AND SPECIFIED VALUE"
1680 FOR I = 1 TO ND
1690 INPUT NT(I), U(I)
1700 PRINT #1, NT(I), U(I): NEXT I
1710 REM ** STIFFNESS MATRIX
1720 DIM BT(2, 3)
1730 FOR I = 1 TO NE
1740 I1 = NOC(I, 1): I2 = NOC(I, 2): I3 = NOC(I, 3)
1750 X32 = X(I3, 1) - X(I2, 1): X13 = X(I1, 1) - X(I3, 1): X21 = X(I2, 1) - X(I1, 1)
1760 Y23 = X(I2, 2) - X(I3, 2): Y31 = X(I3, 2) - X(I1, 2): Y12 = X(I1, 2) - X(I2, 2)
1770 DETJ = X13 * Y23 - X32 * Y31
1780 AREA = .5 * ABS(DETJ)
1790 BT(1, 1) = Y23: BT(1, 2) = Y31: BT(1, 3) = Y12
1800 BT(2, 1) = X32: BT(2, 2) = X13: BT(2, 3) = X21
1810 FOR II = 1 TO 3
1820 FOR JJ = 1 TO 2
1830 BT(JJ, II) = BT(JJ, II) / DETJ
1840 NEXT JJ: NEXT II
1850 FOR II = 1 TO 3
1860 FOR JJ = 1 TO 3
1870 II1 = NOC(I, II): II2 = NOC(I, JJ)
1880 IF II1 > II2 THEN 1930
1890 SUM = 0!
1900 FOR J = 1 TO 2
1910 SUM = SUM + BT(J, II) * BT(J, JJ): NEXT J
1920 S(II1, II2 - II1 + 1) = S(II1, II2 - II1 + 1) + SUM * AREA
1930 NEXT JJ: NEXT II
1940 REM ** LOAD VECTOR
1950 C = 2! * AREA / 3!
1960 F(I1) = F(I1) + C: F(I2) = F(I2) + C: F(I3) = F(I3) + C
1970 NEXT I
1980 SUM = 0!
1990 FOR I = 1 TO NP
2000 SUM = SUM + S(I, 1): NEXT I
2010 CNST = SUM * 10000!
2020 FOR I = 1 TO ND
2030 N = NT(I)
2040 S(N, 1) = S(N, 1) + CNST
2050 F(N) = F(N) + CNST * U(I): NEXT I
2060 REM ** EQUATION SOLVING
2070 N = NP
2080 GOSUB 2510
2090 PRINT #1, "NODE#   STRESS FUNCTION VALUE"
2100 PRINT "NODE#   STRESS FUNCTION VALUE"
```

```
2110 FOR I = 1 TO NP
2120 PRINT #1, USING "###"; I;
2122 PRINT #1, USING "   #.####^^^^"; F(I)
2130 PRINT USING "###"; I;
2132 PRINT USING "   #.####^^^^"; F(I)
2140 NEXT I
2150 REM ** ANGLE OF TWIST, PER UNIT LENGTH
2160 SUM = 0!
2170 FOR I = 1 TO NE
2180 I1 = NOC(I, 1): I2 = NOC(I, 2): I3 = NOC(I, 3)
2190 X32 = X(I3, 1) − X(I2, 1): X13 = X(I1, 1) − X(I3, 1): X21 = X(I2, 1) − X(I1, 1)
2200 Y23 = X(I2, 2) − X(I3, 2): Y31 = X(I3, 2) − X(I1, 2): Y12 = X(I1, 2) − X(I2, 2)
2210 DETJ = X13 * Y23 − X32 * Y31
2220 SUM = SUM + ABS(DETJ) * (F(I1) + F(I2) + F(I3)) / 3!
2230 NEXT I
2240 INPUT "TORQUE = "; TORQUE
2250 INPUT "SHEAR MODULUS = "; SMOD
2260 INPUT "SYMMETRY FACTOR (eg. if 1/4 symmetry, then =4.0) = "; SFAC
2270 ALPHA = TORQUE / (SMOD * SUM * SFAC)
2280 PRINT #1, "TWIST PER UNIT LENGTH = "; ALPHA
2290 PRINT "TWIST PER UNIT LENGTH = "; ALPHA
2300 PRINT #1, " ** SHEARING STRESSES TAUYZ, TAUXZ IN EACH ELEMENT ** "
2302 PRINT " ** SHEARING STRESSES TAUYZ, TAUXZ IN EACH ELEMENT ** "
2310 PRINT #1, "ELEM#   TAUYZ   TAUXZ "
2312 PRINT "ELEM#   TAUYZ   TAUXZ "
2320 FOR I = 1 TO NE
2330 I1 = NOC(I, 1): I2 = NOC(I, 2): I3 = NOC(I, 3)
2340 X32 = X(I3, 1) − X(I2, 1): X13 = X(I1, 1) − X(I3, 1): X21 = X(I2, 1) − X(I1, 1)
2350 Y23 = X(I2, 2) − X(I3, 2): Y31 = X(I3, 2) − X(I1, 2): Y12 = X(I1, 2) − X(I2, 2)
2360 DETJ = X13 * Y23 − X32 * Y31
2370 BT(1, 1) = Y23: BT(1, 2) = Y31: BT(1, 3) = Y12
2380 BT(2, 1) = X32: BT(2, 2) = X13: BT(2, 3) = X21
2390 FOR II = 1 TO 3
2400 FOR JJ = 1 TO 2
2410 BT(JJ, II) = BT(JJ, II) / DETJ
2420 NEXT JJ: NEXT II
2430 TAUYZ = −(BT(1, 1) * F(I1) + BT(1, 2) * F(I2) + BT(1, 3) * F(I3))
2440 TAUXZ = BT(2, 1) * F(I1) + BT(2, 2) * F(I2) + BT(2, 3) * F(I3)
2450 TAUYZ = TAUYZ * SMOD * ALPHA
2460 TAUXZ = TAUXZ * SMOD * ALPHA
2470 PRINT #1, USING "### "; I;
2472 PRINT #1, USING "   #.####^^^^"; TAUYZ; TAUXZ
2480 PRINT USING "### "; I;
2482 PRINT USING "   #.####^^^^"; TAUYZ; TAUXZ
2490 NEXT I
2500 END
2510 REM ** FORWARD ELIMINATION
2520 FOR K = 1 TO N − 1
2530 NBK = N − K + 1
2540 IF N − K + 1 > NBW THEN NBK = NBW
2550 FOR I = K + 1 TO NBK + K − 1
2560 I1 = I − K + 1
2570 C = S(K, I1) / S(K, 1)
2580 FOR J = I TO NBK + K − 1
2590 J1 = J − I + 1
2600 J2 = J − K + 1
2610 S(I, J1) = S(I, J1) − C * S(K, J2): NEXT J
2620 F(I) = F(I) − C * F(K)
```

```
2630 NEXT I: NEXT K
2640 REM ** BACK SUBSTITUTION **
2650 F(N) = F(N) / S(N, 1)
2660 FOR II = 1 TO N - 1
2670 I = N - II
2680 NBI = N - I + 1
2690 IF N - I + 1 > NBW THEN NBI = NBW
2700 SUM = 0!
2710 FOR J = 2 TO NBI
2720 SUM = SUM + S(I, J) * F(I + J - 1): NEXT J
2730 F(I) = (F(I) - SUM) / S(I, 1): NEXT II
2740 REM ** F CONTAINS THE SOLUTION. 'S' IS OVER-WRITTEN"
2750 RETURN
```

PROBLEMS

10.1. Consider a brick wall (Fig. P10.1) of thickness $L = 30$ cm, $k = 0.7$ W/m. °C. The inner surface is at 28°C and the outer surface is exposed to cold air at -15°C. The heat transfer coefficient associated with the outside surface is $h = 40$ W/m². °C. Determine the steady-state temperature distribution within the wall and also the heat flux through the wall. Use a two-element model, and obtain the solution by hand calculations. Assume one-dimensional flow.

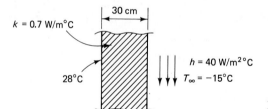

Figure P10.1

10.2. Repeat P10.1 using program HEAT1D. Compare the solution obtained using a two-element model with that obtained from a three-element model.

10.3. Consider a pin fin (Fig. P10.3) having a diameter of $\frac{5}{16}$ in. and length of 5 in. At the root, the temperature is 150°F. The ambient temperature is 80°F and $h = 6$ BTU/(h.ft².°F). Take $k = 24.8$ BTU/(h.ft.°F). Assume that the tip of the fin is insulated. Using a two-element model, determine the temperature distribution and heat loss in the fin (by hand calculations).

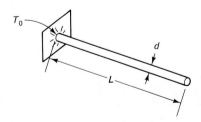

Figure P10.3

10.4. In our derivation using Galerkin's approach for straight rectangular fins, we assume that the fin tip is insulated. Modify the derivation to account for the case when convection occurs from the tip of the fin as well. Repeat Example 10.3 with this type of boundary condition.

10.5. A long steel tube (Fig. P10.5a) with inner radius $r_1 = 3$ cm and outer radius $r_2 = 5$ cm, and $k = 20$ W/m. °C, has its inner surface heated at a rate $q_0 = -100\ 000$ W/m² (the minus sign indicates that heat flows into the body). Heat is dissipated by convection from the outer surface into a fluid at temperature $T_\infty = 120$°C and $h = 400$ W/m². °C. Considering the eight-element, nine-node finite element model shown in Fig. P10.5b, determine:
  (a) The boundary conditions for the model.
  (b) The temperatures T_1, T_2 at the inner and outer surfaces, respectively. Use program HEAT2D.

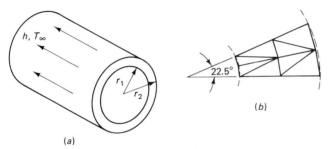

(a)

(b)

Figure P10.5

*10.6. In P10.5, assume that the steel tube is free of stress at a room temperature $T = 30$°C. Determine the thermal stresses in the tube using program AXYSYM. Take $E = 200\ 000$ MPa, $\nu = 0.3$.

10.7. Consider a long pipe with the cross section as shown in Fig. P10.7. If the temperatures at the inner and outer surfaces are $T_i = 500$°C and $T_0 = 100$°C, respectively, determine (using program HEAT2D):
  (a) An appropriate finite element model. Take advantage of symmetry.
  (b) The temperature distribution in the wall.

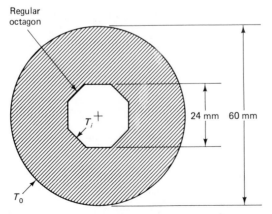

Regular
octagon

T_i

T_0

24 mm   60 mm

Figure P10.7

(c) The total heat flow, q_{total}, through the wall. Take the average of the values at the inner and outer surfaces. Hence, calculate the conduction shape factor S from

$$q_{total} = kS \, \Delta T_{overall}$$

where $\Delta T_{overall} = (500-100)°C = 400°C$ for this problem.

10.8. A large industrial furnace is supported on a long column of fireclay brick, which is 1×1 m on a side (Fig. P10.8). During steady-state operation, installation is such that three surfaces of the column are maintained at 600 K while the remaining surface is exposed to an airstream for which $T_\infty = 300$ K and $h = 12$ W/m^2 K. Determine, using program HEAT2D, the temperature distribution in the column and the heat rate to the airstream per unit length of column. Take $k = 1$ W/m K.

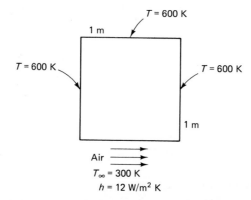

Figure P10.8

*10.9. A thermal diffuser of axisymmetric shape is shown in Fig. P10.9. The thermal diffuser receives a constant thermal flux of magnitude $q_1 = 400\ 000$ W/m^2.°C from a solid-state device on which the diffuser is mounted. At the opposite end, the diffuser is kept at a uniform value of $T = 0°C$ by isothermalizing heatpipes. The lateral sur-

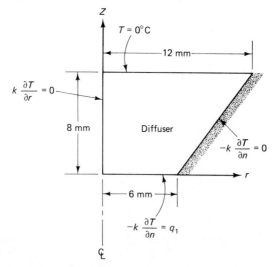

Figure P10.9

face of the diffuser is insulated, and thermal conductivity $k = 200$ W/m.°C. The differential equation is

$$\frac{1}{r}\frac{\partial}{\partial r}\left(r\frac{\partial T}{\partial r}\right) + \frac{\partial^2 T}{\partial z^2} = 0$$

Develop an axisymmetric element to determine the temperature distribution and the outward heat flux at the heatpipes. Refer to Chapter 6 for details on the axisymmetric element.

*10.10. Develop a four-node quadrilateral for heat conduction and solve problem P10.8. Refer to Chapter 7 for details pertaining to the quadrilateral element. Compare your solution with use of three-node triangles.

10.11. The L-shaped beam in Fig. P10.11, which supports a floor slab in a building, is subjected to a twisting moment T in.lb. Determine, using program TORSION:

(a) The angle of twist per unit length, α.

(b) The contribution of each finite element to the total twisting moment.

Leave your answers in terms of torque T and shear modulus G. Verify your answers by refining the finite element grid.

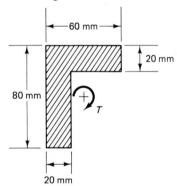

Figure P10.11

10.12. The cross section of the steel beam in Fig. P10.12 is subjected to a torque $T = 5000$ in.lb. Determine, using program TORSION, the angle of twist and the location and magnitude of the maximum shearing stresses.

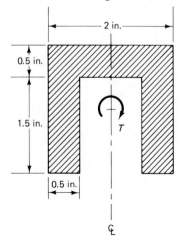

Figure P10.12

10.13. For Fig. 10.14a in the text, let $u^0 = 1$ m/s, $L = 5$ m, $D = 1.5$ m, and $H = 2.0$ m.
Determine the velocity field using a coarse grid and a fine grid (with smaller elements
nearer the cylinder). In particular, determine the maximum velocity in the flow. Comment on the relation of this problem to a stress concentration problem.

10.14. Determine and plot the stream lines for the flow in the venturi meter shown in Fig.
P10.14. The incoming flow has a velocity of 100 cm/s. Also plot the velocity distribution at the waist a–a.

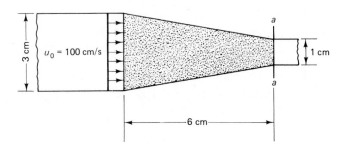

Figure P10.14

10.15. The dam shown in Fig. P10.15 rests on a homogeneous isotropic soil which has
confined impermeable boundaries as shown. The walls and base of the dam are im-
pervious. The dam retains water at a constant height of 5 m, the downstream level being zero. Determine and plot the equipotential lines, and find the quantity of water
seeping underneath the dam per unit width of the dam. Take hydraulic conductivity
$k = 30$ m/day.

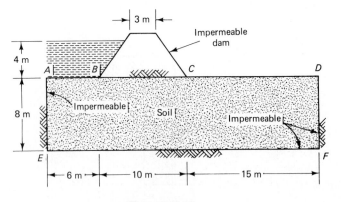

Figure P10.15

*10.16. For the dam section shown in Fig. P10.16, $k = 0.003$ ft/min. Determine:
  (a) The line of seepage.
  (b) The quantity of seepage per 100-ft length of the dam.
  (c) The length of the surface of seepage a.

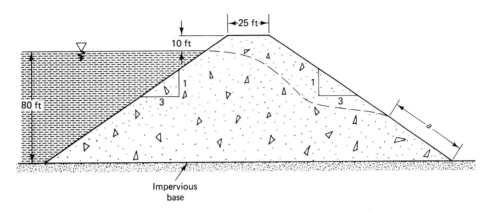

Figure P10.16

*10.17. For the triangular duct shown in Fig. P10.17, obtain the constant C which relates the Fanning friction factor f to the Reynolds number R_e as $f = C/R_e$. Use triangular finite elements. Verify your answer by refining the finite element model. Compare your result for C with that for a square duct having the same perimeter.

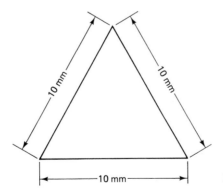

Figure P10.17

10.18. Figure P10.18 shows the cross section of a rectangular coaxial cable. At the inner surface of the dielectric insulator ($\epsilon_R = 3$), a voltage of 100 V is applied. If the voltage at the outer surface is zero, determine the voltage field distribution in the annular space.

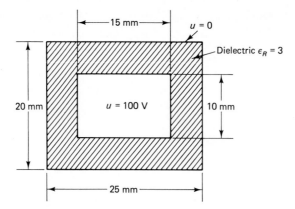

Figure P10.18

10.19. A pair of strip lines, shown in Fig. P10.19, are separated by a dielectric medium $\epsilon_R = 5.4$. The strips are enclosed by a fictitious box 2×1 m with enclosed space having $\epsilon_R = 1$. Assuming $u = 0$ on the boundary of this box, find the voltage field distribution. (Use large elements away from the strip plates.)

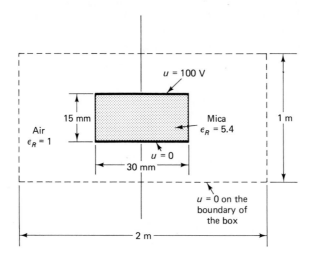

Figure P10.19

10.20. Determine the scalar magnetic potential u for the simplified model of the slot in an electric motor armature shown in Fig. P10.20.

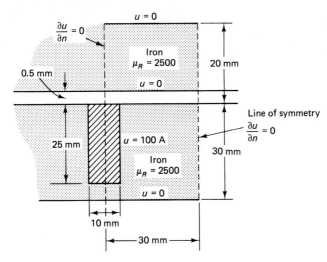

Figure P10.20

chapter eleven

Dynamic Considerations

11.1 INTRODUCTION

In Chapters 3–9, we have discussed the static analysis of structures. Static anaysis holds when the loads are slowly applied. When the loads are suddenly applied, or when the loads are of a variable nature, the mass and acceleration effects come into the picture. If a solid body, such as an engineering structure, is deformed elastically and suddenly released, it tends to vibrate about its equilibrium position. This periodic motion due to the restoring strain energy is called **free vibration**. The number of cycles per unit time is called the **frequency**. The maximum displacement from the equilibrium position is the **amplitude**. In the real world, the vibrations subside with time due to damping action. In the simplest vibration model, the damping effects are neglected. The undamped free vibration model of a structure gives significant information about its dynamic behavior. We present here the considerations needed to apply finite elements to the analysis of undamped free vibrations of structures.

11.2 FORMULATION

We define the Lagrangean L by

$$L = T - \Pi \tag{11.1}$$

where T is the kinetic energy and Π is the potential energy.

Hamilton's principle.   For an arbitrary time interval from t_1 to t_2, the state of motion of a body extremizes the functional

$$I = \int_{t_1}^{t_2} L \, dt \qquad (11.2)$$

If L can be expressed in terms of the generalized variables $(q_1, q_2, \ldots, q_n, \dot{q}_1, \dot{q}_2, \ldots, \dot{q}_n)$ where $\dot{q}_i = dq_i/dt$, then the equations of motion are given by

$$\frac{d}{dt}\left(\frac{\partial L}{\partial \dot{q}_i}\right) - \frac{\partial L}{\partial q_i} = 0 \qquad i = 1 \text{ to } n \qquad (11.3)$$

To illustrate the principle, let us consider two point masses connected by springs. Consideration of distributed masses follows the example.

Example 11.1

Consider the spring–mass system in Fig. E11.1. The kinetic and potential energies are given by

$$T = \tfrac{1}{2}m_1\dot{x}_1^2 + \tfrac{1}{2}m_2\dot{x}_2^2$$

$$\Pi = \tfrac{1}{2}k_1 x_1^2 + \tfrac{1}{2}k_2(x_2 - x_1)^2$$

Using $L = T - \Pi$, we obtain the equations of motion:

$$\frac{d}{dt}\left(\frac{\partial L}{\partial \dot{x}_1}\right) - \frac{\partial L}{\partial x_1} = m_1\ddot{x}_1 + k_1 x_1 - k_2(x_2 - x_1) = 0$$

$$\frac{d}{dt}\left(\frac{\partial L}{\partial \dot{x}_2}\right) - \frac{\partial L}{\partial x_2} = m_2\ddot{x}_2 + k_2(x_2 - x_1) = 0$$

which can be written in the form

$$\begin{bmatrix} m_1 & 0 \\ 0 & m_2 \end{bmatrix}\begin{Bmatrix} \ddot{x}_1 \\ \ddot{x}_2 \end{Bmatrix} + \begin{bmatrix} (k_1 + k_2) & -k_2 \\ -k_2 & k_2 \end{bmatrix}\begin{Bmatrix} x_1 \\ x_2 \end{Bmatrix} = \mathbf{0}$$

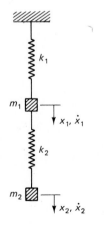

Figure E11.1

which is of the form

$$\mathbf{M\ddot{x}} + \mathbf{Kx} = \mathbf{0} \tag{11.4}$$

where $\mathbf{M}$ is the mass matrix, $\mathbf{K}$ is the stiffness matrix, and $\mathbf{\ddot{x}}$ and $\mathbf{x}$ are vectors representing accelerations and displacements.   ■

Solid Body with Distributed Mass

We now consider a solid body with distributed mass (Fig. 11.1). The potential energy term has already been considered in Chapter 1. The kinetic energy is given by

$$T = \frac{1}{2} \int_V \mathbf{\dot{u}}^T \mathbf{\dot{u}} \rho \, dV \tag{11.5}$$

where ρ is the density (mass per unit volume) of the material and $\mathbf{\dot{u}}$ is the velocity vector of point at $\mathbf{x}$, with components $\dot{u}$, $\dot{v}$, and $\dot{w}$.

$$\mathbf{\dot{u}} = [\dot{u}, \dot{v}, \dot{w}]^T \tag{11.6}$$

In the finite element method, we divide the body into elements, and in each element we express $\mathbf{u}$ in terms of the nodal displacements $\mathbf{q}$, using shape functions $\mathbf{N}$. Thus,

$$\mathbf{u} = \mathbf{Nq} \tag{11.7}$$

In dynamic analysis, the elements of $\mathbf{q}$ are dependent on time, while $\mathbf{N}$ represents (spatial) shape functions defined on a master element. The velocity vector is then given by

$$\mathbf{\dot{u}} = \mathbf{N\dot{q}} \tag{11.8}$$

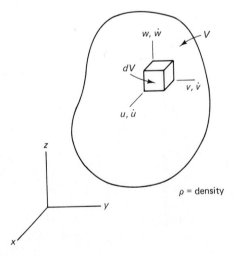

Figure 11.1 Body with distributed mass.

When Eq. 11.8 is substituted into Eq. 11.5, the kinetic energy T_e in element e is

$$T_e = \tfrac{1}{2}\dot{\mathbf{q}}^{\mathsf{T}}\left[\int_e \rho \mathbf{N}^{\mathsf{T}}\mathbf{N}\ dV\right]\dot{\mathbf{q}} \tag{11.9}$$

where the bracketed expression is the element mass matrix

$$\mathbf{m}^e = \int_e \rho \mathbf{N}^{\mathsf{T}}\mathbf{N}\ dV \tag{11.10}$$

This mass matrix is consistent with the shape functions chosen and is called the *consistent mass matrix*. Mass matrices for various elements are given in the next section. On taking summation over all the elements, we get

$$T = \sum_e T_e = \sum_e \tfrac{1}{2}\dot{\mathbf{q}}^{\mathsf{T}}\mathbf{m}^e\dot{\mathbf{q}} = \tfrac{1}{2}\dot{\mathbf{Q}}^{\mathsf{T}}\mathbf{M}\dot{\mathbf{Q}} \tag{11.11}$$

The potential energy is given by

$$\Pi = \tfrac{1}{2}\mathbf{Q}^{\mathsf{T}}\mathbf{K}\mathbf{Q} - \mathbf{Q}^{\mathsf{T}}\mathbf{F} \tag{11.12}$$

Using the Lagrangean $L = T - \Pi$, we obtain the equations of motion:

$$\mathbf{M}\ddot{\mathbf{Q}} + \mathbf{K}\mathbf{Q} = \mathbf{F} \tag{11.13}$$

For free vibrations the force $\mathbf{F}$ is zero. Thus,

$$\mathbf{M}\ddot{\mathbf{Q}} + \mathbf{K}\mathbf{Q} = \mathbf{0} \tag{11.14}$$

For the steady-state condition, starting from the equilibrium state, we set

$$\mathbf{Q} = \mathbf{U}\sin\omega t \tag{11.15}$$

where $\mathbf{U}$ is the vector of nodal amplitudes of vibration and ω (rad/s) is the circular frequency ($=2\pi f$, f = cycles/s or Hz). Introducing Eq. 11.15 into Eq. 11.14, we have

$$\mathbf{K}\mathbf{U} = \omega^2\mathbf{M}\mathbf{U} \tag{11.16}$$

This is the generalized eigenvalue problem

$$\mathbf{K}\mathbf{U} = \lambda\mathbf{M}\mathbf{U} \tag{11.17}$$

where $\mathbf{U}$ is the eigenvector, representing the vibrating mode, corresponding to the eigenvalue λ. The eigenvalue λ is the square of the circular frequency ω. The frequency f in hertz (cycles per second) is obtained from $f = \omega/(2\pi)$.

The above equations can also be obtained by using D'Alembert's principle and the principle of virtual work. Galerkin's approach applied to equations of motion of an elastic body also yields the above set of equations.

11.3 ELEMENT MASS MATRICES

Since the shape functions for various elements have been discussed in detail in the earlier chapters, we now give the element mass matrices. Treating the material density ρ to be constant over the element, we have from Eq. 11.10

$$\mathbf{m}^e = \rho \int_e \mathbf{N}^T \mathbf{N} \, dV \tag{11.18}$$

One-dimensional bar element.   For the bar element shown in Fig. 11.2,

$$\mathbf{q}^T = [q_1 \quad q_2]$$
$$\mathbf{N} = [N_1 \quad N_2] \tag{11.19}$$

where

$$N_1 = \frac{1 - \xi}{2} \qquad N_2 = \frac{1 + \xi}{2}$$

$$\mathbf{m}^e = \rho \int_e \mathbf{N}^T \mathbf{N} A \, dx = \frac{\rho A_e \ell_e}{2} \int_{-1}^{+1} \mathbf{N}^T \mathbf{N} \, d\xi$$

On carrying out the integration of each term in $\mathbf{N}^T \mathbf{N}$, we find

$$\mathbf{m}^e = \frac{\rho A_e \ell_e}{6} \begin{bmatrix} 2 & 1 \\ 1 & 2 \end{bmatrix} \tag{11.20}$$

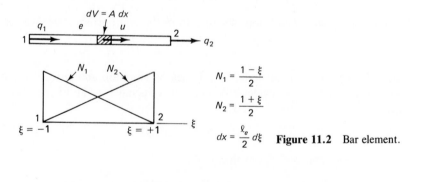

$$N_1 = \frac{1 - \xi}{2}$$
$$N_2 = \frac{1 + \xi}{2}$$
$$dx = \frac{\ell_e}{2} d\xi$$

Figure 11.2   Bar element.

Truss element.   For the truss element shown in Fig. 11.3,

$$\mathbf{u}^T = [u \quad v]$$
$$\mathbf{q}^T = [q_1 \quad q_2 \quad q_3 \quad q_4]$$
$$\mathbf{N} = \begin{bmatrix} N_1 & 0 & N_2 & 0 \\ 0 & N_1 & 0 & N_2 \end{bmatrix} \tag{11.21}$$

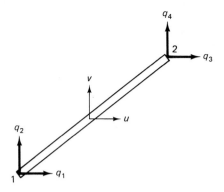

Figure 11.3   Truss element.

where

$$N_1 = \frac{1 - \xi}{2} \qquad N_2 = \frac{1 + \xi}{2}$$

and ξ is defined from -1 to $+1$. Then,

$$\mathbf{m}^e = \frac{\rho A_e \ell_e}{6} \begin{bmatrix} 2 & 0 & 1 & 0 \\ 0 & 2 & 0 & 1 \\ 1 & 0 & 2 & 0 \\ 0 & 1 & 0 & 2 \end{bmatrix} \tag{11.22}$$

CST element.   For the plane stress, plane strain conditions for the CST element shown in Fig. 11.4, we have from Chapter 5

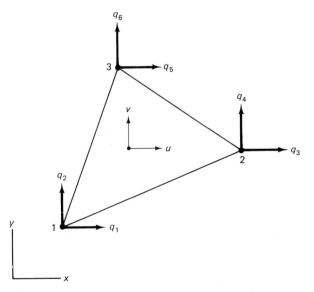

Figure 11.4   CST element.

$$\mathbf{u}^\mathrm{T} = [u \quad v]$$

$$\mathbf{q}^\mathrm{T} = [q_1 \quad q_2 \quad \cdots \quad q_6]$$

$$\mathbf{N} = \begin{bmatrix} N_1 & 0 & N_2 & 0 & N_3 & 0 \\ 0 & N_1 & 0 & N_2 & 0 & N_3 \end{bmatrix}$$

(11.23)

The element mass matrix is then given by

$$\mathbf{m}^e = \rho t_e \int_e \mathbf{N}^\mathrm{T} \mathbf{N} \, dA$$

Noting that $\int_e N_1^2 \, dA = \frac{1}{6} A_e$, $\int_e N_1 N_2 \, dA = \frac{1}{12} A_e$, etc., we have

$$\mathbf{m}^e = \frac{\rho t_e A_e}{12} \begin{bmatrix} 2 & 0 & 1 & 0 & 1 & 0 \\ & 2 & 0 & 1 & 0 & 1 \\ & & 2 & 0 & 1 & 0 \\ & & & 2 & 0 & 1 \\ & \text{Symmetric} & & & 2 & 0 \\ & & & & & 2 \end{bmatrix}$$

(11.24)

Axisymmetric triangular element. For the axisymmetric triangular element, we have

$$\mathbf{u}^\mathrm{T} = [u \quad w]$$

where u and w are the radial and axial displacements, respectively. The vectors $\mathbf{q}$ and $\mathbf{N}$ are similar to those for the triangular element given in Eq. 11.23. We have

$$m^e = \int_e \rho \mathbf{N}^\mathrm{T} \mathbf{N} \, dV = \int_e \rho \mathbf{N}^\mathrm{T} \mathbf{N} 2\pi r \, dA$$

(11.25)

Since $r = N_1 r_1 + N_2 r_2 + N_3 r_3$, we have

$$m^e = 2\pi \rho \int_e (N_1 r_1 + N_2 r_2 + N_3 r_3) \mathbf{N}^\mathrm{T} \mathbf{N} \, dA$$

Noting that

$$\int_e N_1^3 \, dA = \frac{2A_e}{20}, \int_e N_1^2 N_2 \, dA = \frac{2A_e}{60}, \int_e N_1 N_2 N_3 \, dA = \frac{2A_e}{120}, \text{ etc.}$$

we get

$$\mathbf{m}_e = \frac{\pi \rho A_e}{10} \begin{bmatrix} \frac{4}{3}r_1 + 2\bar{r} & 0 & 2\bar{r} - \frac{r_3}{3} & 0 & 2\bar{r} - \frac{r_2}{3} & 0 \\ & \frac{4}{3}r_1 + 2\bar{r} & 0 & 2\bar{r} - \frac{r_3}{3} & 0 & 2\bar{r} - \frac{r_2}{3} \\ & & \frac{4}{3}r_2 + 2\bar{r} & 0 & 2\bar{r} - \frac{r_1}{3} & 0 \\ & & & \frac{4}{3}r_2 + 2\bar{r} & 0 & 2\bar{r} - \frac{r_1}{3} \\ & \text{Symmetric} & & & \frac{4}{3}r_3 + 2\bar{r} & 0 \\ & & & & & \frac{4}{3}r_3 + 2\bar{r} \end{bmatrix}$$

$$(11.26)$$

where

$$\bar{r} = \frac{r_1 + r_2 + r_3}{3} \tag{11.27}$$

Quadrilateral element.   For the quadrilateral element for plane stress and plane strain,

$$\mathbf{u}^{\mathrm{T}} = [u \quad v]$$

$$\mathbf{q}^{\mathrm{T}} = [q_1 \quad q_2 \quad \cdots \quad q_8] \tag{11.28}$$

$$\mathbf{N} = \begin{bmatrix} N_1 & 0 & N_2 & 0 & N_3 & 0 & N_4 & 0 \\ 0 & N_1 & 0 & N_2 & 0 & N_3 & 0 & N_4 \end{bmatrix}$$

The mass matrix is then given by

$$\mathbf{m}^e = \rho t_e \int_{-1}^{1} \int_{-1}^{1} \mathbf{N}^{\mathrm{T}} \mathbf{N} \det \mathbf{J} \, d\xi \, d\eta \tag{11.29}$$

The integral above needs to be evaluated by numerical integration.

Beam element.   For the beam element shown in Fig. 11.5, we use the Hermite shape functions given in Chapter 8. We have

$$v = \mathbf{Hq}$$

$$\mathbf{m}^e = \int_{-1}^{+1} \mathbf{H}^{\mathrm{T}} \mathbf{H} \rho A_e \frac{\ell_e}{2} \, d\xi \tag{11.30}$$

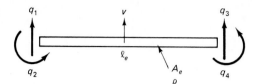

Figure 11.5 Beam element.

On integrating, we get

$$\mathbf{m}^e = \frac{\rho A_e \ell_e}{420} \begin{bmatrix} 156 & 22\ell_e & 54 & -13\ell_e \\ & 4\ell_e^2 & 13\ell_e & -3\ell_e^2 \\ \text{Symmetric} & & 156 & -22\ell_e \\ & & & 4\ell_e^2 \end{bmatrix} \tag{11.31}$$

Frame element. We refer to Fig. 8.9, showing the frame element. In the body coordinate system x', y', the mass matrix can be seen as a combination of bar element and beam element. Thus, the mass matrix in the prime system is given by

$$\mathbf{m}^{e'} = \begin{bmatrix} 2a & 0 & 0 & a & 0 & 0 \\ & 156b & 22\ell_e b & 0 & 54b & -13\ell_e b \\ & & 4\ell_e^2 b & 0 & 13\ell_e b & -3\ell_e^2 b \\ & & & 2a & 0 & 0 \\ \text{Symmetric} & & & & 156b & -22\ell_e b \\ & & & & & 4\ell_e^2 b \end{bmatrix} \tag{11.32}$$

where

$$a = \frac{\rho A_e \ell_e}{6} \quad \text{and} \quad b = \frac{\rho A_e \ell_e}{420}$$

Using the transformation matrix $\mathbf{L}$ given by Eq. 8.48, we now obtain the mass matrix $\mathbf{m}^e$ in the global system:

$$\mathbf{m}^e = \mathbf{L}^{\mathsf{T}} \mathbf{m}^{e'} \mathbf{L} \tag{11.33}$$

Tetrahedral element. For the tetrahedral element presented in Chapter 9,

$$\mathbf{u}^{\mathsf{T}} = [u \quad v \quad w]$$

$$\mathbf{N} = \begin{bmatrix} N_1 & 0 & 0 & N_2 & 0 & 0 & N_3 & 0 & 0 & N_4 & 0 & 0 \\ 0 & N_1 & 0 & 0 & N_2 & 0 & 0 & N_3 & 0 & 0 & N_4 & 0 \\ 0 & 0 & N_1 & 0 & 0 & N_2 & 0 & 0 & N_3 & 0 & 0 & N_4 \end{bmatrix} \tag{11.34}$$

The mass matrix of the element is then given by

$$\mathbf{m}^e = \frac{\rho V_e}{20} \begin{bmatrix} 2 & 0 & 0 & 1 & 0 & 0 & 1 & 0 & 0 & 1 & 0 & 0 \\ & 2 & 0 & 0 & 1 & 0 & 0 & 1 & 0 & 0 & 1 & 0 \\ & & 2 & 0 & 0 & 1 & 0 & 0 & 1 & 0 & 0 & 1 \\ & & & 2 & 0 & 0 & 1 & 0 & 0 & 1 & 0 & 0 \\ & & & & 2 & 0 & 0 & 1 & 0 & 0 & 1 & 0 \\ & \text{Symmetric} & & & & 2 & 0 & 0 & 1 & 0 & 0 & 1 \\ & & & & & & 2 & 0 & 0 & 1 & 0 & 0 \\ & & & & & & & 2 & 0 & 0 & 1 & 0 \\ & & & & & & & & 2 & 0 & 0 & 1 \\ & & & & & & & & & 2 & 0 & 0 \\ & & & & & & & & & & 2 & 0 \\ & & & & & & & & & & & 2 \end{bmatrix} \qquad (11.35)$$

Lumped mass matrices.   Consistent mass matrices have been presented above. Practicing engineers also used lumped mass techniques, where the total element mass in each direction is distributed equally to the nodes of the element and the masses are associated with translational degrees of freedom only. For the truss element, the lumped mass approach gives a mass matrix of

$$\mathbf{m}^e = \frac{\rho A_e \ell_e}{2} \begin{bmatrix} 1 & 0 & 0 & 0 \\ & 1 & 0 & 0 \\ & & 1 & 0 \\ \text{Symmetric} & & & 1 \end{bmatrix} \qquad (11.36)$$

For the beam element, the lumped element mass matrix is

$$\mathbf{m}^e = \frac{\rho A_e \ell_e}{2} \begin{bmatrix} 1 & 0 & 0 & 0 \\ & 0 & 0 & 0 \\ & & 1 & 0 \\ \text{Symmetric} & & & 0 \end{bmatrix} \qquad (11.37)$$

Consistent mass matrices yield more accurate results for flexural elements such as beams. The lumped mass technique is easier to handle since only diagonal elements are involved. The natural frequencies obtained from lumped mass techniques are lower than the exact values. In our presentation we discuss techniques for determining the eigenvalues and eigenvectors with consistent mass formulations. The programs presented can be used for lumped mass cases also.

11.4 EVALUATION OF EIGENVALUES AND EIGENVECTORS

The generalized problem in free vibration is that of evaluating an eigenvalue λ $(=\omega^2)$, which is a measure of the frequency of vibration together with the corresponding eigenvector **U** indicating the mode shape, as in Eq. 11.17, which is re-stated here:

$$KU = \lambda MU \tag{11.38}$$

We observe here that **K** and **M** are symmetric matrices. Further, **K** is positive definite for properly constrained problems.

Properties of Eigenvectors

For a positive definite symmetric stiffness matrix of size n, there are n real eigenvalues and corresponding eigenvectors satisfying Eq. 11.38. The eigenvalues may be arranged in ascending order:

$$0 \le \lambda_1 \le \lambda_2 \le \cdots \le \lambda_n \tag{11.39}$$

If $U_1, U_2, \ldots, U_n$ are the corresponding eigenvectors, we have

$$KU_i = \lambda_i MU_i \tag{11.40}$$

The eigenvectors possess the property of being orthogonal with respect to both the stiffness and mass matrices:

$$U_i^T MU_j = 0 \qquad \text{if } i \ne j \tag{11.41a}$$

$$U_i^T KU_j = 0 \qquad \text{if } i \ne j \tag{11.41b}$$

The lengths of eigenvectors are generally normalized so that

$$U_i^T MU_i = 1 \tag{11.42a}$$

The above normalization of the eigenvectors leads to the relation

$$U_i^T KU_i = \lambda_i \tag{11.42b}$$

In many codes, other normalization schemes are also used. The length of an eigenvector may be fixed by setting its largest component to a preset value, say unity.

Eigenvalue–Eigenvector Evaluation

The eigenvalue–eigenvector evaluation procedures fall into the following basic categories:

1. Characteristic polynomial technique
2. Vector iteration methods
3. Transformation methods

Characteristic polynomial.   From Eq. 11.38, we have

$$(K - \lambda M)U = 0 \tag{11.43}$$

If the eigenvector is to be nontrivial, the required condition is

$$\det(\mathbf{K} - \lambda\mathbf{M}) = 0 \tag{11.44}$$

This represents the characteristic polynomial in λ.

Example 11.2

Determine the eigenvalues and eigenvectors for the stepped bar shown in Fig. E11.2a.

Solution   Gathering the stiffness and mass values corresponding to the degrees of freedom Q_2 and Q_3, we get the eigenvalue problem:

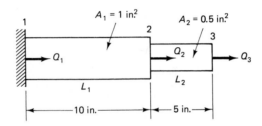

$$E\begin{bmatrix} \left(\dfrac{A_1}{L_1} + \dfrac{A_2}{L_2}\right) & -\dfrac{A_2}{L_2} \\[2mm] -\dfrac{A_2}{L_2} & \dfrac{A_2}{L_2} \end{bmatrix} \begin{Bmatrix} U_2 \\ U_3 \end{Bmatrix} = \lambda\frac{\rho}{6}\begin{bmatrix} 2(A_1 L_1 + A_2 L_2) & A_2 L_2 \\ A_2 L_2 & 2 A_2 L_2 \end{bmatrix}\begin{Bmatrix} U_2 \\ U_3 \end{Bmatrix}$$

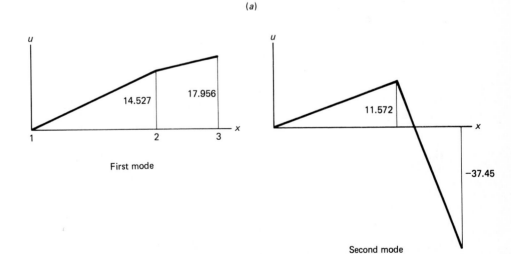

$A_1 = 1$ in.^{2}

$A_2 = 0.5$ in.^{2}

L_1

L_2

10 in.

5 in.

$E = 30 \times 10^6$ psi

Specific weight $f = 0.283$ lb/in.^{3}

(a)

14.527

17.956

First mode

11.572

−37.45

Second mode

(b)

Figure E11.2

We note here that the density

$$\rho = \frac{f}{g} = \frac{0.283}{32.2 \times 12} = 7.324 \times 10^{-4} \text{ lbs}^2/\text{in.}^4$$

Substituting the values we get

$$30 \times 10^6 \begin{bmatrix} 0.2 & -0.1 \\ -0.1 & 0.1 \end{bmatrix} \begin{Bmatrix} U_2 \\ U_3 \end{Bmatrix} = \lambda\, 1.22 \times 10^{-4} \begin{bmatrix} 25 & 2.5 \\ 2.5 & 5 \end{bmatrix} \begin{Bmatrix} U_2 \\ U_3 \end{Bmatrix}$$

The characteristic equation is

$$\det \begin{bmatrix} (6 \times 10^6 - 30.5 \times 10^{-4}\lambda) & (-3 \times 10^6 - 3.05 \times 10^{-4}\lambda) \\ (-3 \times 10^6 - 3.05 \times 10^{-4}\lambda) & (3 \times 10^6 - 6.1 \times 10^{-4}\lambda) \end{bmatrix} = 0$$

which simplifies to

$$1.77 \times 10^{-6}\lambda^2 - 1.465 \times 10^4\lambda + 9 \times 10^{12} = 0$$

The eigenvalues are

$$\lambda_1 = 6.684 \times 10^8$$

$$\lambda_2 = 7.61 \times 10^9$$

Note that $\lambda = \omega^2$, where ω is the circular frequency given by $2\pi f$, f = frequency in hertz (cycles/s).

The above frequencies are

$$f_1 = 4115 \text{ Hz}$$

$$f_2 = 13\ 884 \text{ Hz}$$

The eigenvector for λ_1 is found from

$$(\mathbf{K} - \lambda_1\mathbf{M})\mathbf{U}_1 = \mathbf{0}$$

which gives

$$10^6 \begin{bmatrix} 3.96 & -3.204 \\ -3.204 & 2.592 \end{bmatrix} \begin{Bmatrix} U_2 \\ U_3 \end{Bmatrix}_1 = \mathbf{0}$$

The two equations above are not independent since the determinant of the matrix is zero. This gives

$$3.96U_2 = 3.204U_3$$

Thus,

$$\mathbf{U}_1^\mathsf{T} = [U_2,\ 1.236U_2]$$

For normalization, we set

$$\mathbf{U}_1^\mathsf{T}\mathbf{M}\mathbf{U}_1 = 1$$

On substituting for $\mathbf{U}_1$, we get

$$\mathbf{U}_1^\mathsf{T} = [14.527 \quad 17.956]$$

The eigenvector corresponding to the second eigenvalue is similarly found to be

$$\mathbf{U}_2^{\mathsf{T}} = [11.572 \quad -37.45]$$

The mode shapes are shown in Fig. E11.2*b*.   ∎

Implementation of characteristic polynomial approach on computers is rather tedious and requires further mathematical considerations. We now discuss the other two categories.

Vector iteration methods.   Various vector iteration methods use the properties of the *Rayleigh quotient*. For the generalized eigenvalue problem given by Eq. 11.38, we define the Rayleigh quotient $Q(\mathbf{v})$,

$$Q(\mathbf{v}) = \frac{\mathbf{v}^{\mathsf{T}}\mathbf{K}\mathbf{v}}{\mathbf{v}^{\mathsf{T}}\mathbf{M}\mathbf{v}} \tag{11.45}$$

where $\mathbf{v}$ is arbitrary vector. A fundamental property of the Rayleigh quotient is that it lies between the smallest and the largest eigenvalue,

$$\lambda_1 \le Q(\mathbf{v}) \le \lambda_n \tag{11.46}$$

Power iteration, inverse iteration, and subspace iteration methods use this property. Power iteration leads to evaluation of the largest eigenvalue. Subspace iteration technique is suitable for large-scale problems and is used in several codes. The inverse iteration scheme can be used in evaluating the lowest eigenvalues. We present here the inverse iteration scheme and give a computer program for banded matrices.

Inverse Iteration Method.   In the inverse iteration scheme, we start with a trial vector $\mathbf{u}^0$. The iterative solution proceeds as follows:

Step 0. Estimate an initial trial vector $\mathbf{u}^0$. Set iteration index $k = 0$.

Step 1. Set $k = k + 1$

Step 2. Determine right side $\mathbf{v}^{k-1} = \mathbf{M}\mathbf{u}^{k-1}$

Step 3. Solve equations $\mathbf{K}\hat{\mathbf{u}}^k = \mathbf{v}^{k-1}$

Step 4. Denote $\hat{\mathbf{v}}^k = \mathbf{M}\hat{\mathbf{u}}^k$

Step 5. Estimate eigenvalue $\lambda^k = \dfrac{\hat{\mathbf{u}}^{k\mathsf{T}}\mathbf{v}^{k-1}}{\hat{\mathbf{u}}^{k\mathsf{T}}\hat{\mathbf{v}}^k}$ (11.47)

Step 6. Normalize eigenvector $\mathbf{u}^k = \dfrac{\hat{\mathbf{u}}^k}{(\hat{\mathbf{u}}^{k\mathsf{T}}\hat{\mathbf{v}}^k)^{1/2}}$

Step 7. Check for tolerance $\left| \dfrac{\lambda^k - \lambda^{k-1}}{\lambda^k} \right| \le$ tolerance

   If satisfied, denote the eigenvector $\mathbf{u}^k$ as $\mathbf{U}$ and exit. Otherwise, go to step 1.

The above algorithm converges to the lowest eigenvalue, provided the trial vector does not coincide with one of the eigenvectors. Other eigenvalues can be evaluated by shifting, or by taking the trial vector from a space that is M orthogonal to the calculated eigenvectors.

Shifting.   We define a shifted stiffness matrix as

$$\mathbf{K}_s = \mathbf{K} + s\mathbf{M} \tag{11.48}$$

where $\mathbf{K}_s$ is the shifted matrix. Now the shifted eigenvalue problem is

$$\mathbf{K}_s\mathbf{U} = \lambda_s\mathbf{M}\mathbf{U} \tag{11.49}$$

We state here without proof that the eigenvectors of the shifted problem are the same as those of the original problem. The eigenvalues get shifted by s.

$$\lambda_s = \lambda + s \tag{11.50}$$

By shifting nearer to one of the eigenvalues, the inverse iteration scheme can be made to converge to that eigenvalue.

Orthogonal Space.   Higher eigenvalues can be obtained by the inverse iteration method by choosing the trial vector from a space M orthogonal to the calculated eigenvectors. This can be done effectively by the Gram–Schmidt process. Let U_1, $U_2, \ldots, U_m$ be the first m eigenvectors already determined. The trial vector for each iteration is then taken as

$$\mathbf{u}^{k-1} = \mathbf{u}^{k-1} - (\mathbf{u}^{k-1^{\mathrm{T}}}\mathbf{M}\mathbf{U}_1)\mathbf{U}_1 - (\mathbf{u}^{k-1^{\mathrm{T}}}\mathbf{M}\mathbf{U}_2)\mathbf{U}_2 - \cdots - (\mathbf{u}^{k-1^{\mathrm{T}}}\mathbf{M}\mathbf{U}_m)\mathbf{U}_m \tag{11.51}$$

This is the Gram–Schmidt process, which results in the evaluation of λ_{m+1} and $\mathbf{U}_{m+1}$. This is used in the program given in this chapter. Equation 11.51 is implemented after step 1 in the algorithm given above.

Example 11.3.

Evaluate the lowest eigenvalue and the corresponding eigenmode for the beam shown in Fig. E11.3a.

Solution   Using only the degrees of freedom Q_3, Q_4, Q_5, and Q_6, we obtain the stiffness and mass matrices:

$$\mathbf{K} = 10^3 \begin{bmatrix} 355.56 & 0 & -177.78 & 26.67 \\ & 10.67 & -26.67 & 2.667 \\ \text{Symmetric} & & 177.78 & -26.67 \\ & & & 5.33 \end{bmatrix}$$

$$\mathbf{M} = \begin{bmatrix} 0.4193 & 0 & 0.0726 & -.0052 \\ & .000967 & .0052 & -.00036 \\ \text{Symmetric} & & 0.2097 & -.0089 \\ & & & .00048 \end{bmatrix}$$

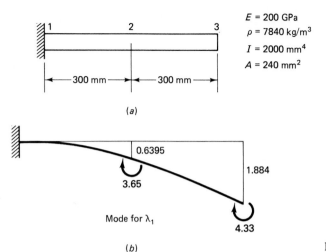

$E = 200$ GPa
$\rho = 7840$ kg/m³
$I = 2000$ mm⁴
$A = 240$ mm²

(a)

0.6395

1.884

3.65

Mode for λ_1

4.33

(b)   **Figure E11.3**

The inverse iteration program requires the creation of a data file. The format of the file for the above problem is as follows:

Data File

```
              4                 4
Banded Stiffness Matrix
      3.556E5              0          -1.778E5      2.667E4
      1.067E4         -2.667E4         2.667E3         0
      1.778E5         -2.667E4            0            0
      5.333E3              0              0            0
Banded Mass Matrix
      0.4193               0           0.0726      -0.0052
      0.000967         0.0052        -0.00036         0
      0.2097          -0.0089            0            0
      0.00048              0              0            0
```

The first line of data contains the values of n = dimension of the matrices, and nbw = half-band width. This is followed by **K** and **M** matrices in banded form (see Chapter 2). The two titles are part of the data file. Though these data were created by hand calculations, it is possible to write a program, as discussed at the end of this chapter.

Feeding the above data file into the inverse iteration program, INVITR, gives the lowest eigenvalue

$$\lambda_1 = 2.03 \times 10^4$$

and the corresponding eigenvector or mode shape

$$\mathbf{U}_1^T = [0.64, 3.65, 1.88, 4.32]$$

λ_1 corresponds to a circular frequency of 142.48 rad/s or 22.7 Hz ($=142.48/2\pi$). The mode shape is shown in Fig. E11.3b.   ∎

Transformation methods.   The basic approach here is to transform the matrices to a simpler form and then determine the eigenvalues and eigenvectors. The major methods in this category are the generalized Jacobi method and the QR method. These methods are suitable for large-scale problems. In the QR method, the matrices are first reduced to tridiagonal form using Householder matrices. The generalized Jacobi method uses the transformation to simultaneously diagonalize the stiffness and mass matrices. This method needs the full matrix locations and is quite efficient for calculating all eigenvalues and eigenvectors for small problems. We present here the generalized Jacobi method as an illustration of the transformation approach.

If all the eigenvectors are arranged as columns of a square matrix $\mathbb{U}$, and all eigenvalues as the diagonal elements of a square matrix Λ, then the generalized eigenvalue problem can be written in the form

$$\mathbf{K}\mathbb{U} = \mathbf{M}\mathbb{U}\Lambda \tag{11.52}$$

where

$$\mathbb{U} = [\mathbf{U}_1, \mathbf{U}_2, \ldots, \mathbf{U}_n] \tag{11.53}$$

$$\Lambda = \begin{bmatrix} \lambda_1 & & & \\ & \lambda_2 & & 0 \\ 0 & & \ddots & \\ & & & \lambda_n \end{bmatrix} \tag{11.54}$$

Using the M-orthonormality of the eigenvectors, we have

$$\mathbb{U}^\mathsf{T}\mathbf{K}\mathbb{U} = \Lambda \tag{11.55a}$$

and

$$\mathbb{U}^\mathsf{T}\mathbf{M}\mathbb{U} = \mathbf{I} \tag{11.55b}$$

where $\mathbf{I}$ is the identity matrix.

Generalized Jacobi Method

In the generalized Jacobi method, a series of transformations $\mathbf{P}_1, \mathbf{P}_2, \ldots, \mathbf{P}_\ell$ are used such that if $\mathbf{P}$ represents the product

$$\mathbf{P} = \mathbf{P}_1\mathbf{P}_2 \cdots \mathbf{P}_\ell \tag{11.56}$$

then the off-diagonal terms of $\mathbf{P}^\mathsf{T}\mathbf{K}\mathbf{P}$ and $\mathbf{P}^\mathsf{T}\mathbf{M}\mathbf{P}$ are zero. In practice, the off-diagonal terms are set to be less than a small tolerance. If we denote these diagonal matrices as

$$\hat{\mathbf{K}} = \mathbf{P}^\mathsf{T}\mathbf{K}\mathbf{P} \tag{11.57a}$$

and

$$\hat{\mathbf{M}} = \mathbf{P}^{\mathrm{T}}\mathbf{M}\mathbf{P} \tag{11.57b}$$

then the eigenvectors are given by

$$\mathbb{U} = \hat{\mathbf{M}}^{-1/2}\mathbf{P} \tag{11.58}$$

and

$$\mathbf{\Lambda} = \hat{\mathbf{M}}^{-1}\hat{\mathbf{K}} \tag{11.59}$$

where

$$\hat{\mathbf{M}}^{-1} = \begin{bmatrix} \hat{M}_{11}^{-1} & & & 0 \\ & \hat{M}_{22}^{-1} & & \\ & & \ddots & \\ 0 & & & \hat{M}_{nn}^{-1} \end{bmatrix} \tag{11.60}$$

$$\hat{\mathbf{M}}^{-1/2} = \begin{bmatrix} \hat{M}_{11}^{-1/2} & & & 0 \\ & \hat{M}_{22}^{-1/2} & & \\ & & \ddots & \\ 0 & & & \hat{M}_{nn}^{-1/2} \end{bmatrix} \tag{11.61}$$

Computationally, (11.58) indicates that each row of $\mathbf{P}$ is divided by the square root of the diagonal element of $\hat{\mathbf{M}}$, and Eq. 11.59 indicates that each diagonal element of $\hat{\mathbf{K}}$ is divided by the diagonal element of $\hat{\mathbf{M}}$.

We mentioned that the diagonalization follows in several steps. At step k, we choose a transformation matrix $\mathbf{P}_k$ given by

$$\begin{array}{c} \text{Column} \to \quad i \qquad\qquad j \\ \mathbf{P}_k = \begin{array}{c} \text{Row} \\ \downarrow \\ \\ i \\ \\ \\ \\ j \end{array} \begin{bmatrix} 1 & & & & & & & & \\ & 1 & & & & & & & \\ & & 1 & & & & & & \\ & & & 1 & \cdots & \cdots & \alpha & \cdots & \\ & & & & 1 & & & & \\ & & & & & 1 & & & \\ & & & & & & 1 & & \\ & & \beta & \cdots & \cdots & & 1 & & \\ & & & & & & & 1 & \\ & & & & & & & & 1 \end{bmatrix} \end{array} \tag{11.62}$$

$\mathbf{P}_k$ has all diagonal elements equal to 1, a value of α at row i and column j and β at row j and column i, and has all other elements equal to zero. The scalars α and β are chosen so that the ij locations of $\mathbf{P}_k^{\mathrm{T}}\mathbf{K}\mathbf{P}_k$ and $\mathbf{P}_k^{\mathrm{T}}\mathbf{M}\mathbf{P}_k$ are simultaneously zero.

This is represented by

$$\alpha K_{ii} + (1 + \alpha\beta)K_{ij} + \beta K_{jj} = 0 \tag{11.63}$$

$$\alpha M_{ii} + (1 + \alpha\beta)M_{ij} + \beta M_{jj} = 0 \tag{11.64}$$

where K_{ii}, K_{ij}, . . . , M_{ii}, . . . are elements of the stiffness and mass matrices. The solution of these equations is as follows:

Denote

$$A = K_{ii}M_{ij} - M_{ii}K_{ij}$$

$$B = K_{jj}M_{ij} - M_{jj}K_{ij}$$

$$C = K_{ii}M_{jj} - M_{ii}K_{jj}$$

then α and β are given by

$A \neq 0, B \neq 0$: $\qquad \alpha = \dfrac{-0.5C + \text{sgn}(C)\sqrt{0.25C^2 + AB}}{A}$

$$\beta = -\dfrac{A\alpha}{B}$$

$A = 0$: $\qquad \beta = 0$

$$\alpha = -\dfrac{K_{ij}}{K_{ii}}$$

$B = 0$: $\qquad \alpha = 0$

$$\beta = -\dfrac{K_{ij}}{K_{jj}}$$

When both A and B are zero, any one of the above two values can be chosen. (*Note:* There is no summation on repeated indices in the above expressions.)

In the generalized Jacobi program given at the end of the chapter, the elements of **K** and **M** are zeroed out in the order indicated in Fig. 11.6. Once $\mathbf{P}_k$ is defined by determining α and β, $\mathbf{P}_j^T[\ ]\mathbf{P}_k$ can be performed on **K** and **M** as shown in Fig. 11.7. Also by starting with $\mathbf{P} = \mathbf{I}$, the product $\mathbf{PP}_k$ is computed after each step. When all elements are covered as shown in Fig. 11.6, one pass is completed. After the operations at step k, some of the previously zeroed elements are altered. Another pass is conducted to check for the value of diagonal elements. The transformation is performed if the element at ij is larger than a tolerance value. A tolerance of $10^6 \times$ smallest K_{ii} is used for stiffness and $10^{-6} \times$ largest M_{ii} is used for the mass. The tolerance can be redefined for higher accuracy. The process stops when all off-diagonal elements are less than the tolerance.

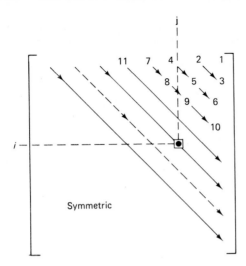

Figure 11.6 Diagonalization.

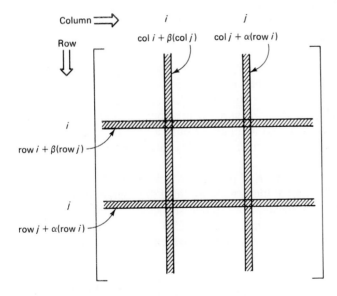

Figure 11.7 Multiplication $\mathbf{P}_{\overline{i}}^T[\ ]\mathbf{P}$

If the diagonal masses are less than the tolerance, the diagonal value is re-placed by the tolerance value, and thus a large eigenvalue will be obtained. In this method, **K** need not be positive definite.

Example 11.4

Determine all the eigenvalues and eigenvectors for the beam discussed in Example 11.3 using program JACOBI.

Solution   The input data for JACOBI is same as that for INVITR. However, the program converts to full matrices in calculations. Convergence occurs at the fourth sweep.

$$\lambda_1 = 2.0304 \times 10^4 \quad (22.7 \text{ Hz})$$

$$U_1^T = [0.64, 3.65, 1.88, 4.32]$$

$$\lambda_2 = 8.0987 \times 10^5 \quad (143.2 \text{ Hz})$$

$$U_2^T = [-1.37, 1.39, 1.901, 15.27]$$

$$\lambda_3 = 9.2651 \times 10^6 \quad (484.4 \text{ Hz})$$

$$U_3^T = [-0.20, 27.16, -2.12, -33.84]$$

$$\lambda_4 = 7.7974 \times 10^7 \quad (1405.4 \text{ Hz})$$

$$U_4^T = [0.8986, 30.89, 3.546, 119.15]$$

Note that the eigenvalues are arranged in ascending order after they are evaluated.   ∎

11.5 INTERFACING WITH PREVIOUS FINITE ELEMENT PROGRAMS AND A PROGRAM FOR DETERMINING CRITICAL SPEEDS OF SHAFTS

Once the stiffness matrix **K** and mass matrix **M** for a structure are known, then the inverse iteration or Jacobi programs that are provided can be used to determine the natural frequencies and mode shapes. The finite element programs for rod, truss, beam, and elasticity problems that we used in previous chapters can be readily modified to output the banded **K** and **M** matrices onto a file. This file is then input into the inverse iteration program which gives the natural frequencies and mode shapes.

We have provided the BEAM_KM program which outputs the banded **K** and **M** matrices for a beam. This output file is then provided to program INVITR, which calculates the eigenvalues and eigenvectors (mode shapes). The following example illustrates the use of these two programs. Program CST_KM, which outputs **K** and **M** matrices for the CST element, has also been provided.

Example 11.5

Determine the lowest critical speed (or transverse natural frequency) of the shaft shown in Fig. E11.5. The shaft has two lumped weights, W_1 and W_2, representing flywheels,

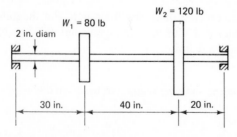

Figure E11.5

as shown. Take $E = 30 \times 10^6$ psi and mass density of shaft $\rho = 0.0007324$ lb.s^2/in.^{4} (=0.283 lb/in^3).

Solution   The lumped weights W_1 and W_2 correspond to lumped masses W_1/g and W_2/g, respectively, where $g = 386$ in./s^2. Program BEAM_KM is executed, followed by program INVITR. The input data and solution are given at the end of the next section.

Now, we can obtain the critical speed in rpm from the eigenvalue 4042 as

$$n = \sqrt{\lambda} \times \frac{60}{2\pi} \text{ rpm}$$

$$= \sqrt{4042} \times \frac{60}{2\pi}$$

$$= 607 \text{ rpm}$$

This example illustrates how the inverse iteration and Jacobi programs given in this chapter can be interfaced with other programs for vibration analysis.   ■

11.6 CONCLUSION

In this chapter, the application of finite elements for free vibrations is discussed in a general setting using consistent mass matrices. Solution techniques and computer programs are given. These programs can be integrated with static analysis programs to get dynamic behavior of structures. Natural frequencies and mode shapes of structures give us the needed data concerning what excitation frequencies should be avoided.

Use of Programs BEAM_KM and INVITR for Example 11.5

```
NUMBER OF ELEMENTS=?  3
NUMBER OF CONSTRAINED DOF=?  2
GIVE NUMBER OF DIFFERENT MATERIALS=?  1
GIVE YOUNG'S MODULUS AND MASS DENSITY FOR EACH MATERIAL
?  30E6,.0007324
GIVE NODE NUMBER AND ITS COORDINATE
?  1,0
?  2,30
?  3,70
?  4,90
GIVE EACH ELEMENT #,MATERIAL #,AREA,AND ITS MOMENT OF INERTIA
?  1,1,3.1416,.7854
?  2,1,3.1416,.7854
?  3,1,3.1416,.7854
GIVE DOF NO, KNOWN DISPLACEMENT
?  1,0
?  7,0
```

```
NUMBER OF SPRINGS =? 0
NUMBER OF LUMPED (POINT) MASSES=? 2
GIVE DOF #, LUMPED MASS=? 3,.2072538
GIVE DOF #, LUMPED MASS=? 5,.310881
GIVE NAME OF OUTPUT FILE=? EG11_5.DAT
```

```
*****  Inverse Iteration Method    *****
** for eigenvalues and eigenvectors ***
*****     Searching in Subspace     *****
*****       for Banded Matrices      *****
```

```
GIVE NAME OF INPUT FILE? EG11_5.DAT
DEFAULT TOLERANCE ISTARTING VECTOR
NUMBER OF EIGENVALUES DESIRED ? 1
                    1. 1 1 1 ...
                    2. Vector of Your Choice ?
                    Your Choice < 1 or 2 >? 1
ITERATION NUMBER   3
EIGENVALUE NUMBER     1
EIGENVALUE IS   4.042E+03
EIGENVECTOR
 1.716E-08   5.527E-02   1.378E+00   2.758E-02   1.050E+00
 -4.219E-02   2.121E-08   -5.
765E-02
```

```
1000 CLS : DEFINT I-N
1010 REM   *****          INVITR          *****
1020 PRINT "***** Inverse Iteration Method   *****"
1030 PRINT "** for eigenvalues and eigenvectors ***"
1040 PRINT "*****     Searching in Subspace    *****"
1050 PRINT "*****      for Banded Matrices      *****"
1060 PRINT "-------------------------------------------"
1070 PRINT "        Program written by         "
1080 PRINT "  T.R.Chandrupatla and A.D.Belegundu   "
1090 PRINT "======================================="
1100 INPUT "GIVE NAME OF INPUT FILE"; FILE1$
1102 INPUT "GIVE NAME OF OUTPUT FILE"; FILE2$
1110 CLS
1120 OPEN FILE1$ FOR INPUT AS #1
1122 OPEN FILE2$ FOR OUTPUT AS #2
1130 REM *** Read in Number of Equations ***
1140 INPUT #1, NQ, NBW
1150 DIM S(NQ, NBW), EM(NQ, NBW), EV1(NQ), EV2(NQ)
1152 DIM EVT(NQ), EVS(NQ), ST(NQ)
1160 TOL = .000001: ITMAX = 50: SH = 0: NEV = 0
1170 INPUT "Tolerance default  < 1E-6>"; A$
1172 IF VAL(A$) > 0 THEN TOL = VAL(A$)
1180 INPUT "Number of Eigenvalues Desired"; NEV
1190 DIM EVC(NQ, NEV)
1200 REM *** Read in Stiffness Matrix ***
1210 LINE INPUT #1, TITLE$
1220 FOR I = 1 TO NQ: FOR J = 1 TO NBW
1230 INPUT #1, S(I, J): NEXT J: NEXT I
```

```
1240 REM *** Read in Mass Matrix ***
1250 LINE INPUT #1, TITLE$
1260 FOR I = 1 TO NQ: FOR J = 1 TO NBW
1270 INPUT #1, EM(I, J): NEXT J: NEXT I
1272 CLOSE #1
1280 LOCATE 8, 20: PRINT "STARTING VECTOR"
1290 LOCATE 10, 20: PRINT "1. 1 1 1 ..."
1300 LOCATE 11, 20: PRINT "2. Vector of Your Choice ?"
1310 LOCATE 12, 20: INPUT "Your Choice < 1 or 2 >"; NSV
1320 FOR I = 1 TO NQ
1330 IF NSV = 1 GOTO 1360
1340 LOCATE 15, 16 + 4 * I: INPUT ST(I)
1350 GOTO 1380
1360 ST(I) = 1!
1380 NEXT I
1390 GOSUB 5000: REM <----Stiffness to Upper Triangle
1400 FOR NV = 1 TO NEV
1410 REM *** Starting Value for Eigenvector ***
1420 FOR I = 1 TO NQ
1430 EV1(I) = ST(I): NEXT I
1440 EL2 = 0: ITER = 0
1450 EL1 = EL2
1460 ITER = ITER + 1
1470 IF ITER > ITMAX GOTO 2140
1480 IF NV = 1 GOTO 1630
1490 REM ---- Starting Vector Orthogonal to ----
1492 REM ----        Evaluated Vectors         ----
1500 FOR I = 1 TO NV - 1
1510 CV = 0
1520 FOR K = 1 TO NQ
1530 KA = K - NBW + 1: KZ = K + NBW - 1
1540 IF KA < 1 THEN KA = 1
1550 IF KZ > NQ THEN KZ = NQ
1560 FOR L = KA TO KZ
1570 IF L < K GOTO 1590
1580 K1 = K: L1 = L - K + 1: GOTO 1600
1590 K1 = L: L1 = K - L + 1
1600 CV = CV + EVS(K) * EM(K1, L1) * EVC(L, I): NEXT L: NEXT K
1610 FOR K = 1 TO NQ
1620 EV1(K) = EV1(K) - CV * EVC(K, I): NEXT K: NEXT I
1630 FOR I = 1 TO NQ
1640 IA = I - NBW + 1: IZ = I + NBW - 1: EVT(I) = 0
1650 IF IA < 1 THEN IA = 1
1660 IF IZ > NQ THEN IZ = NQ
1670 FOR K = IA TO IZ
1680 IF K < I GOTO 1700
1690 I1 = I: K1 = K - I + 1: GOTO 1710
1700 I1 = K: K1 = I - K + 1
1710 EVT(I) = EVT(I) + EM(I1, K1) * EV1(K): NEXT K
1720 EV2(I) = EVT(I)
1730 NEXT I
1740 GOSUB 5130: REM <---Reduce Right Side and Solve
1750 C1 = 0: C2 = 0
1760 FOR I = 1 TO NQ
1770 C1 = C1 + EV2(I) * EVT(I): NEXT I
1780 FOR I = 1 TO NQ
1790 IA = I - NBW + 1: IZ = I + NBW - 1: EVT(I) = 0
1800 IF IA < 1 THEN IA = 1
1810 IF IZ > NQ THEN IZ = NQ
1820 FOR K = IA TO IZ
1830 IF K < I GOTO 1850
1840 I1 = I: K1 = K - I + 1: GOTO 1860
1850 I1 = K: K1 = I - K + 1
1860 EVT(I) = EVT(I) + EM(I1, K1) * EV2(K): NEXT K: NEXT I
1870 FOR I = 1 TO NQ
```

```
1880 C2 = C2 + EV2(I) * EVT(I): NEXT I
1890 EL2 = C1 / C2
1900 C2 = SQR(C2)
1910 FOR I = 1 TO NQ
1920 EV1(I) = EV2(I) / C2
1930 EVS(I) = EV1(I): NEXT I
1940 IF ABS(EL2 - EL1) / ABS(EL2) > TOL GOTO 1450
1950 FOR I = 1 TO NQ
1960 EVC(I, NV) = EV1(I): NEXT I
1970 PRINT "Iteration Number "; ITER
1972 PRINT #2, "Iteration Number "; ITER
1980 EL2 = EL2 + SH: EVL(NV) = EL2
1990 PRINT "Eigenvalue Number  "; NV
1992 PRINT #2, "Eigenvalue Number  "; NV
2000 PRINT USING "Eigenvalue is ##.###^^^^"; EL2
2002 PRINT #2, USING "Eigenvalue is ##.####^^^^"; EL2
2010 PRINT "Eigenvector "
2020 FOR I = 1 TO NQ
2030 PRINT USING "##.###^^^^ "; EVC(I, NV),
2032 PRINT #2, USING "##.###^^^^ "; EVC(I, NV),
2034 IFL = I - 7 * INT(I / 7)
2036 IF IFL = 0 THEN PRINT : PRINT #2,
2038 NEXT I
2040 PRINT : PRINT #2, : IF NV = NEV GOTO 2120
2042 PRINT "Default Shift Value "; SH
2050 INPUT "Shift default < return for default > "; A$
2060 IF A$ = "" GOTO 2120
2062 SH = VAL(A$)
2070 OPEN FILE1$ FOR INPUT AS #1
2080 INPUT #1, NQ, NBW
2090 INPUT #1, TITLE$
2100 FOR I = 1 TO NQ: FOR J = 1 TO NBW
2110 INPUT #1, S(I, J)
2112 S(I, J) = S(I, J) - SH * EM(I, J): NEXT J: NEXT I
2114 GOSUB 5000
2116 CLOSE #1
2120 NEXT NV
2130 GOTO 2150
2140 PRINT "No Convergence for "; ITER; " Iterations"
2150 CLOSE #2
2160 END
5000 REM    **** Gauss Elimination LDU Approach ****
5010 REM  --- REDUCTION TO UPPER TRIANGULAR MATRIX ---
5020 FOR K = 1 TO NQ - 1
5030 NK = NQ - K + 1
5040 IF NK > NBW THEN NK = NBW
5050 FOR I = 2 TO NK
5060 C1 = S(K, I) / S(K, 1)
5070 I1 = K + I - 1
5080 FOR J = I TO NK
5090 J1 = J - I + 1
5100 S(I1, J1) = S(I1, J1) - C1 * S(K, J)
5110 NEXT J: NEXT I: NEXT K
5120 RETURN
5130 REM    **** Reduction of the right hand side ****
5140 FOR K = 1 TO NQ - 1
5150 NK = NQ - K + 1
5160 IF NK > NBW THEN NK = NBW
5170 FOR I = 2 TO NK: I1 = K + I - 1
5180 C1 = 1 / S(K, 1)
5190 EV2(I1) = EV2(I1) - C1 * S(K, I) * EV2(K)
5200 NEXT I: NEXT K
5210 REM    **** Back Substitution ****
5220 EV2(NQ) = EV2(NQ) / S(NQ, 1)
5230 FOR I1 = 1 TO NQ - 1
```

```
5240 I = NQ - II: C1 = 1 / S(I, 1)
5250 NI = NQ - I + 1
5260 IF NI > NBW THEN NI = NBW
5270 EV2(I) = C1 * EV2(I)
5280 FOR K = 2 TO NI
5290 EV2(I) = EV2(I) - C1 * S(I, K) * EV2(I + K - 1)
5300 NEXT K: NEXT II
5310 RETURN
5320 END


1000 CLS
1010 PRINT "======================================="
1020 PRINT "***** Generalized Jacobi's Method *****"
1030 PRINT "*****   for symmetric matrices    *****"
1040 PRINT "---------------------------------------"
1070 PRINT "         Program written by         "
1080 PRINT " T.R.Chandrupatla and A.D.Belegundu   "
1090 PRINT "======================================="
1100 INPUT "GIVE NAME OF INPUT FILE"; FILE1$
1102 INPUT "GIVE NAME OF OUTPUT FILE"; FILE2$
1104 OPEN FILE1$ FOR INPUT AS #1
1106 OPEN FILE2$ FOR OUTPUT AS #2
1108 REM *** Read in Number of Equations and Bandwidth ***
1110 INPUT #1, NQ, NBW
1120 DIM S(NQ, NQ), EM(NQ, NQ), EVL(NQ), EVC(NQ, NQ), NORD(NQ)
1122 REM NORD( ) is for ascending order of eigenvalues
1124 FOR I = 1 TO NQ: NORD(I) = I: NEXT I
1130 TOL = .000001#
1140 PRINT "DEFAULT TOLERANCE IS  1E-6"
1150 REM *** BANDED STIFFNESS MATRIX INTO S(NQ,NQ) ***
1152 LINE INPUT #1, TITLE$
1160 FOR I = 1 TO NQ: FOR JN = 1 TO NBW
1162 INPUT #1, STIFF: J = I + JN - 1
1164 IF J > NQ GOTO 1172
1170 S(I, J) = STIFF: S(J, I) = STIFF
1172 NEXT JN: NEXT I
1180 REM *** READ IN MASS MATRIX ***
1182 LINE INPUT #1, TITLE$
1190 FOR I = 1 TO NQ: FOR JN = 1 TO NBW
1192 INPUT #1, EMASS: J = I + JN - 1
1194 IF J > NQ GOTO 1202
1200 EM(I, J) = EMASS: EM(J, I) = EMASS
1202 NEXT JN: NEXT I
1204 CLOSE #1
1210 REM *** INITIALIZE EIGENVECTOR MATRIX ***
1220 FOR I = 1 TO NQ: FOR J = 1 TO NQ
1230 EVC(I, J) = 0: NEXT J
1240 EVC(I, I) = 1: NEXT I
1250 C1 = S(1, 1): C2 = EM(1, 1)
1260 FOR I = 1 TO NQ
1270 IF C1 > S(I, I) THEN C1 = S(I, I)
1280 IF C2 < EM(I, I) THEN C2 = EM(I, I)
1290 NEXT I
1300 TOLS = TOL * C1: TOLM = TOL * C2
1310 INPUT "MAX. NUMBER OF SWEEPS "; NSWMAX
1320 CLS
1330 REM ***** GENERALIZED JACOBI'S METHOD *****
1340 K1 = 1: I1 = 1: NSW = 0
1350 NSW = NSW + 1
1360 IF NSW > NSWMAX THEN GOTO 2130
1370 PRINT "******  SWEEP NUMBER    "; NSW
1380 FOR K = K1 TO NQ - 1
1390 FOR I = I1 TO K
1400 J = NQ - K + I
1410 IF ABS(S(I, J)) <= TOLS AND ABS(EM(I, J)) <= TOLM THEN GOTO 1860
```

```
1420 AA = S(I, I) * EM(I, J) – EM(I, I) * S(I, J)
1430 BB = S(J, J) * EM(I, J) – EM(J, J) * S(I, J)
1440 CC = S(I, I) * EM(J, J) – EM(I, I) * S(J, J)
1450 CAB = .25 * CC * CC + AA * BB
1460 IF CAB < 0 GOTO 2150
1470 IF AA = 0 GOTO 1530
1480 IF BB = 0 GOTO 1550
1490 SQC = SQR(CAB): IF CC < 0 THEN SQC = –SQC
1500 ALP = (–.5 * CC + SQC) / AA
1510 BET = –AA * ALP / BB
1520 GOTO 1560
1530 BET = 0: ALP = –S(I, J) / S(I, I)
1540 GOTO 1560
1550 ALP = 0: BET = –S(I, J) / S(J, J)
1560 REM * ONLY UPPER TRIANGULAR PART IS USED IN DIAGONALIZATION *
1570 IF I = 1 THEN GOTO 1630
1580 FOR N = 1 TO I – 1
1590 SI = S(N, I): SJ = S(N, J): EMI = EM(N, I): EMJ = EM(N, J)
1600 S(N, I) = SI + BET * SJ: S(N, J) = SJ + ALP * SI
1610 EM(N, I) = EMI + BET * EMJ: EM(N, J) = EMJ + ALP * EMI
1620 NEXT N
1630 IF J = NQ THEN GOTO 1690
1640 FOR N = J + 1 TO NQ
1650 SI = S(I, N): SJ = S(J, N): EMI = EM(I, N): EMJ = EM(J, N)
1660 S(I, N) = SI + BET * SJ: S(J, N) = SJ + ALP * SI
1670 EM(I, N) = EMI + BET * EMJ: EM(J, N) = EMJ + ALP * EMI
1680 NEXT N
1690 IF J = I + 1 THEN GOTO 1750
1700 FOR N = I + 1 TO J – 1
1710 SI = S(I, N): SJ = S(N, J): EMI = EM(I, N): EMJ = EM(N, J)
1720 S(I, N) = SI + BET * SJ: S(N, J) = SJ + ALP * SI
1730 EM(I, N) = EMI + BET * EMJ: EM(N, J) = EMJ + ALP * EMI
1740 NEXT N
1750 SII = S(I, I): SIJ = S(I, J): SJJ = S(J, J)
1760 S(I, J) = 0: S(I, I) = SII + 2 * BET * SIJ + BET * BET * SJJ
1770 S(J, J) = SJJ + 2 * ALP * SIJ + ALP * ALP * SII
1780 EII = EM(I, I): EIJ = EM(I, J): EJJ = EM(J, J)
1790 EM(I, J) = 0: EM(I, I) = EII + 2 * BET * EIJ + BET * BET * EJJ
1800 EM(J, J) = EJJ + 2 * ALP * EIJ + ALP * ALP * EII
1810 REM *** EIGENVECTORS ***
1820 FOR N = 1 TO NQ
1830 EVI = EVC(N, I): EVJ = EVC(N, J)
1840 EVC(N, I) = EVI + BET * EVJ: EVC(N, J) = EVJ + ALP * EVI
1850 NEXT N
1860 NEXT I: NEXT K
1870 FOR K = 1 TO NQ – 1
1880 FOR I = 1 TO K
1890 J = NQ – K + I
1900 IF ABS(S(I, J)) <= TOLS AND ABS(EM(I, J)) <= TOLM THEN GOTO 1930
1910 K1 = K: I1 = I
1920 GOTO 1350
1930 NEXT I: NEXT K
1940 REM *** CALCULATION OF EIGENVALUES ***
1950 FOR I = 1 TO NQ
1960 IF ABS(EM(I, I)) < TOLM THEN EM(I, I) = TOLM
1970 EVL(I) = S(I, I) / EM(I, I): NEXT I
1980 REM *** SCALING OF EIGENVECTORS ***
1990 FOR I = 1 TO NQ
2000 EM2 = SQR(ABS(EM(I, I)))
2010 FOR J = 1 TO NQ
2020 EVC(J, I) = EVC(J, I) / EM2: NEXT J: NEXT I
2022 REM *****   RESULTS   ******
2024 REM Ascending Order of Eigenvalues
2026 FOR I = 1 TO NQ
```

```
2028 II = NORD(I): I1 = II
2030 C1 = EVL(II): J1 = I
2032 FOR J = I TO NQ
2034 IJ = NORD(J)
2036 IF C1 <= EVL(IJ) GOTO 2040
2038 C1 = EVL(IJ): I1 = IJ: J1 = J
2040 NEXT J
2042 IF I1 = II GOTO 2044
2043 NORD(I) = I1: NORD(J1) = II
2044 NEXT I
2046 FOR I = 1 TO NQ
2048 II = NORD(I)
2050 PRINT "Eigenvalue Number  "; I
2060 PRINT #2, "Eigenvalue Number  "; I
2070 PRINT USING "Eigenvalue is ##.###^^^^"; EVL(II)
2080 PRINT #2, USING "Eigenvalue is ##.####^^^^"; EVL(II)
2090 PRINT "Eigenvector "
2092 FOR J = 1 TO NQ
2094 PRINT USING "##.###^^^^ "; EVC(J, II),
2096 PRINT #2, USING "##.###^^^^ "; EVC(J, II),
2098 IFL = J - 7 * INT(J / 7)
2100 IF IFL = 0 THEN PRINT : PRINT #2,
2102 NEXT J
2104 PRINT : PRINT #2,
2110 NEXT I
2120 GOTO 2160
2130 PRINT "NO CONVERGENCE": PRINT #2, "No Convergence"
2140 GOTO 2160
2150 PRINT "SQUARE ROOT OF NEGATIVE TERM -- CHECK MATRICES"
2152 PRINT #2, "Square Root of Negative Term -- Check Matrices"
2160 CLOSE #2
2180 END
1000 PRINT "********          BEAM_KM          **********"
1010 PRINT "Beam Mass and Stiffness Matrices calculated in banded form"
1020 PRINT " Output file is then the input file for program INVITR "
1030 PRINT "===================================="
1040 PRINT "        Program written by             "
1050 PRINT " T.R.Chandrupatla and A.D.Belegundu   "
1060 PRINT "===================================="
1070 INPUT "NUMBER OF ELEMENTS="; NE
1080 INPUT "NUMBER OF CONSTRAINED DOF="; ND
1090 INPUT "GIVE NUMBER OF DIFFERENT MATERIALS="; NMG
1100 DIM YM(NMG), RO(NMG)
1110 PRINT "GIVE YOUNG'S MODULUS AND MASS DENSITY FOR EACH MATERIAL"
1120 FOR I = 1 TO NMG
1130 INPUT YM(I), RO(I): NEXT I
1140 NN = NE + 1
1150 NBW = 4
1160 REM ** THE TOTAL DOF IS "NQ" **
1170 NQ = 2 * NN
1180 DIM X(NN), DB(ND), U(ND), MI(NE), AREA(NE)
1190 DIM S(NQ, 4), EM(NQ, 4)
1200 PRINT "GIVE NODE NUMBER AND ITS COORDINATE"
1210 FOR I = 1 TO NN
1220 INPUT N, X(N): NEXT I
1230 PRINT "GIVE EACH ELEMENT #,MATERIAL #,AREA,AND ITS MOMENT OF INERTIA"
1240 DIM MG(NE)
1250 FOR I = 1 TO NE
1260 INPUT N, MG(I), AREA(N), MI(N): NEXT I
1270 FOR I = 1 TO NQ
1280 FOR J = 1 TO NBW
```

```
1290 S(I, J) = 0: NEXT J: NEXT I
1300 PRINT "GIVE DOF NO, KNOWN DISPLACEMENT"
1310 FOR I = 1 TO ND
1320 INPUT DB(I), U(I): NEXT I
1330 REM *** GLOBAL STIFFNESS MATRIX ***
1340 FOR I = 1 TO NE
1350 M = MG(I)
1360 YMM = YM(M)
1370 I1 = 2 * I - 1
1380 EL = ABS(X(I + 1) - X(I))
1390 EIL = YMM * MI(I) / EL ^ 3
1400 S(I1, 1) = S(I1, 1) + 12 * EIL
1410 S(I1, 2) = S(I1, 2) + EIL * 6 * EL
1420 S(I1, 3) = S(I1, 3) - 12 * EIL
1430 S(I1, 4) = S(I1, 4) + EIL * 6 * EL
1440 S(I1 + 1, 1) = S(I1 + 1, 1) + EIL * 4 * EL * EL
1450 S(I1 + 1, 2) = S(I1 + 1, 2) - EIL * 6 * EL
1460 S(I1 + 1, 3) = S(I1 + 1, 3) + EIL * 2 * EL * EL
1470 S(I1 + 2, 1) = S(I1 + 2, 1) + EIL * 12
1480 S(I1 + 2, 2) = S(I1 + 2, 2) - EIL * 6 * EL
1490 S(I1 + 3, 1) = S(I1 + 3, 1) + EIL * 4 * EL * EL
1500 NEXT I
1510 REM *** MODIFY FOR BOUNDARY CONDITIONS ***
1520 SUM = 0!
1530 FOR J = 1 TO NN
1540 SUM = SUM + S(J, 1): NEXT J
1550 CNST = SUM * 10000!
1560 FOR I = 1 TO ND
1570 N = DB(I)
1580 S(N, 1) = S(N, 1) + CNST: NEXT I
1590 REM *** SPRINGS, IF ANY
1600 INPUT "NUMBER OF SPRINGS ="; NSPR
1610 IF NSPR = 0 THEN 1670
1620 PRINT "GIVE DOF NUMBER AND SPRING CONSTANT FOR EACH SPRING"
1630 FOR I = 1 TO NSPR
1640 INPUT N, SCONST
1650 S(N, 1) = S(N, 1) + SCONST: NEXT I
1660 REM ** MASS MATRIX **
1670 FOR I = 1 TO NQ
1680 FOR J = 1 TO NBW
1690 EM(I, J) = 0!: NEXT J: NEXT I
1700 FOR I = 1 TO NE
1710 M = MG(I): ROO = RO(M): AR = AREA(I)
1740 ELL = ABS(X(I + 1) - X(I))
1750 C1 = ROO * AR * ELL / 420!
1760 I1 = 2 * I - 1
1770 EM(I1, 1) = EM(I1, 1) + 156! * C1
1780 EM(I1, 2) = EM(I1, 2) + 22! * ELL * C1
1790 EM(I1, 3) = EM(I1, 3) + 54! * C1
1800 EM(I1, 4) = EM(I1, 4) - 13! * ELL * C1
1810 EM(I1 + 1, 1) = EM(I1 + 1, 1) + 4! * ELL * ELL * C1
1820 EM(I1 + 1, 2) = EM(I1 + 1, 2) + 13! * ELL * C1
1830 EM(I1 + 1, 3) = EM(I1 + 1, 3) - 3! * ELL * ELL * C1
1840 EM(I1 + 2, 1) = EM(I1 + 2, 1) + 156! * C1
1850 EM(I1 + 2, 2) = EM(I1 + 2, 2) - 22! * ELL * C1
1860 EM(I1 + 3, 1) = EM(I1 + 3, 1) + 4! * ELL * ELL * C1
1870 NEXT I
1880 INPUT "NUMBER OF LUMPED(POINT) MASSES="; NLM
1890 IF NLM = 0 THEN 1940
```

```
1900 FOR I = 1 TO NLM
1910 INPUT "GIVE DOF #, LUMPED MASS="; N, XM
1920 EM(N, 1) = EM(N, 1) + XM
1930 NEXT I
1940 INPUT "GIVE NAME OF OUTPUT FILE="; FILENAME$
1950 OPEN FILENAME$ FOR OUTPUT AS #1
1960 REM ** PRINT INFO IN OUTPUT FILE
1970 PRINT #1, NQ, NBW
1980 PRINT #1, "BANDED STIFFNESS MATRIX"
1990 FOR I = 1 TO NQ
2000 FOR J = 1 TO NBW
2010 PRINT #1, S(I, J);
2020 NEXT J: PRINT #1, : NEXT I
2030 PRINT #1, "BANDED MASS MATRIX"
2040 FOR I = 1 TO NQ
2050 FOR J = 1 TO NBW
2060 PRINT #1, EM(I, J);
2070 NEXT J: PRINT #1, : NEXT I
2080 CLOSE #1
2090 END
1000 REM ************* CST_KM  *************
1010 REM ** Stiffness and Mass Matrices for CST Element**
1020 CLS
1030 PRINT TAB(22); "======================================="
1040 PRINT TAB(22); "       Program written by       "
1050 PRINT TAB(22); " T.R.Chandrupatla and A.D.Belegundu   "
1060 PRINT TAB(22); "======================================="
1061 DEFINT I-N
1062 LOCATE 8, 28: PRINT "1. Interactive Data Input"
1064 LOCATE 9, 28: PRINT "2. Data Input from a File"
1066 LOCATE 12, 28: INPUT "  Your Choice <1 or 2> "; IFL1
1070 CLS: IF IFL1 = 2 GOTO 1122
1080 INPUT "Number of Elements ="; NE
1090 INPUT "Number of Nodes ="; NN
1100 INPUT "Number of Constrained DOF ="; ND
1120 INPUT "Number of Materials ="; NM
1121 GOTO 1130
1122 INPUT "File Name for Input Data "; FILE1$
1124 OPEN FILE1$ FOR INPUT AS #1
1126 INPUT #1, NN, NE, ND, NL, NM, NDIM, NEN
1130 INPUT "Thickness ="; TH
1160 REM TOTAL DOF IS "NQ"
1170 NQ = 2 * NN
1180 DIM X(NN, 2), NOC(NE, 3), MAT(NE), PM(NM, 3), NU(ND), U(ND)
1190 DIM D(3, 3), B(3, 6), DB(3, 6), SE(6, 6), EME(6, 6)
1192 IF IFL1 = 1 GOTO 1200
1194 GOSUB 2510: GOTO 1242
1200 REM BANDWIDTH NBW FROM CONNECTIVITY"NOC"
1210 NBW = 0
1220 PRINT "Element#, 3 Nodes, Material# "
1230 FOR I = 1 TO NE
1240 INPUT N, NOC(N, 1), NOC(N, 2), NOC(N, 3), MAT(N): NEXT I
1242 FOR I = 1 TO NE
1250 CMIN = NN + 1: CMAX = 0
1260 FOR J = 1 TO 3
1270 IF CMIN > NOC(I, J) THEN CMIN = NOC(I, J)
1280 IF CMAX < NOC(I, J) THEN CMAX = NOC(I, J)
1290 NEXT J
1300 C = 2 * (CMAX - CMIN + 1)
```

```
1310 IF NBW < C THEN NBW = C
1320 NEXT I
1330 PRINT "The Bandwidth is"; NBW
1340 REM *** INITIALIZATION ***
1350 DIM S(NQ, NBW), EM(NQ, NBW)
1360 FOR I = 1 TO NQ: FOR J = 1 TO NBW
1390 S(I, J) = 0: NEXT J: NEXT I
1392 IF IFL1 = 2 GOTO 1520
1400 PRINT "Node#,  X-Coord,  Y-Coord"
1410 FOR I = 1 TO NN
1420 INPUT N, X(N, 1), X(N, 2): NEXT I
1430 PRINT "Material#,  E,  Poissons Ratio, Density"
1440 FOR I = 1 TO NM
1450 INPUT N, PM(N, 1), PM(N, 2), PM(N, 3): NEXT I
1460 PRINT "DOF#,  Displacement"
1470 FOR I = 1 TO ND
1480 INPUT NU(I), U(I): NEXT I
1520 REM *** GLOBAL STIFFNESS MATRIX **
1530 FOR N = 1 TO NE
1540 PRINT "Forming Stiffness Matrix of Element "; N
1542 GOSUB 3000
1560 PRINT ".... Placing in Global Locations"
1570 FOR II = 1 TO 3
1580 NRT = 2 * (NOC(N, II) - 1)
1590 FOR IT = 1 TO 2
1600 NR = NRT + IT
1610 I = 2 * (II - 1) + IT
1620 FOR JJ = 1 TO 3
1630 NCT = 2 * (NOC(N, JJ) - 1)
1640 FOR JT = 1 TO 2
1650 J = 2 * (JJ - 1) + JT
1660 NC = NCT + JT - NR + 1
1670 IF NC <= 0 GOTO 1690
1680 S(NR, NC) = S(NR, NC) + SE(I, J)
1682 EM(NR, NC) = EM(NR, NC) + EME(I, J)
1690 NEXT JT: NEXT JJ: NEXT IT: NEXT II
1700 NEXT N
1710 REM  *** MODIFY FOR BOUNDARY CONDITIONS ***
1720 CNST = S(1, 1) * 10000
1730 FOR I = 1 TO ND
1740 N = NU(I)
1750 S(N, 1) = S(N, 1) + CNST
1760 NEXT I
1770 INPUT "NUMBER OF LUMPED(POINT) MASSES="; NLM
1780 IF NLM = 0 THEN 1830
1790 FOR I = 1 TO NLM
1800 INPUT "GIVE DOF #, LUMPED MASS="; N, XM
1810 EM(N, 1) = EM(N, 1) + XM
1820 NEXT I
1830 INPUT "Give Name of Output File ="; FILE2$
1840 OPEN FILE2$ FOR OUTPUT AS #2
1850 REM ** PRINT INFO IN OUTPUT FILE **
1860 PRINT #2, NQ, NBW
1870 PRINT #2, "BANDED STIFFNESS MATRIX"
1880 FOR I = 1 TO NQ
1890 FOR J = 1 TO NBW
1900 PRINT #2, S(I, J);
1910 NEXT J: PRINT #2, : NEXT I
```

```
1920 PRINT #2, "BANDED MASS MATRIX"
1930 FOR I = 1 TO NQ
1940 FOR J = 1 TO NBW
1950 PRINT #2, EM(I, J);
1960 NEXT J: PRINT #2, : NEXT I
1970 CLOSE #2
1980 END
2500 '=============== READ DATA  ====================
2510 FOR I = 1 TO NN: FOR J = 1 TO NDIM
2520 INPUT #1, X(I, J): NEXT J: NEXT I
2530 FOR I = 1 TO NE: FOR J = 1 TO NEN
2540 INPUT #1, NOC(I, J): NEXT J: NEXT I
2550 FOR I = 1 TO NE: INPUT #1, MAT(I): NEXT I
2570 FOR I = 1 TO ND: INPUT #1, NU(I), U(I): NEXT I
2580 FOR I = 1 TO NL: INPUT #1, N, F: NEXT I
2590 FOR I = 1 TO NM: FOR J = 1 TO 3
2600 INPUT #1, PM(I, J): NEXT J: NEXT I
2610 CLOSE #1
2620 RETURN
3000 'ELEMENT STIFFNESS FORMATION
3010 '***FIRST THE D-MATRIX***
3020 M = MAT(N): E = PM(M, 1): PNU = PM(M, 2): RHO = PM(M, 3)
3030 C1 = E / (1 - PNU ^ 2): C2 = C1 * PNU: C3 = .5 * E / (1 + PNU)
3040 D(1, 1) = C1: D(1, 2) = C2: D(1, 3) = 0
3050 D(2, 1) = C2: D(2, 2) = C1: D(2, 3) = 0
3060 D(3, 1) = O: D(3, 2) = 0: D(3, 3) = C3
3070 '***STRAIN-DISPLACEMENT MATRIX***
3080 I1 = NOC(N, 1): I2 = NOC(N, 2): I3 = NOC(N, 3)
3090 X1 = X(I1, 1): Y1 = X(I1, 2)
3100 X2 = X(I2, 1): Y2 = X(I2, 2)
3110 X3 = X(I3, 1): Y3 = X(I3, 2)
3120 X21 = X2 - X1: X32 = X3 - X2: X13 = X1 - X3
3130 Y12 = Y1 - Y2: Y23 = Y2 - Y3: Y31 = Y3 - Y1
3140 DJ = X13 * Y23 - X32 * Y31'DETERMINANT OF JACOBIAN
3150 '***FORMATION OF B MATRIX***
3160 B(1, 1) = Y23 / DJ: B(2, 1) = 0: B(3, 1) = X32 / DJ
3170 B(1, 2) = 0: B(2, 2) = X32 / DJ: B(3, 2) = Y23 / DJ
3180 B(1, 3) = Y31 / DJ: B(2, 3) = 0: B(3, 3) = X13 / DJ
3190 B(1, 4) = 0: B(2, 4) = X13 / DJ: B(3, 4) = Y31 / DJ
3200 B(1, 5) = Y12 / DJ: B(2, 5) = 0: B(3, 5) = X21 / DJ
3210 B(1, 6) = 0: B(2, 6) = X21 / DJ: B(3, 6) = Y12 / DJ
3220 '***DB MATRIX DB=D*B***
3230 FOR I = 1 TO 3
3240 FOR J = 1 TO 6
3250 DB(I, J) = 0
3260 FOR K = 1 TO 3
3270 DB(I, J) = DB(I, J) + D(I, K) * B(K, J)
3280 NEXT K: NEXT J: NEXT I
3290 FOR I = 1 TO 6
3300 FOR J = 1 TO 6
3310 SE(I, J) = 0
3320 FOR K = 1 TO 3
3330 SE(I, J) = SE(I, J) + .5 * ABS(DJ) * B(K, I) * DB(K, J) * TH
3340 NEXT K: NEXT J: NEXT I
3350 '*** ELEMENT MASS MATRIX EME( , ) ***
3360 CM = RHO * TH * .5 * ABS(DJ) / 12
3370 FOR I = 1 TO 6: FOR J = 1 TO 6: EME(I, J) = 0: NEXT J: NEXT I
3380 '+Non-zero element mass matrix values are defined+
```

```
3390 EME(1, 1) = 2 * CM: EME(1, 3) = CM: EME(1, 5) = CM
3400 EME(2, 2) = 2 * CM: EME(2, 4) = CM: EME(2, 6) = CM
3410 EME(3, 1) = CM: EME(3, 3) = 2 * CM: EME(3, 5) = CM
3420 EME(4, 2) = CM: EME(4, 4) = 2 * CM: EME(4, 6) = CM
3430 EME(5, 1) = CM: EME(5, 3) = CM: EME(5, 5) = 2 * CM
3440 EME(6, 2) = CM: EME(6, 4) = CM: EME(6, 6) = 2 * CM
3450 RETURN
7320 ITER = ITER + 1
7330 IF ITER > ITMAX GOTO 7940
7340 IF NV = 1 GOTO 7490
7350 REM *** STARTING VECTOR IN SUBSPACE ***
7360 FOR I = 1 TO NV - 1
7370 CV = 0
7380 FOR K = 1 TO NQ
7390 KA = K - NBW + 1: KZ = K + NBW - 1
7400 IF KA < 1 THEN KA = 1
7410 IF KZ > NQ THEN KZ = NQ
7420 FOR L = KA TO KZ
7430 IF L < K GOTO 7450
7440 K1 = K: L1 = L - K + 1: GOTO 7460
7450 K1 = L: L1 = K - L + 1
7460 CV = CV + EVS(K) * EM(K1, L1) * EVC(L, I): NEXT L: NEXT K
7470 FOR K = 1 TO NQ
7480 EV1(K) = EV1(K) - CV * EVC(K, I): NEXT K: NEXT I
7490 FOR I = 1 TO NQ
7500 IA = I - NBW + 1: IZ = I + NBW - 1: EVT(I) = 0
7510 IF IA < 1 THEN IA = 1
7520 IF IZ > NQ THEN IZ = NQ
7530 FOR K = IA TO IZ
7540 IF K < I GOTO 7560
7550 I1 = I: K1 = K - I + 1: GOTO 7570
7560 I1 = K: K1 = I - K + 1
7570 EVT(I) = EVT(I) + EM(I1, K1) * EV1(K): NEXT K
7580 EV2(I) = EVT(I)
7590 NEXT I
7600 GOSUB 8090
7610 C1 = 0: C2 = 0
7620 FOR I = 1 TO NQ
7630 C1 = C1 + EV2(I) * EVT(I): NEXT I
7640 FOR I = 1 TO NQ
7650 IA = I - NBW + 1: IZ = I + NBW - 1: EVT(I) = 0
7660 IF IA < 1 THEN IA = 1
7670 IF IZ > NQ THEN IZ = NQ
7680 FOR K = IA TO IZ
7690 IF K < I GOTO 7710
7700 I1 = I: K1 = K - I + 1: GOTO 7720
7710 I1 = K: K1 = I - K + 1
7720 EVT(I) = EVT(I) + EM(I1, K1) * EV2(K): NEXT K: NEXT I
7730 FOR I = 1 TO NQ
7740 C2 = C2 + EV2(I) * EVT(I): NEXT I
7750 EL2 = C1 / C2
7760 C2 = SQR(C2)
7770 FOR I = 1 TO NQ
7780 EV1(I) = EV2(I) / C2
7790 EVS(I) = EV1(I): NEXT I
7800 IF ABS(EL2 - EL1) / ABS(EL2) > TOL GOTO 7310
7810 FOR I = 1 TO NQ
7820 EVC(I, NV) = EV1(I): NEXT I
7830 PRINT "ITERATION NUMBER "; ITER
7840 EL2 = EL2 + SH: EVL(NV) = EL2
7850 PRINT "EIGENVALUE NUMBER  "; NV
7860 PRINT USING "EIGENVALUE IS ##.###^^^^"; EL2
7870 PRINT "EIGENVECTOR "
```

```
7880 FOR I = 1 TO NQ
7890 PRINT USING "##.###^^^^ "; EVC(I, NV), : NEXT I
7900 PRINT : PRINT : IF NV = NEV GOTO 7920
7910 INPUT "SHIFT VALUE SH= "; SH
7920 NEXT NV
7930 GOTO 7950
7940 PRINT "NO CONVERGENCE FOR "; ITER; " ITERATIONS"
7950 END
7960 REM    **** Gauss Elimination LDU Approach ****
7970 REM  --- REDUCTION TO UPPER TRIANGULAR MATRIX ---
7980 FOR K = 1 TO NQ - 1
7990 NK = NQ - K + 1
8000 IF NK > NBW THEN NK = NBW
8010 FOR I = 2 TO NK
8020 C1 = S(K, I) / S(K, 1)
8030 I1 = K + I - 1
8040 FOR J = I TO NK
8050 J1 = J - I + 1
8060 S(I1, J1) = S(I1, J1) - C1 * S(K, J)
8070 NEXT J: NEXT I: NEXT K
8080 RETURN
8090 REM    **** Reduction of the right hand side ****
8100 FOR K = 1 TO NQ - 1
8110 NK = NQ - K + 1
8120 IF NK > NBW THEN NK = NBW
8130 FOR I = 2 TO NK: I1 = K + I - 1
8140 C1 = 1 / S(K, 1)
8150 EV2(I1) = EV2(I1) - C1 * S(K, I) * EV2(K)
8160 NEXT I: NEXT K
8170 REM    **** Back Substitution ****
8180 EV2(NQ) = EV2(NQ) / S(NQ, 1)
8190 FOR II = 1 TO NQ - 1
8200 I = NQ - II: C1 = 1 / S(I, 1)
8210 NI = NQ - I + 1
8220 IF NI > NBW THEN NI = NBW
8230 EV2(I) = C1 * EV2(I)
8240 FOR K = 2 TO NI
8250 EV2(I) = EV2(I) - C1 * S(I, K) * EV2(I + K - 1)
8260 NEXT K: NEXT II
8270 RETURN
```

PROBLEMS

11.1. Consider *axial* vibration of the steel bar shown in Fig. P11.1.
   (a) Develop the global stiffness and mass matrices.
   (b) By hand calculations, determine the lowest natural frequency and mode shape using the inverse iteration algorithm.

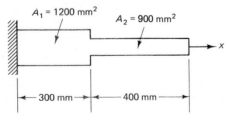

Steel bar

Figure P11.1   Steel bar.

(c) Verify your results in (b) using programs INVITR and JACOBI.

(d) Verify the properties in Eqs. 11.41a and 11.41b.

11.2. By hand calculations, determine the natural frequencies and mode shapes for the rod in P11.1 using the characteristic polynomial technique.

11.3. Use a lumped mass model for the rod in P11.1, and compare the results obtained with the consistent mass model. Use program INVITR or JACOBI.

11.4. Determine all natural frequencies of the simply supported beam shown in Fig. P11.4. Compare the results obtained using

(a) A one-element model.

(b) A two-element model.

Use either programs INVITR or JACOBI.

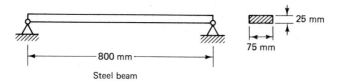

Figure P11.4   Steel beam.

11.5. Determine, with the help of program BEAM _ KM, the two lowest natural frequencies (critical speeds) of the steel shaft shown in Fig. P11.5, considering the following cases:

(a) The three journals are like simple supports.

(b) Each journal bearing is like a spring of stiffness equal to 25 000 lb/in.

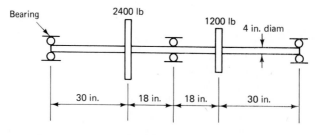

Figure P11.5

11.6. The existence of a crack renders an overall reduction in the stiffness of a structure. A crack in a bending member, such as a beam, suggests a slope discontinuity at the section containing the crack, even though the displacement is still continuous there. Thus, the effect of a fracture at a section may be represented by torsional spring connecting two elements, whose torsional stiffness k may be determined analytically or experimentally.

Consider the cracked cantilever beam shown in Fig. P11.6.

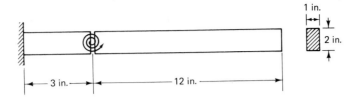

Figure P11.6

(a) Discuss how you will model this using beam elements. Write down the boundary conditions at the cracked section, and the resulting modifications to the stiffness matrix.

(b) Determine the first three natural frequencies and mode shapes, and compare these with those of an uncracked beam of same dimensions. Take $k = 8 \times 10^6$ in.-lb and $E = 30 \times 10^6$ psi.

11.7. A simplified model of a steel turbine blade is shown in Fig. P11.7. We want to determine the lowest resonant frequency with motion in x direction and corresponding mode shape. It is important that we do not excite this resonant frequency to avoid contact of the blades with the casing. The outer ring connecting all the blades is represented as a lumped mass. Use programs CST_KM and INVITR.

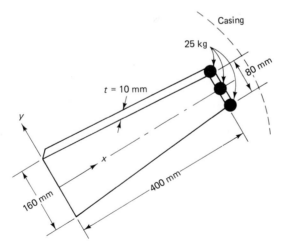

Figure P11.7

*11.8. Figure P11.8 shows a beam modeled using 4-node quadrilateral elements. Develop a program that will generate the banded **K** and **M** matrices. Then use program INVITR to determine the two lowest natural frequencies and mode shapes. Compare your results with those obtained using beam elements.

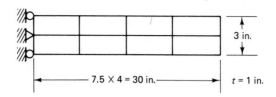

Figure P11.8   Steel beam.

*11.9. Determine the two lowest natural frequencies and mode shapes for the one-bay, two-story planar steel frame shown in Fig. P11.9. You need to develop a program, analogous to BEAM_KM, that will generate the banded **K** and **M** matrices, and then use program INVITR.

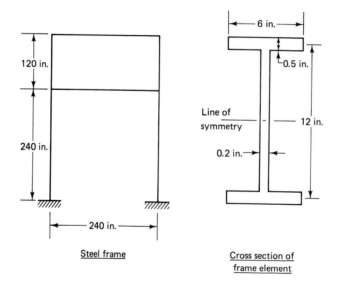

Steel frame

Cross section of frame element

Figure P11.9

*11.10. For the signal pole arrangement shown in Fig. P11.10, a 2-D frame, determine the natural frequencies and mode shapes. (*Note:* Develop a program in line with BEAM_KM to write stiffness and mass matrices to a file. Then an eigenvalue routine like INVITR can be run.)

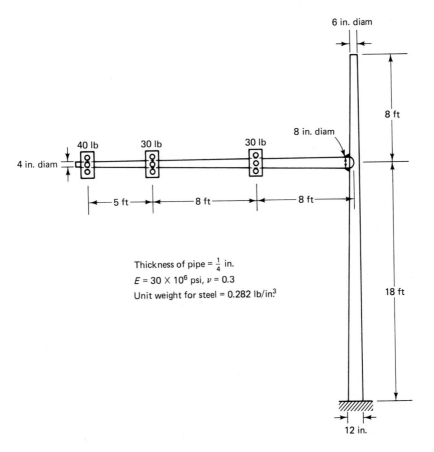

Thickness of pipe = $\frac{1}{4}$ in.
$E = 30 \times 10^6$ psi, $\nu = 0.3$
Unit weight for steel = 0.282 lb/in.³

Figure P11.10

chapter twelve

Preprocessing and Postprocessing

12.1 INTRODUCTION

Finite element analysis involves three stages of activity: preprocessing, processing, and postprocessing. Preprocessing involves the preparation of data, such as nodal coordinates, connectivity, boundary conditions, and loading and material information. The processing stage involves stiffness generation, stiffness modification, and solution of equations, resulting in the evaluation of nodal variables. Other derived quantities, such as gradients or stresses, may be evaluated at this stage. The processing stage is presented in detail in earlier chapters, where the data were prepared in an interactive manner. The postprocessing stage deals with the presentation of results. Typically, the deformed configuration, mode shapes, temperature, and stress distribution are computed and displayed at this stage. A complete finite element analysis is a logical interaction of the three stages. The preparation of data and postprocessing require considerable effort if all data are to be handled manually. The tedium of handling the data and the possibility of errors creeping in as the number of elements increase are discouraging factors for the finite element analyst. In the following sections, we present a systematic development of preprocessing and postprocessing considerations. This should make finite element analysis a more interesting computational tool. We first present a general purpose mesh generation scheme for two-dimensional plane problems.

12.2 MESH GENERATION

Region and Block Representation

The basic idea of a mesh generation scheme is to generate element connectivity data and nodal coordinate data by reading in input data for a few key points. We present here the theory and computer implementation of a mesh generation scheme suggested by Zienkiewicz and Philips.* In this scheme, a complex region is divided into eight-noded quadrilaterals, which are then viewed in the form of a rectangular block pattern. Consider the region shown in Fig. 12.1. The full rectangular block pattern is convenient for node numbering. To match the pattern in the region, the block number 4 is to be treated as void and the two hatched edges need to be merged. In general, a complex region is viewed as a rectangle, composed of rectangular blocks, with some blocks left as void and some edges identified to be merged.

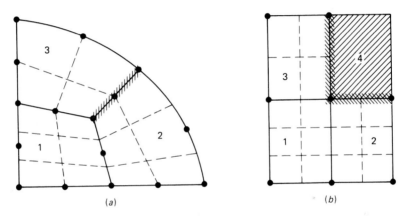

Figure 12.1   (*a*) Region and (*b*) block diagram.

Block Corner Nodes, Sides, and Subdivisions

A general configuration of the full rectangle composed of blocks is shown in Fig. 12.2. We represent the sides of the rectangle as S and W, with respective numbers of spans of NS and NW. For consistent coordinate mapping, S, W, and the third coordinate direction Z must form a right-hand system. For mesh generation, each span is subdivided. Spans KS and KW are divided into $NSD(KS)$ and $NWD(KW)$ divisions respectively. Since the node numbering will be carried out in the S direction first and incremented in the W direction next, the bandwidth of resulting matrices

*Zienkiewicz, O. C., and D. V. Philips, An automatic mesh generation scheme for plane and curved surfaces by "isoparametric" coordinates. *International Journal for Numerical Methods in Engineering* 3: 519–528 (1971).

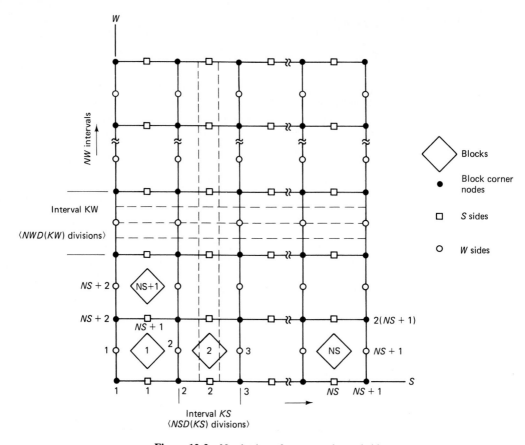

Figure 12.2   Numbering of corner nodes and sides.

will be small if total number of divisions in the S direction is less than the total number in the W direction. S and W are chosen to represent short and wide directions in this sense. In this scheme, the bandwidth is a minimum when there are no void blocks and there is no side merging. We note here that the total number of nodes in the S and W directions are

$$NNS = 1 + \sum_{KS=1}^{NS} NSD(KS)$$

$$NNW = 1 + \sum_{KW=1}^{NW} NWD(KW)$$

(12.1)

The maximum possible nodes for quadrilateral or triangular division is taken as $NNT\,(= NNS \times NNW)$. We define an array $NNAR(NNT)$ to define the nodes in the problem. We also define a block identifier array $IDBLK(NSW)$, which stores the material number in the location representing the block. A zero is stored in the location corresponding to a void block. The x, y coordinates of all valid block corner nodes are read into $XB(NGN, 2)$. The program is given for planar regions. By introducing the z coordinate, three-dimensional surfaces can be modeled. Two arrays, $SR(NSR, 2)$ and $WR(NWR, 2)$, are used for storing the coordinates of the nodes on the corresponding sides. First we generate the nodes for all sides, assuming that the side is a straight line and the node is at the midpoint between the corner nodes. This represents the default configuration. Then, for sides that are curved and for those straight sides with nodes not located at physical midpoints, the x, y coordinates are read into $SR(.,2)$, $WR(.,2)$ at appropriate locations. The sides to be merged are identified by the end node numbers of the sides. We now discuss the node numbering and coordinate generation schemes.

Generation of node numbers.   We present the node numbering strategy by means of an example. Consider the region and block representation shown in Fig. 12.1. The node numbering scheme is shown in Fig. 12.3. We have two blocks in the S direction and two in the W direction. Block 4 is void. Array $NNAR(30)$ has all the locations defined. Edges 18–20 and 18–28 are to be merged. We first initialize the array $NNAR(30)$ by putting -1 at each of its locations. We then cover each of the void blocks and put *zero* where nodes do not exist. Existence of neighboring blocks is checked in implementing this process. For side merging, at the node locations of the side with higher node numbers, the location numbers of the correspond-

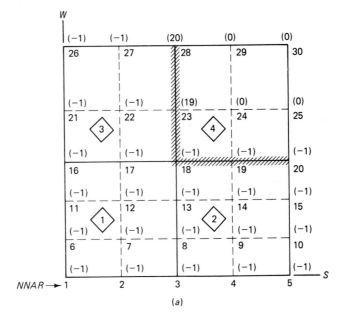

Figure 12.3 Node numbering.

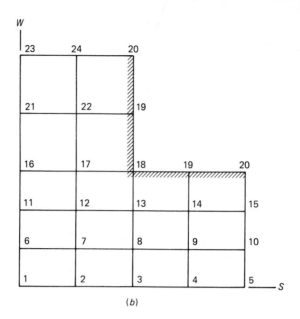

(b)

Figure 12.3 *(cont.)*

ing nodes of the merging side are entered. The final node numbering is a simple process. We sweep along S and then increment along W. The node numbers are incremented by 1 whenever the location has a negative value. When the value is zero it is skipped. If the location has a positive value, it indicates side merging and the corresponding node number from the location indicated by the value is inserted. The scheme is simple and nodal coordinate checking is not necessary in this process.

Generation of coordinates and connectivity.   Here we use the shape functions for isoparametric mapping for an eight-noded quadrilateral developed in Chapter 7. We refer to Fig. 12.4, which establishes the relationships for the master block or ξ-η block, the S-W block, and the region block or x-y block. The first step is one of extracting the global coordinates of corner and midside nodes of the block under consideration. For a general node $N1$, the ξ, η coordinates are obtained using the number of divisions. The coordinates of $N1$ are given by

$$x = \sum_{I=1}^{8} SH(I) \cdot X(I)$$

$$y = \sum_{I=1}^{8} SH(I) \cdot Y(I)$$

(12.2)

where $SH(\ )$ are shape functions and $X(\ )$, $Y(\ )$ are corner node coordinates. For the small rectangular shaded division with lower left corner $N1$, shown in Fig. 12.4, the other three nodes $N2$, $N3$, $N4$ are computed. For quadrilateral elements, we use $N1$–$N2$–$N3$–$N4$ as the element, with the first element of the block starting at the

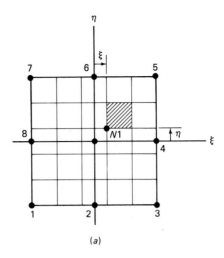

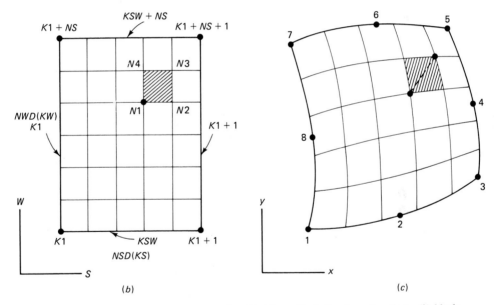

Figure 12.4 Coordinates and connectivity. (*a*) Master block for shape functions, (*b*) block for node numbers, (*c*) block in region.

lower left corner. The element numbers for the next block start after the last number of the previous block. For triangular element division, each rectangle is divided into two triangles, $N1-N2-N3$ and $N3-N4-N1$. The triangular division is readjusted to connect the diagonal of shorter length. The process of coordinate and connectivity generation is skipped for void blocks.

This is a general purpose mesh generation scheme with the capability to model complex problems. This scheme can be readily generalized to model three-dimensional surfaces by introducing the z coordinate. To illustrate the use of the program, we consider a few examples.

Examples of mesh generation. In the first example shown in Fig. 12.5, there are four blocks. The default material number for all blocks is 1. Material number for block 4 is read in as zero to represent void space. S spans 1 and 2 are divided into four and two divisions, respectively. W spans 1 and 2 are each divided into three divisions. The coordinates of corner nodes 1–8 and the coordinates of midpoints of curved sides $W1$ and $W4$ are read in. The generated mesh with node numbers is also shown in Fig. 12.5. If triangular mesh is desired, the shorter diagonal of each quadrilateral will be joined.

In the second example, shown in Fig. 12.6, we model a full annular region. To achieve a minimum bandwidth, the block diagram shown in Fig. 12.6a is suggested. Blocks 2 and 5 are void space. The side 1–2 merges with 4–3, and side 9–10 merges with 12–11. Coordinates of all corner nodes and the midpoints of $W1$, $W2$, . . . , $W8$ of the block diagram need to be given. The resulting mesh for the span divisions shown in the block diagram is given in Fig. 12.6c.

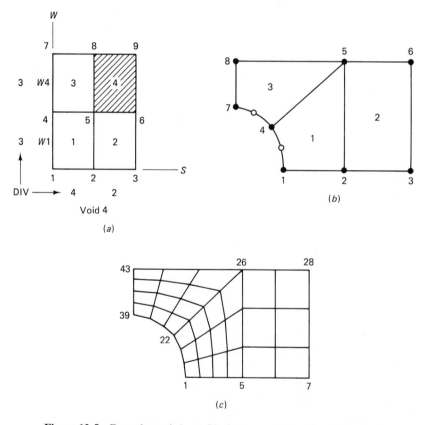

Figure 12.5 Example mesh 1. (a) Block diagram, (b) region, (c) mesh.

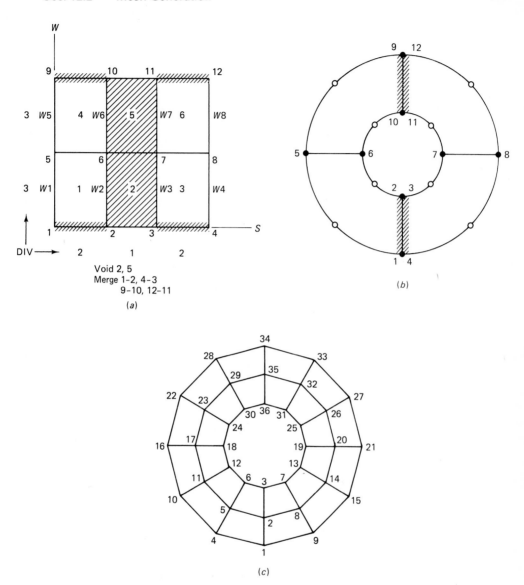

Figure 12.6   Example mesh 2. (*a*) Block diagram, (*b*) region, (*c*) mesh.

Figure 12.7 shows an eyelet. The full geometric shape is modeled. The block diagram shows void blocks and span divisions. Merging sides are indicated. Coordinates of all corner points of the block diagram are to be read in. The coordinates of midpoints of curved sides $W1$, $W2$, $W4$, $W7$, $W10$, $W13$, $W16$, and $W17$ have to be input. The mesh is shown for quadrilateral elements.

Division of a region and making a block diagram form the first step in the preparation of data for mesh generation.

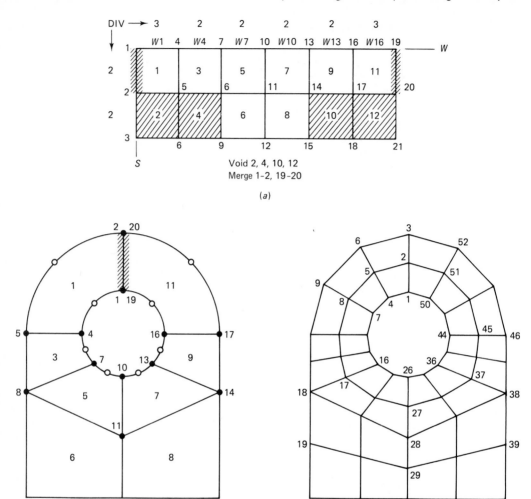

Figure 12.7   Example mesh 3. (*a*) Block diagram, (*b*) region, (*c*) mesh.

Mesh plotting.   The generated data are saved in a file. The convenient way of reviewing the coordinate and connectivity data is by plotting it using the computer. The plots will quickly reveal if there are any errors. Points to be readjusted can easily be identified. The program PLOT2D can be used for plotting 2-D meshes on the screen. In mesh plotting, we scan each element and draw the element boundaries using the connectivity information. The coordinate bounds must first be adjusted for the screen resolution and size.

Data handling and editing.   In simple problems with small number of elements and nodes, it is convenient to prepare the data directly. Data obtained from mesh generation need to be edited. Often some badly placed node points must be adjusted. Boundary condition data, loading data, and material property data need to be added. To handle all these situations, and also to edit previously created data, we include the program DATAFEM.

The initial screen has three choices:

1. Create New Data.
2. Edit Data.
3. Edit Data from MESHGEN.

For choice 1 we get screens for coordinate data, connectivity data, boundary data, load data, and material data. We then get into the Edit mode. For choice 3 the data file generated by MESHGEN is first read. Then, screens for boundary data, load data, and material data appear in succession. We then enter the Edit mode. For choice 2 the previously created data file is read and then directly goes to the Edit mode. The Edit screen looks as shown below:

Edit Menu

1. Coordinate Data
2. Connectivity/Material
3. Boundary Data
4. Load Data
5. Material Properties
6. ⟨SAVE DATA and EXIT⟩
7. EXIT without Saving

When new data are created, the node numbers, element numbers, and other data are incremented sequentially in ascending order. In the edit mode, the specific data number can be chosen and the appropriate value can be changed. The program PLOT2D can read the data file created and plot two-dimensional meshes.

The data thus created can be processed by the finite element programs presented in earlier chapters. Some minor changes may be needed in the programs to change from interactive mode of data input to that of reading from a data file. The finite element program processes the data and calculates nodal variable quantities such as displacements and temperatures, and element quantities such as stresses and gradients. The stage is now set for the discussion of postprocessing.

12.3 POSTPROCESSING

We discuss here the aspects of plotting a displaced configuration, plotting nodal data in the form of contour plots, such as isotherms and isobars, and conversion of element-oriented data into best fitting nodal values. We restrict our discussion here to two-dimensional problems; however, the ideas can be extended to three-dimensional problems with some additional effort.

Deformed Configuration and Mode Shape

Plotting a deformed or displaced shape is a simple extension of PLOT2D. If the displacements or components of the eigenvector are read into the matrix $U(NN, 2)$ and the coordinates are stored in $X(NN, 2)$, we can define the displaced position matrix $XP(NN, 2)$. NN represents the number of nodes.

$$XP(I, J) = X(I, J) + \alpha U(I, J) \qquad J = 1, 2$$
$$I = 1, \ldots, NN \tag{12.3}$$

where α is a magnification factor so chosen that the largest component of $\alpha U(I, J)$ is of reasonable proportion in relation to the body size. One may try this largest component to be about 10% of the body size parameter. In the program PLOT2D, we need to make changes to read displacements $U(NN, 2)$, decide the value of α, and replace $\mathbf{X}$ by $\mathbf{X} + \alpha\mathbf{U}$.

Contour Plotting

Contour plotting of a scalar nodal variable such as temperature is straightforward for three-noded triangular elements. We consider the variable f on one triangular element shown in Fig. 12.8. The nodal values are f_1, f_2, and f_3 at the three nodes 1, 2, and 3, respectively. The function f is interpolated using the linear shape functions used for the constant strain triangle. f represents a plane surface with values f_1, f_2, and f_3 at the three nodes. We check for each desired level. Say $\hat{f}$ represents a typical level for contour map. If $\hat{f}$ lies in the interval f_2–f_3, it also lies in one of the intervals f_1–f_2 or f_1–f_3. Say it lies in the interval f_1–f_3, as shown in Fig. 12.8. Then f has the value of $\hat{f}$ at points A and B and is constant along the line AB. Determination of the coordinates of points A and B will give us the contour line AB. Coordinates of point A can be obtained from

$$\xi = \frac{\hat{f} - f_2}{f_3 - f_2}$$
$$x_A = \xi x_3 + (1 - \xi)x_2 \tag{12.4}$$
$$y_A = \xi y_3 + (1 - \xi)y_2$$

The coordinates of point B can be obtained by replacing the indices 2, 3 by 1 and 3, respectively.

The program CONTOUR plots the variable FF represented by its nodal values. The coordinate, connectivity, and function data are read in from data files. In the first part of the program, the boundary limits are set on the screen. The function limits are found and the number of contour levels is read in. The boundary of the

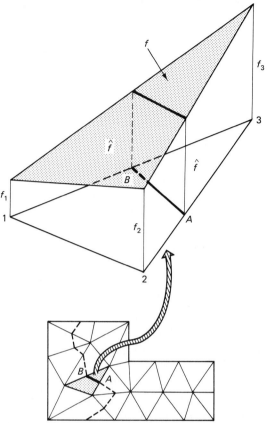

Figure 12.8 Constant level of variation f.

region is plotted on the screen. Each element is then scanned for the function levels and the constant value lines are drawn. The result is a contour map.

There are also some quantities, such as stresses, temperature, and velocity gradients, which are constant over triangular elements. For these, the contour mapping requires the evaluation of nodal values. We present here the procedure for evaluating the nodal values for least-squares fit. The procedure discussed below is useful in diverse situations, such as smoothing data obtained in image processing.

Nodal Values from Known Constant Element Values

We evaluate the nodal values that minimize the least-squares error. We consider here triangular elements with constant function values. A triangular element having function value f_e is shown in Fig. 12.9. Let f_1, f_2, f_3 be the local nodal values. The interpolated function is given by

$$f_e = \mathbf{Nf} \tag{12.5}$$

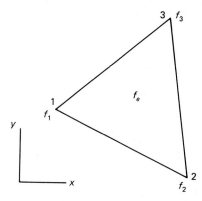

Figure 12.9 Triangular element for least-squares fit study.

where $\mathbf{N}$ is the vector of shape functions

$$\mathbf{N} = [N_1, N_2, N_3] \qquad (12.6)$$

and

$$\mathbf{f} = [f_1, f_2, f_3]^T \qquad (12.7)$$

The squared error may be represented by

$$E = \sum_e \frac{1}{2} \int_e (f - f_e)^2 \, dA \qquad (12.8)$$

On expanding and substituting from (12.5), we get

$$E = \sum_e \left[\tfrac{1}{2}\mathbf{f}^T \left(\int_e \mathbf{N}^T \mathbf{N} \, dA \right) \mathbf{f} - \mathbf{f}^T \left(f_e \int_A \mathbf{N}^T \, dA \right) + \tfrac{1}{2} f_e^2 A \right] \qquad (12.9)$$

Noting that the last term is a constant, we write the equation in the form

$$E = \sum_e \left[\tfrac{1}{2}\mathbf{f}^T \mathbf{W}^e \mathbf{f} - \mathbf{f}^T \mathbf{R}^e \right] + \text{constant} \qquad (12.10)$$

where

$$\mathbf{W}^e = \int_e \mathbf{N}^T \mathbf{N} \, dA = \frac{A_e}{12} \begin{bmatrix} 2 & 1 & 1 \\ 1 & 2 & 1 \\ 1 & 1 & 2 \end{bmatrix} \qquad (12.11)$$

$$\mathbf{R}^e = f_e \int_e \mathbf{N}^T \, dA = \frac{f_e A_e}{3} \begin{Bmatrix} 1 \\ 1 \\ 1 \end{Bmatrix} \qquad (12.12)$$

$\int_e \mathbf{N}^T \mathbf{N} \, dA$ is similar to the evaluation of mass matrix for a triangle in Chapter 11. On assembling the stiffness $\mathbf{W}^e$ and load vector from $\mathbf{R}^e$, we get

$$E = \tfrac{1}{2}\mathbf{F}^T \mathbf{W} \mathbf{F} - \mathbf{F}^T \mathbf{R} + \text{constant} \qquad (12.13)$$

where **F** is the global nodal value vector given by

$$\mathbf{F} = [F_1, F_2, \ldots, F_{NN}]^T \tag{12.14}$$

For least-squares error, setting the derivatives of E with respect to each F_i to be zero, we get

$$\mathbf{WF} = \mathbf{R} \tag{12.15}$$

Here **W** is a banded symmetric matrix. The set of equations is solved using the equation solving techniques used in other finite element programs. The program BESTFIT takes the mesh data and element value data $FS(NE)$ and evaluates the nodal data $F(NN)$.

Element quantities such as maximum shear stress, Von Mises stress, and temperature gradient can be converted to nodal values and then contour plotting can be done.

12.4 CONCLUSION

Preprocessing and postprocessing are integral parts of finite element analysis. The general purpose mesh generation scheme can model a variety of complex regions. One needs to use some imagination in preparing the block representation of the region. Definition of void blocks and merging of sides enables one to model multiple-connected regions. The node numbering gives sparse matrices and in many cases should give minimum bandwidth by proper block representation. Mesh plotting shows the element layout. The data handling program is a dedicated routine for finite element data preparation and data editing. Ideas for the plotting of deformed configuration and mode shapes can be readily implemented into the programs included here. Contour plotting for triangular elements has been presented and a program is included. The computation of nodal values that best fit the element values takes some of the very same steps used in the development of finite elements in earlier chapters.

Finite element analysis involves solution of a wide variety of problems in solid mechanics, fluid mechanics, heat transfer, electrical and magnetic fields, and other areas. Problem solving involves large amounts of data which must be systematically handled and clearly presented. The ideas developed in this chapter should make preparation and handling of input and output data an interesting endeavor rather than a tedious task.

Example 12.1

The quadrant shown in Fig. 12.1 is meshed using program MESHGEN. The input data given below is constructed from the display in Fig. E12.1. Connectivity and nodal coordinate data are contained in the output file, and a plot of the mesh can be obtained by running program PLOT2D.

```
**Type of Elements   <1>Triangle   <2>Quadrilateral ? 1
```

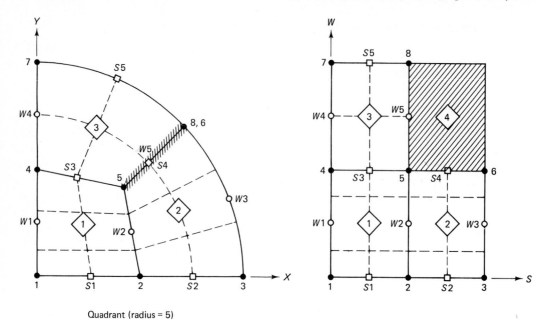

Quadrant (radius = 5)

Figure E12.1

```
**Number of S — Spans ? 2
**Number of W — Spans ? 2
**Input Block# For Those Blocks With Material#
  Not Equal To 1 ( =0 Completes Block Data)
    Block# ? 4
>Enter Material# For The Above Block (=1 By Default)
       ( = 0, For Void Block  )
       ( = n, For Material# n )
     Material# ? 0
    Block# ? 0
**S—Span # 1  Number of Divisions ? 2
**S—Span # 2  Number of Divisions ? 2
**W—Span # 1  Number of Divisions ? 3
**W—Span # 2  Number of Divisions ? 2
**Input Corner Node# ( = 0 Completes Corner Data),
  Followed By the X,Y Coordinates Of The Corner Node
    Corner Node# ? 1
      X,Y Coordinates ? 0,0
    Corner Node# ? 2
      X,Y Coordinates ? 2.5,0
    Corner Node# ? 3
      X,Y Coordinates ? 5,0
    Corner Node# ? 4
      X,Y Coordinates ? 0,2.5
    Corner Node# ? 5
```

```
     X,Y Coordinates ? 1.8,1.8
  Corner Node# ? 6
     X,Y Coordinates ? 3.536,3.536
  Corner Node# ? 7
     X,Y Coordinates ? 0,5
  Corner Node# ? 8
     X,Y Coordinates ? 3.536,3.536
  Corner Node# ? 0
**Input Mid-Side Node Data
  — only for those sides which are curved or those
    straight sides with displaced mid-points for grading
 >Enter S-Side# ( = 0 Completes S-Side Data )
    S-Side# ? 5
     X,Y Coordinates ? 1.913,4.619
    S-Side# ? 0
 >Enter W-Side# ( = 0 Completes W-Side Data )
    W-Side# ? 3
     X,Y Coordinates ? 4.619,1.913
    W-Side# ? 0
**Number of Pairs of Sides to be Merged ? 1
 >Side 1 is the one with Lower Corner Node Numbers
    End nodes of Side 1 ? 5,6
    Corresponding nodes of Merging Side ? 5,8
File Name for Saving Data <eg A:FEM.DAT>? eg12_1.dat
# Constrained DOF, # Component Loads? 0,0
OK                                                          ■
```

Use of Programs *CONTOUR* and *BESTFIT*

CONTOUR.   Input to this program requires two files. One contains the mesh data, such as the output file resulting from use of the MESHGEN program (using three noded triangles). The other input file contains the values of the variable at each node of the mesh. The variable can represent temperature or a nodal stress component. The reader may, for example, execute program HEAT2D which allows the temperature solution to be directed to a specified file. This together with the MESHGEN output file can be input into CONTOUR to generate isotherms.

BESTFIT.   Input to this program also requires two files. One is the mesh data such as the data file resulting from use of program MESHGEN (with three noded triangles). The other file contains stress (or heat flux) values in each element. The output consists of the nodal values of the variable. The nodal stress output is important in itself because of greater accuracy. Also, it permits use of program CONTOUR for plotting contour maps. The element stress data file can be created by either editing the output file generated by the analysis program (FE2CST), or by inserting appropriate print statements within the analysis program for the stress component of interest.

```
1000 '    **********  MESHGEN  ***************
1010 PRINT "===================================="
1020 PRINT "        Program written by        "
1030 PRINT " T.R.Chandrupatla and A.D.Belegundu  "
1040 PRINT "===================================="
1050 DEFINT I-N
1060 CLS : NDIM = 2
1070 INPUT "**Type of Elements  <1>Triangle  <2>Quadrilateral "; NTMP
1080 IF NTMP = 1 THEN NEN = 3 ELSE NEN = 4
1090 INPUT "**Number of S - Spans "; NS
1100 INPUT "**Number of W - Spans "; NW
1110 NSW = NS * NW: NGN = (NS + 1) * (NW + 1): NM = 1
1120 DIM IDBLK(NSW), NSD(NS), NWD(NW), NGCN(NGN)
1130 '-------- Block Identifier / Material# --------
1140 FOR I = 1 TO NSW: IDBLK(I) = 1: NEXT I
1150 PRINT "**Input Block# For Those Blocks With Material#"
1160 PRINT " Not Equal To 1 ( =0 Completes Block Data)"
1170 INPUT "   Block# "; NTMP
1180 IF NTMP = 0 GOTO 1260
1190 PRINT ">Enter Material# For The Above Block ( =1 by Default)"
1200 PRINT "      ( = 0 For a Void Block )"
1210 PRINT "      ( = n For Material# n )"
1220 INPUT "     Material# "; IDBLK(NTMP)
1230 IF NM < IDBLK(NTMP) THEN NM = IDBLK(NTMP)
1240 GOTO 1170
1250 '------------- Span Divisions ---------------
1260 NNS = 1: NNW = 1
1270 FOR KS = 1 TO NS
1280 PRINT "**S-Span #"; KS;
1290 INPUT " Number of Divisions "; NSD(KS)
1300 NNS = NNS + NSD(KS): NEXT KS
1310 FOR KW = 1 TO NW
1320 PRINT "**W-Span #"; KW;
1330 INPUT " Number of Divisions "; NWD(KW)
1340 NNW = NNW + NWD(KW): NEXT KW
1350 '------------ Block Corner and Side Data ---------------
1360 NSR = NS * (NW + 1): NWR = NW * (NS + 1)
1370 DIM XB(NGN, 2), SR(NSR, 2), WR(NWR, 2)
1380 PRINT "**Enter Corner Node# ( =0 Completes Corner Data)"
1390 PRINT " Followed By The X,Y Coordinates of the Corner Node"
1400 INPUT "   Corner Node# "; NTMP
1410 IF NTMP = 0 GOTO 1450
1420 INPUT "    X, Y Coordinates "; XB(NTMP, 1), XB(NTMP, 2)
1430 GOTO 1400
1440 '---------- Evaluate Mid-points of S-Sides -------------
1450 FOR I = 1 TO NW + 1: FOR J = 1 TO NS
1460 IJ = (I - 1) * NS + J
1470 SR(IJ, 1) = .5 * (XB(IJ + I - 1, 1) + XB(IJ + I, 1))
1480 SR(IJ, 2) = .5 * (XB(IJ + I - 1, 2) + XB(IJ + I, 2))
1490 NEXT J: NEXT I
1500 '---------- Evaluate Mid-points of W-Sides -------------
1510 FOR I = 1 TO NW: FOR J = 1 TO NS + 1
1520 IJ = (I - 1) * (NS + 1) + J
1530 WR(IJ, 1) = .5 * (XB(IJ, 1) + XB(IJ + NS + 1, 1))
1540 WR(IJ, 2) = .5 * (XB(IJ, 2) + XB(IJ + NS + 1, 2))
1550 NEXT J: NEXT I
1560 '------ Input Locations of Points on Sides ------
1570 PRINT "**Input Mid-Side Node Data"
1580 PRINT "--only for those sides which are curved or those"
1590 PRINT " straight sides with displaced mid-points for grading"
1600 PRINT "Enter S-Side# ( =0 Completes S-Side Data)"
1610 INPUT "   S-Side# "; NTMP
1620 IF NTMP = 0 GOTO 1650
1630 INPUT "    X, Y Coordinates "; SR(NTMP, 1), SR(NTMP, 2)
1640 GOTO 1610
```

```
1650 PRINT "Enter W-Side# ( =0 Completes W-Side Data)"
1660 INPUT "    W-Side# "; NTMP
1670 IF NTMP = 0 GOTO 1700
1680 INPUT "      X, Y Coordinates "; WR(NTMP, 1), WR(NTMP, 2)
1690 GOTO 1660
1700 '------------------- Merging Sides ----------------------
1710 INPUT "**Number of Pairs of Sides to be Merged "; NSJ
1720 IF NSJ = 0 GOTO 1840
1730 DIM MERG(NSJ, 4)
1740 PRINT " Side 1 is the one with lower corner node numbers"
1750 FOR I = 1 TO NSJ
1760 INPUT "   End nodes of Side 1 "; L1, L2
1770 I1 = L1: I2 = L2: GOSUB 2830: II1 = IDIV
1780 INPUT "   Corresponding nodes of merging side "; L3, L4
1790 I1 = L3: I2 = L4: GOSUB 2830: II2 = IDIV
1800 IF II1 <> II2 THEN PRINT "#Div don't match, Check Merge data ": END
1810 MERG(I, 1) = L1: MERG(I, 2) = L2: MERG(I, 3) = L3: MERG(I, 4) = L4
1820 NEXT I
1830 '------- Global Node Locations of Corner Nodes ---------
1840 NTMPI = 1
1850 FOR I = 1 TO NW + 1
1860 IF I = 1 THEN IINC = 0 ELSE IINC = NNS * NWD(I - 1)
1870 NTMPI = NTMPI + IINC: NTMPJ = 0
1880 FOR J = 1 TO NS + 1
1890 IJ = (NS + 1) * (I - 1) + J
1900 IF J = 1 THEN JINC = 0 ELSE JINC = NSD(J - 1)
1910 NTMPJ = NTMPJ + JINC
1920 NGCN(IJ) = NTMPI + NTMPJ: NEXT J: NEXT I
1930 '---------------- Node Point Array --------------------
1940 NNT = NNS * NNW
1950 DIM NNAR(NNT)
1960 FOR I = 1 TO NNT
1970 NNAR(I) = -1: NEXT I
1980 '--------- Zero Non-Existing Node Locations ---------
1990 FOR KW = 1 TO NW: FOR KS = 1 TO NS
2000 KSW = NS * (KW - 1) + KS
2010 IF IDBLK(KSW) > 0 GOTO 2160
2020 '-------- Operation within an Empty Block --------
2030 K1 = (KW - 1) * (NS + 1) + KS: N1 = NGCN(K1)
2040 NS1 = 2: IF KS = 1 THEN NS1 = 1
2050 NW1 = 2: IF KW = 1 THEN NW1 = 1
2060 NS2 = NSD(KS) + 1: IF KS = NS GOTO 2080
2070 IF IDBLK(KSW + 1) > 0 THEN NS2 = NSD(KS)
2080 NW2 = NWD(KW) + 1: IF KW = NW GOTO 2100
2090 IF IDBLK(KSW + NS) > 0 THEN NW2 = NWD(KW)
2100 FOR I = NW1 TO NW2: IN1 = N1 + (I - 1) * NNS
2110 FOR J = NS1 TO NS2: IJ = IN1 + J - 1
2120 NNAR(IJ) = 0: NEXT J: NEXT I
2130 IF NS2 = NSD(KS) OR NW2 = NWD(KW) GOTO 2160
2140 IF KS = NS OR KW = NW GOTO 2160
2150 IF IDBLK(KSW + NS + 1) > 0 THEN NNAR(IJ) = -1
2160 NEXT KS: NEXT KW
2170 '-------- Node Identification for Side Merging ------
2180 IF NSJ = 0 GOTO 2300
2190 FOR I = 1 TO NSJ
2200 I1 = MERG(I, 1): I2 = MERG(I, 2): GOSUB 2830
2210 IA1 = NGCN(I1): IA2 = NGCN(I2): IASTP = (IA2 - IA1) / IDIV
2220 I1 = MERG(I, 3): I2 = MERG(I, 4): GOSUB 2830
2230 IB1 = NGCN(I1): IB2 = NGCN(I2): IBSTP = (IB2 - IB1) / IDIV
2240 IAA = IA1 - IASTP
2250 FOR IBB = IB1 TO IB2 STEP IBSTP
2260 IAA = IAA + IASTP
2270 IF IBB = IAA THEN NNAR(IAA) = -1 ELSE NNAR(IBB) = IAA
2280 NEXT IBB: NEXT I
2290 '---------- Final Node Numbers in the Array --------
```

```
2300 NODE = 0
2310 FOR I = 1 TO NNT
2320 IF NNAR(I) = 0 GOTO 2360
2330 IF NNAR(I) > 0 GOTO 2350
2340 NODE = NODE + 1: NNAR(I) = NODE: GOTO 2360
2350 II = NNAR(I): NNAR(I) = NNAR(II)
2360 NEXT I
2370 '------------ Nodal Coordinates ---------------
2380 NN = NODE: NELM = 0
2390 DIM X(NN, 2), XP(8, 2), NOC(2 * NNT, NEN), MAT(2 * NNT)
2400 FOR KW = 1 TO NW: FOR KS = 1 TO NS
2410 KSW = NS * (KW - 1) + KS
2420 IF IDBLK(KSW) = 0 GOTO 2690
2430 '--------- Extraction of Block Data ----------
2440 NODW = NGCN(KSW + KW - 1) - NNS - 1
2450 FOR JW = 1 TO NWD(KW) + 1
2460 ETA = -1 + 2 * (JW - 1) / NWD(KW)
2470 NODW = NODW + NNS: NODS = NODW
2480 FOR JS = 1 TO NSD(KS) + 1
2490 XI = -1 + 2 * (JS - 1) / NSD(KS)
2500 NODS = NODS + 1: NODE = NNAR(NODS)
2510 GOSUB 2880: GOSUB 2990
2520 FOR J = 1 TO 2: C1 = 0: FOR I = 1 TO 8
2530 C1 = C1 + SH(I) * XP(I, J): NEXT I
2540 X(NODE, J) = C1: NEXT J
2550 '-------------------- Connectivity ----------------------
2560 IF JS = NSD(KS) + 1 OR JW = NWD(KW) + 1 GOTO 2680
2570 N1 = NODE: N2 = NNAR(NODS + 1)
2580 N4 = NNAR(NODS + NNS): N3 = NNAR(NODS + NNS + 1)
2590 NELM = NELM + 1: IF NEN = 3 GOTO 2640
2600 '-------------------- Quadrilateral Elements ----------------
2610 NOC(NELM, 1) = N1: NOC(NELM, 2) = N2: MAT(NELM) = IDBLK(KSW)
2620 NOC(NELM, 3) = N3: NOC(NELM, 4) = N4: GOTO 2680
2630 '-------------------- Triangular Elements -------------------
2640 NOC(NELM, 1) = N1: NOC(NELM, 2) = N2
2650 NOC(NELM, 3) = N3: MAT(NELM) = IDBLK(KSW)
2660 NELM = NELM + 1: NOC(NELM, 1) = N3: NOC(NELM, 2) = N4
2670 NOC(NELM, 3) = N1: MAT(NELM) = IDBLK(KSW)
2680 NEXT JS: NEXT JW
2690 NEXT KS: NEXT KW
2700 NE = NELM: IF NEN = 4 GOTO 2800
2710 '--------- Readjustment for Triangle Connectivity ----------
2720 NE2 = NE / 2: FOR I = 1 TO NE2
2730 I1 = 2 * I - 1: N1 = NOC(I1, 1): N2 = NOC(I1, 2)
2740 N3 = NOC(I1, 3): N4 = NOC(2 * I, 2)
2750 X13 = X(N1, 1) - X(N3, 1): Y13 = X(N1, 2) - X(N3, 2)
2760 X24 = X(N2, 1) - X(N4, 1): Y24 = X(N2, 2) - X(N4, 2)
2770 IF (X13 * X13 + Y13 * Y13) <= 1.1 * (X24 * X24 + Y24 * Y24) GOTO 2790
2780 NOC(I1, 3) = N4: NOC(2 * I, 3) = N2
2790 NEXT I
2800 GOSUB 3090
2810 END
2820 '========== Number of Divisions for Side I1,I2  ===========
2830 IMIN = I1: IMAX = I2: IF IMIN > I2 THEN IMIN = I2: IMAX = I1
2840 IF (IMAX - IMIN) = 1 THEN IDIV = NGCN(IMAX) - NGCN(IMIN): RETURN
2850 IDIV = (NGCN(IMAX) - NGCN(IMIN)) / NNS
2860 RETURN
2870 '====== Coordinates of 8-Nodes of the Block  ======
2880 N1 = KSW + KW - 1
2890 XP(1, 1) = XB(N1, 1): XP(1, 2) = XB(N1, 2)
2900 XP(3, 1) = XB(N1 + 1, 1): XP(3, 2) = XB(N1 + 1, 2)
2910 XP(5, 1) = XB(N1 + NS + 2, 1): XP(5, 2) = XB(N1 + NS + 2, 2)
2920 XP(7, 1) = XB(N1 + NS + 1, 1): XP(7, 2) = XB(N1 + NS + 1, 2)
2930 XP(2, 1) = SR(KSW, 1): XP(2, 2) = SR(KSW, 2)
2940 XP(6, 1) = SR(KSW + NS, 1): XP(6, 2) = SR(KSW + NS, 2)
```

```
2950 XP(8, 1) = WR(N1, 1): XP(8, 2) = WR(N1, 2)
2960 XP(4, 1) = WR(N1 + 1, 1): XP(4, 2) = WR(N1 + 1, 2)
2970 RETURN
2980 '=============== Shape Functions ================
2990 SH(1) = -(1 - XI) * (1 - ETA) * (1 + XI + ETA) / 4
3000 SH(2) = (1 - XI * XI) * (1 - ETA) / 2
3010 SH(3) = -(1 + XI) * (1 - ETA) * (1 - XI + ETA) / 4
3020 SH(4) = (1 - ETA * ETA) * (1 + XI) / 2
3030 SH(5) = -(1 + XI) * (1 + ETA) * (1 - XI - ETA) / 4
3040 SH(6) = (1 - XI * XI) * (1 + ETA) / 2
3050 SH(7) = -(1 - XI) * (1 + ETA) * (1 + XI - ETA) / 4
3060 SH(8) = (1 - ETA * ETA) * (1 - XI) / 2
3070 RETURN
3080 '=============== SAVE DATA ====================
3090 INPUT "File Name for Saving Data <eg A:FEM.DAT>"; FILE$
3100 OPEN FILE$ FOR OUTPUT AS #1
3110 INPUT "# Constrained DOF, # Component Loads"; ND, NL
3120 PRINT #1, NN; NE; ND; NL; NM; NDIM; NEN;
3130 FOR I = 1 TO NN: FOR J = 1 TO NDIM
3140 PRINT #1, X(I, J); : NEXT J: PRINT #1, : NEXT I
3150 FOR I = 1 TO NE: FOR J = 1 TO NEN
3160 PRINT #1, NOC(I, J); : NEXT J: PRINT #1, : NEXT I
3170 FOR I = 1 TO NE: PRINT #1, MAT(I); : NEXT I: PRINT #1,
3180 CLOSE #1
3190 RETURN


1000 '    ************** PLOT2D ***************
1010 PRINT "====================================="
1020 PRINT "      Program written by         "
1030 PRINT "  T.R.Chandrupatla and A.D.Belegundu  "
1040 PRINT "====================================="
1050 '======== Screen for Graphics =========
1060 KEY OFF: DEFINT I-N
1070 SCREEN 2: F$ = "####.##"
1080 ASP = .46 '*** Change Aspect Ratio
1090 '        *** if a square is not a square
1100 LOCATE 16, 25
1110 INPUT "Plot File Name <path fn.ext> ", FILE$
1120 OPEN FILE$ FOR INPUT AS #1
1130 INPUT #1, NN, NE, ND,NL,NM,NDIM, NEN
1140 DIM X(NN, NDIM), NOC(NE, NEN)
1150 '============= READ DATA ===============
1160 FOR I = 1 TO NN: FOR J = 1 TO NDIM
1170 INPUT #1, X(I, J): NEXT J: NEXT I
1180 FOR I = 1 TO NE: FOR J = 1 TO NEN
1190 INPUT #1, NOC(I, J): NEXT J: NEXT I
1200 XMAX = X(1, 1): YMAX = Y(1, 2): XMIN = X(1, 1): YMIN = X(1, 2)
1210 FOR I = 2 TO NN
1220 IF XMAX < X(I, 1) THEN XMAX = X(I, 1)
1230 IF YMAX < X(I, 2) THEN YMAX = X(I, 2)
1240 IF XMIN > X(I, 1) THEN XMIN = X(I, 1)
1250 IF YMIN > X(I, 2) THEN YMIN = X(I, 2)
1260 NEXT I
1270 CLS : XL = (XMAX - XMIN): YL = (YMAX - YMIN)
1280 X0 = XMIN - XL / 10: Y0 = YMIN - YL / 10
1290 LINE (100, 1)-(100, 169)
1300 LINE (100, 169)-(639, 169)
1310 VIEW (101, 1)-(639, 168)
1320 AA = 538 * ASP / 167
1330 IF XL / YL > AA THEN YL = XL / AA
1340 IF XL / YL < AA THEN XL = YL * AA
```

```
1350 XMAX = X0 + 1.2 * XL: YMAX = Y0 + 1.2 * YL
1360 WINDOW (X0, Y0)-(XMAX, YMAX)
1370 LOCATE 1, 5: PRINT USING F$; YMAX
1380 LOCATE 23, 73: PRINT USING F$; XMAX
1390 LOCATE 23, 10: PRINT USING F$; X0
1400 LOCATE 21, 5: PRINT USING F$; Y0
1410 '=========== Draw Elements ================
1420 CLS : FOR IE = 1 TO NE
1430 FOR II = 1 TO NEN
1440 X1 = X(NOC(IE, II), 1): Y1 = X(NOC(IE, II), 2)
1450 IF II = NEN GOTO 1490
1460 X2 = X(NOC(IE, II + 1), 1): Y2 = X(NOC(IE, II + 1), 2)
1470 LINE (X1, Y1)-(X2, Y2)
1480 GOTO 1510
1490 X2 = X(NOC(IE, 1), 1): Y2 = X(NOC(IE, 1), 2)
1500 LINE (X1, Y1)-(X2, Y2)
1510 NEXT II: NEXT IE
1520 '================ Node Numbers ==================
1530 LOCATE 24, 20: INPUT ; " Do You want Node Numbers < Y/N > ", A$
1540 IF A$ = "N" OR A$ = "n" GOTO 1610
1550 FOR I = 1 TO NN
1560 ICOL = (100 + (538 * (X(I, 1) - X0) / (1.2 * XL))) / 8
1570 IROW = (168 - 167 * (X(I, 2) - Y0) / (1.2 * YL)) / 8
1580 LOCATE IROW, ICOL: PRINT I; : NEXT I: LOCATE 24, 20
1590 INPUT ; "Plot without Node Numbers < Y/N > ", A$
1600 IF A$ = "Y" OR A$ = "y" GOTO 1420
1610 END


1000 ' *************** DATAFEM *****************
1010 PRINT "====================================="
1020 PRINT "       Program written by        "
1030 PRINT " T.R.Chandrupatla and A.D.Belegundu   "
1040 PRINT "====================================="
1050 DEFINT I-N
1060 CLS
1070 LOCATE 10, 30: PRINT "1. Create New Data"
1080 LOCATE 11, 30: PRINT "2. Edit Data"
1090 LOCATE 12, 30: PRINT "3. Edit Data from MESHGEN"
1100 LOCATE 14, 25: INPUT " Your Choice <1,2,3> ", IFL1
1110 ON IFL1 GOTO 1120, 1200, 1200
1120 CLS : LOCATE 10, 25: INPUT "Number of Nodes ", NN
1130 LOCATE 11, 25: INPUT "Number of Elements ", NE
1140 LOCATE 12, 25: INPUT "Number of Constrained DOF ", ND
1150 LOCATE 13, 25: INPUT "Number of Component Loads ", NL
1160 LOCATE 14, 25: INPUT "Number of Materials ", NM
1170 LOCATE 15, 25: INPUT "Dimension of the Problem <1,2,3> ", NDIM
1180 LOCATE 16, 25: INPUT "Number of Nodes per Element ", NEN
1190 GOTO 1230
1200 LOCATE 16, 25: INPUT "Edit File Name <path fn.ext> ", FILE$
1210 OPEN FILE$ FOR INPUT AS #1
1220 INPUT #1, NN, NE, ND, NL, NM, NDIM, NEN
1230 DIM X(NN, NDIM), NOC(NE, NEN), MAT(NE)
1240 DIM NU(ND), U(ND), NF(NL), F(NL), PM(NM, 3)
1250 GOSUB 1500
1260 ON IFL1 GOTO 1270, 1320, 1320
1270 FOR I = 1 TO NE: MAT(I) = 1: NEXT I
1280 IF IFL1 = 1 THEN IL1 = 1 ELSE IL1 = 3
1290 FOR IFL2 = IL1 TO 5: GOSUB 1590: GOSUB 2020: NEXT IFL2
```

```
1300 IFL1 = 2: GOTO 1340
1310 '++++++++++ Read Data  ++++++++++++++++
1320 GOSUB 2930: IF IFL1 = 3 GOTO 1280
1330 '++++++++++ Edit Data  ++++++++++++++++
1340 CLS : LOCATE 7, 30: PRINT "EDIT MENU"
1350 LOCATE 9, 25: PRINT "1.  Coordinate Data"
1360 LOCATE 10, 25: PRINT "2.  Connectivity / Material"
1370 LOCATE 11, 25: PRINT "3.  Boundary Data"
1380 LOCATE 12, 25: PRINT "4.  Load Data"
1390 LOCATE 13, 25: PRINT "5.  Material Properties"
1400 LOCATE 15, 25: PRINT "6.< SAVE DATA and EXIT >"
1410 LOCATE 16, 25: PRINT "7.  EXIT without Saving"
1420 LOCATE 18, 25: INPUT "Your Choice <1-7>  ", IFL2
1430 IF IFL2 < 1 OR IFL2 > 7 THEN GOTO 1440 ELSE GOTO 1450
1440 LOCATE 18, 44: PRINT "     ": GOTO 1420
1450 IF IFL2 = 7 GOTO 1480
1455 IF IFL2 = 6 GOTO 1470
1460 GOSUB 1590: GOSUB 2020: GOTO 1340
1470 GOSUB 3060
1480 END
1490 '=============== READ SCREEN & FORMAT DATA  ================
1500 DIM ANE$(9, 5), JCOL(5), NWID(5, 5)
1510 BLNK1$ = "        ": BLNK2$ = "             "
1520 F1$ = " #######.### ": F2$ = "  ##.###^^^^ "
1530 F3$ = "  ####   ": F4$ = "    #      "
1540 FOR I = 1 TO 9: FOR J = 1 TO 5: READ ANE$(I, J): NEXT J: NEXT I
1550 JCOL(1) = NDIM: JCOL(2) = NEN + 1: JCOL(3) = 2: JCOL(4) = 2: JCOL(5) = 3
1560 FOR I = 1 TO 5: FOR J = 1 TO 5: READ NWID(I, J): NEXT J: NEXT I
1570 RETURN
1580 '=============== SCREEN FOR DATA HANDLING  =================
1590 CLS : LOCATE 2, 1: FOR I = 1 TO 77: PRINT CHR$(196); : NEXT I
1600 LOCATE 23, 1: FOR I = 1 TO 77: PRINT CHR$(196); : NEXT I
1610 LOCATE 9, 1: FOR I = 1 TO 23: PRINT CHR$(196); : NEXT I
1620 LOCATE 15, 1: FOR I = 1 TO 23: PRINT CHR$(196); : NEXT I
1630 LOCATE 13, 3: ON IFL1 GOTO 1640, 1650
1640 PRINT "Entry for ": GOTO 1660
1650 PRINT " Number >"
1660 LOCATE 22, 3: PRINT "< RETURN >  to Skip"
1670 FOR I = 1 TO 23
1680 LOCATE I, 1: PRINT CHR$(179)
1690 LOCATE I, 78: PRINT CHR$(179)
1700 LOCATE I, 24: PRINT CHR$(179)
1710 LOCATE I, 33: PRINT CHR$(179)
1720 JWID = 33
1730 FOR J = 1 TO JCOL(IFL2)
1740 JWID = JWID + NWID(IFL2, J)
1750 LOCATE I, JWID: PRINT CHR$(179): NEXT J
1760 NEXT I
1770 LOCATE 1, 4: PRINT ANE$(1, IFL2): LOCATE 5, 3: PRINT ANE$(2, IFL2)
1780 LOCATE 7, 10: GOSUB 1950
1790 PRINT NGEN: LOCATE 13, 14: PRINT NGEN
1800 LOCATE 14, 3: PRINT ANE$(3, IFL2)
1810 '================= RESET SCREEN  =========================
1820 LOCATE 11, 3: PRINT ANE$(4, IFL2); : PRINT BLNK1$
1830 LOCATE 1, 25: COLOR 0, 7: PRINT ANE$(4, IFL2): COLOR 7, 0
1840 FOR J = 1 TO JCOL(IFL2)
1850 JCNST = 1: IF NWID(IFL2, J) > 9 THEN JCNST = 4
1860 J4 = J + 4: IF IFL2 = 2 AND J = JCOL(IFL2) THEN J4 = 9
```

```
1870 IF J > 1 GOTO 1890
1880 JWID = 34: GOTO 1900
1890 JWID = JWID + NWID(IFL2, J - 1)
1900 LOCATE 1, JWID + JCNST: PRINT ANE$(J4, IFL2)
1910 LOCATE 16 + J, 3: PRINT ANE$(J4, IFL2);
1920 PRINT BLNK2$: NEXT J
1930 RETURN
1940 '============= PICK THE RIGHT NUMBER  =============
1950 ON IFL2 GOTO 1960, 1970, 1980, 1990, 2000
1960 NGEN = NN: RETURN
1970 NGEN = NE: RETURN
1980 NGEN = ND: RETURN
1990 NGEN = NL: RETURN
2000 NGEN = NM: RETURN
2010 '============ INPUT OR CHANGE NODE DATA  ============
2020 INN = 0: INNA = 0: NEDM = 0
2030 GOSUB 2370
2040 LOCATE 11, 11: ON IFL1 GOTO 2050, 2060, 2050
2050 NEDM = NEDM + 1: PRINT NEDM: GOTO 2070
2060 INPUT " ", NEDM
2070 IF NEDM > NGEN GOTO 2350
2080 IF NEDM > 0 GOTO 2100
2090 LOCATE 11, 12: PRINT "       ": GOTO 2040
2100 INN = INT((10 * NEDM - 1) / 200): NNN = NEDM - INN * 20
2110 IF INN = INNA GOTO 2130
2120 GOSUB 2370
2130 LOCATE 1, 25: PRINT ANE$(4, IFL2): COLOR 0, 7
2140 LOCATE NNN + 2, 25: PRINT USING F3$; NEDM
2150 FOR J = 1 TO JCOL(IFL2)
2160 J4 = J + 4: IF IFL2 = 2 AND J = JCOL(IFL2) THEN J4 = 9
2170 JCNST = 1: IF NWID(IFL2, J) > 9 THEN JCNST = 4
2180 IF J > 1 GOTO 2200
2190 JWID = 34: GOTO 2210
2200 JWID = JWID + NWID(IFL2, J - 1)
2210 LOCATE 1, JWID + JCNST
2220 COLOR 0, 7: PRINT ANE$(J4, IFL2): COLOR 7, 0
2230 LOCATE 16 + J, 9: INPUT " ", DAT$
2240 IF LEFT$(DAT$, IFL2) = "" GOTO 2280
2250 LOCATE NNN + 2, JWID: IDMY = NEDM: JDMY = J
2260 VDAT = VAL(DAT$)
2270 ON IFL2 GOSUB 2530, 2600, 2680, 2780, 2880
2280 LOCATE 1, JWID + JCNST: PRINT ANE$(J4, IFL2)
2290 NEXT J
2300 LOCATE NNN + 2, 25: PRINT USING F3$; NEDM
2310 LOCATE 11, 12: PRINT "
2320 GOSUB 1820
2330 INNA = INN
2340 GOTO 2040
2350 RETURN
2360 '=================== PRINTING DATA ON SCREEN  ====================
2370 FOR I = 1 TO 20
2380 LOCATE I + 2, 25: INN2 = INN * 20 + I
2390 IF INN2 > NGEN THEN PRINT BLNK1$: GOTO 2410
2400 PRINT USING F3$; INN2
2410 NEXT I
2420 FOR J = 1 TO JCOL(IFL2)
2430 IF J > 1 GOTO 2450
2440 JWID = 34: GOTO 2460
```

```
2450 JWID = JWID + NWID(IFL2, J - 1)
2460 FOR I = 1 TO 20
2470 INN2 = INN * 20 + I: LOCATE I + 2, JWID
2480 IDMY = INN2: JDMY = J
2490 ON IFL2 GOSUB 2540, 2620, 2700, 2800, 2890
2500 NEXT I: NEXT J
2510 RETURN
2520 '=============== PRINT COORDINATES  ==================
2530 X(IDMY, JDMY) = VDAT: GOTO 2550
2540 IF IDMY > NGEN THEN PRINT BLNK2$: RETURN
2550 ABX = ABS(X(IDMY, JDMY)): IF X(IDMY, JDMY) = 0 GOTO 2580
2560 IF ABX > 1000000! OR ABX < .01 THEN FF$ = F2$ ELSE FF$ = F1$
2570 PRINT USING FF$; X(IDMY, JDMY): RETURN
2580 PRINT USING F4$; X(IDMY, JDMY): RETURN
2590 '=============== PRINT CONNECTIVITY  ================
2600 IF JDMY > NEN THEN MAT(IDMY) = VDAT: GOTO 2660
2610 NOC(IDMY, JDMY) = VDAT
2620 IF IDMY > NGEN THEN PRINT BLNK1$: RETURN
2630 IF JDMY > NEN GOTO 2660
2640 IF NOC(IDMY, JDMY) = 0 THEN PRINT BLNK1$: RETURN
2650 PRINT USING F3$; NOC(IDMY, JDMY): RETURN
2660 PRINT USING F3$; MAT(IDMY): RETURN
2670 '=============== PRINT BOUNDARY DATA  =================
2680 IF JDMY > 1 THEN U(IDMY) = VDAT: GOTO 2750
2690 NU(IDMY) = VDAT: GOTO 2720
2700 IF JDMY > 1 GOTO 2740
2710 IF IDMY > NGEN THEN PRINT BLNK1$: RETURN
2720 IF NU(IDMY) = 0 THEN PRINT BLNK1$: RETURN
2730 PRINT USING F3$; NU(IDMY): RETURN
2740 IF IDMY > NGEN THEN PRINT BLNK2$: RETURN
2750 IF U(IDMY) = 0 THEN FF$ = F4$ ELSE FF$ = F2$
2760 PRINT USING FF$; U(IDMY): RETURN
2770 '=============== PRINT LOAD DATA  =================
2780 IF JDMY > 1 THEN F(IDMY) = VDAT: GOTO 2850
2790 NF(IDMY) = VDAT: GOTO 2820
2800 IF JDMY > 1 GOTO 2840
2810 IF IDMY > NGEN THEN PRINT BLNK1$: RETURN
2820 IF NF(IDMY) = 0 THEN PRINT BLNK1$: RETURN
2830 PRINT USING F3$; NF(IDMY): RETURN
2840 IF IDMY > NGEN THEN PRINT BLNK2$: RETURN
2850 IF F(IDMY) = 0 THEN FF$ = F4$ ELSE FF$ = F2$
2860 PRINT USING FF$; F(IDMY): RETURN
2870 '============= PRINT MATERIAL DATA  ==============
2880 PM(IDMY, JDMY) = VDAT: GOTO 2900
2890 IF IDMY > NGEN THEN PRINT BLNK2$: RETURN
2900 IF PM(IDMY, JDMY) = 0 THEN FF$ = F4$ ELSE FF$ = F2$
2910 PRINT USING FF$; PM(IDMY, JDMY): RETURN
2920 '=============== READ DATA  ====================
2930 FOR I = 1 TO NN: FOR J = 1 TO NDIM
2940 INPUT #1, X(I, J): NEXT J: NEXT I
2950 FOR I = 1 TO NE: FOR J = 1 TO NEN
2960 INPUT #1, NOC(I, J): NEXT J: NEXT I
2970 FOR I = 1 TO NE: INPUT #1, MAT(I): NEXT I
2980 IF IFL1 = 3 GOTO 3030
2990 FOR I = 1 TO ND: INPUT #1, NU(I), U(I): NEXT I
3000 FOR I = 1 TO NL: INPUT #1, NF(I), F(I): NEXT I
3010 FOR I = 1 TO NM: FOR J = 1 TO 3
3020 INPUT #1, PM(I, J): NEXT J: NEXT I
```

```
3030 CLOSE #1
3040 RETURN
3050 '=============== SAVE DATA ===================
3060 INPUT "File Name for Saving Data <path fn.ext>"; FILE$
3070 OPEN FILE$ FOR OUTPUT AS #1
3080 PRINT #1, NN; NE; ND; NL; NM; NDIM; NEN
3090 FOR I = 1 TO NN: FOR J = 1 TO NDIM
3100 PRINT #1, X(I, J); : NEXT J: PRINT #1, : NEXT I
3110 FOR I = 1 TO NE: FOR J = 1 TO NEN
3120 PRINT #1, NOC(I, J); : NEXT J: PRINT #1, : NEXT I
3130 FOR I = 1 TO NE: PRINT #1, MAT(I); : NEXT I: PRINT #1,
3140 FOR I = 1 TO ND: PRINT #1, NU(I); U(I): NEXT I
3150 FOR I = 1 TO NL: PRINT #1, NF(I); F(I): NEXT I
3160 FOR I = 1 TO NM: FOR J = 1 TO 3
3170 PRINT #1, PM(I, J); : NEXT J: PRINT #1, : NEXT I
3180 CLOSE #1
3190 RETURN
3200 '=============== DATA FOR SCREENS ===============
3210 DATA "1. COORDINATES","2. CONNECTIVITY","3. BOUNDARY DATA"
3220 DATA            " 4. LOADS","5. PROPERTIES"
3230 DATA "NUM OF NODES","NUM OF ELEMENTS","NUM OF CONSTR. DOF"
3240 DATA         "NUM OF COMP. LOADS","NUM OF MATERIALS"
3250 DATA "Ends Node Data","Ends Element Data","Ends Boundary Data"
3260 DATA         "Ends Load Data","Ends Material Data"
3270 DATA "NODE NO","ELEM NO","SER NO","SER NO","MAT NO"
3280 DATA " X ","NODE1","DOF NO","DOF NO","PROP1"
3290 DATA " Y ","NODE2","DISPL.","LOAD","PROP2"
3300 DATA " Z ","NODE3"," "," ","PROP3"
3310 DATA " ","NODE4"," "," "," "
3320 DATA " ","MAT NO"," "," "," "
3330 DATA 15,15,15,0,0
3340 DATA 9,9,9,9,9
3350 DATA 9,15,0,0,0
3360 DATA 9,15,0,0,0
3370 DATA 15,15,15,0,0
3380 '_____

1000 PRINT "************ CONTOUR ***************"
1010 PRINT " Contour Plotting for Known Nodal Values "
1020 PRINT "====================================="
1030 PRINT "     Program written by       "
1040 PRINT " T.R.Chandrupatla and A.D.Belegundu   "
1050 PRINT "====================================="
1060 '======= Screen for Graphics =========
1070 KEY OFF: DEFINT I-N
1080 SCREEN 2: F$ = "####.##"
1090 ASP = .46 '*** Change Aspect Ratio
1100 '    *** if a square is not a square
1110 LOCATE 16, 25
1120 INPUT "Mesh Data File Name <path fn.ext> ", FILE1$
1130 INPUT "File Name Containing Nodal (FF) Values"; FILE2$
1140 OPEN FILE1$ FOR INPUT AS #1
1150 OPEN FILE2$ FOR INPUT AS #2
1160 INPUT #1, NN, NE, ND, NL, NM, NDIM, NEN
1170 DIM X(NN, NDIM), NOC(NE, NEN), NCON(NE, NEN), FF(NN)
1180 '============= READ DATA ===============
1190 FOR I = 1 TO NN: FOR J = 1 TO NDIM
1200 INPUT #1, X(I, J): NEXT J: NEXT I
1210 FOR I = 1 TO NE: FOR J = 1 TO NEN
```

```
1220 INPUT #1, NOC(I, J): NEXT J: NEXT I
1230 FOR I = 1 TO NN
1240 INPUT #2, FF(I): NEXT I
1250 CLOSE #1: CLOSE #2
1260 XMAX = X(1, 1): YMAX = Y(1, 2): XMIN = X(1, 1): YMIN = X(1, 2)
1270 FOR I = 2 TO NN
1280 IF XMAX < X(I, 1) THEN XMAX = X(I, 1)
1290 IF YMAX < X(I, 2) THEN YMAX = X(I, 2)
1300 IF XMIN > X(I, 1) THEN XMIN = X(I, 1)
1310 IF YMIN > X(I, 2) THEN YMIN = X(I, 2)
1320 IF FMAX < FF(I) THEN FMAX = FF(I)
1330 IF FMIN > FF(I) THEN FMIN = FF(I)
1340 NEXT I
1350 INPUT "Number of Contour Lines"; NCL
1360 STP = (FMAX - FMIN) / (NCL - 1)
1370 CLS : XL = (XMAX - XMIN): YL = (YMAX - YMIN)
1380 X0 = XMIN - XL / 10: Y0 = YMIN - YL / 10
1390 LINE (100, 1)-(100, 169)
1400 LINE (100, 169)-(639, 169)
1410 VIEW (101, 1)-(639, 168)
1420 AA = 538 * ASP / 167
1430 IF XL / YL > AA THEN YL = XL / AA
1440 IF XL / YL < AA THEN XL = YL * AA
1450 XMAX = X0 + 1.2 * XL: YMAX = Y0 + 1.2 * YL
1460 WINDOW (X0, Y0)-(XMAX, YMAX)
1470 LOCATE 1, 5: PRINT USING F$; YMAX
1480 LOCATE 23, 73: PRINT USING F$; XMAX
1490 LOCATE 23, 10: PRINT USING F$; X0
1500 LOCATE 21, 5: PRINT USING F$; Y0
1510 '============= Find Boundary Lines ===============
1520 'Edges defined by nodes in NOC to nodes in NCON
1530 FOR IE = 1 TO NE
1540 FOR I = 1 TO NEN
1550 I1 = I + 1: IF I1 > NEN THEN I1 = 1
1560 NCON(IE, I) = NOC(IE, I1): NEXT I: NEXT IE
1570 FOR IE = 1 TO NE
1580 FOR I = 1 TO NEN
1590 I1 = NCON(IE, I): I2 = NOC(IE, I)
1600 INDX = 0
1610 FOR JE = IE + 1 TO NE
1620 FOR J = 1 TO NEN
1630 IF NCON(JE, J) = 0 GOTO 1670
1640 IF I1 <> NCON(JE, J) AND I1 <> NOC(JE, J) GOTO 1670
1650 IF I2 <> NCON(JE, J) AND I2 <> NOC(JE, J) GOTO 1670
1660 NCON(JE, J) = 0: INDX = INDX + 1
1670 NEXT J: NEXT JE
1680 IF INDX > 0 THEN NCON(IE, I) = 0
1690 NEXT I: NEXT IE
1700 '============= Draw Boundary ==============
1710 CLS : FOR IE = 1 TO NE
1720 FOR I = 1 TO NEN
1730 IF NCON(IE, I) = 0 GOTO 1760
1740 I1 = NCON(IE, I): I2 = NOC(IE, I)
1750 LINE (X(I1, 1), X(I1, 2))-(X(I2, 1), X(I2, 2))
1760 NEXT I: NEXT IE
1770 '=========== Contour Plotting ===========
1780 FOR IE = 1 TO NE
1790 FOR I = 1 TO NCL
```

```
1800 FI = FMIN + (I - 1) * STP: II = 0
1810 FOR J = 1 TO NEN
1820 J1 = J: J2 = J + 1: IF J = NEN THEN J2 = 1
1830 N1 = NOC(IE, J1): N2 = NOC(IE, J2)
1840 IF FI > FF(N1) AND FI <= FF(N2) THEN GOTO 1870
1850 IF FI < FF(N1) AND FI >= FF(N2) THEN GOTO 1870
1860 GOTO 1930
1870 II = II + 1
1880 XI = (FI - FF(N1)) / (FF(N2) - FF(N1))
1890 XX(II) = XI * X(N2, 1) + (1 - XI) * X(N1, 1)
1900 YY(II) = XI * X(N2, 2) + (1 - XI) * X(N1, 2)
1910 IF II <> 2 GOTO 1930
1920 LINE (XX(1), YY(1))-(XX(2), YY(2))
1930 NEXT J: NEXT I: NEXT IE
1940 END


1000 PRINT "************** BESTFIT ***************"
1010 PRINT "Find Nodal Values Given Constant Element Values "
1020 PRINT "====================================="
1030 PRINT "       Program written by          "
1040 PRINT "  T.R.Chandrupatla and A.D.Belegundu   "
1050 PRINT "====================================="
1060 DEFINT I-N: CLS
1070 INPUT "Mesh Data File Name <path fn.ext> ", FILE1$
1080 INPUT "File Name Containing Element Values"; FILE3$
1090 OPEN FILE3$ FOR INPUT AS #3
1100 INPUT "Output Data File Name <path fn.ext>  ", FILE2$
1110 OPEN FILE1$ FOR INPUT AS #1
1120 OPEN FILE2$ FOR OUTPUT AS #2
1130 INPUT #1, NN, NE, ND, NL, NM, NDIM, NEN
1140 DIM X(NN, 2), NOC(NE, 3), FS(NE)
1150 FOR I = 1 TO NN: FOR J = 1 TO NDIM
1160 INPUT #1, X(I, J): NEXT J: NEXT I
1170 FOR I = 1 TO NE: FOR J = 1 TO NEN
1180 INPUT #1, NOC(I, J): NEXT J: NEXT I
1190 FOR I = 1 TO NE
1200 INPUT #3, FS(I): NEXT I
1210 CLOSE #1,#3
1220 ' BANDWIDTH NBW FROM CONNECTIVITY  "NOC"
1230 NQ = NN: NBW = 0
1240 FOR I = 1 TO NE
1250 CMIN = NN + 1: CMAX = 0
1260 FOR J = 1 TO 3
1270 IF CMIN > NOC(I, J) THEN CMIN = NOC(I, J)
1280 IF CMAX < NOC(I, J) THEN CMAX = NOC(I, J)
1290 NEXT J
1300 C = (CMAX - CMIN + 1)
1310 IF NBW < C THEN NBW = C
1320 NEXT I
1330 PRINT "THE BANDWIDTH IS"; NBW
1340 ' *** INITIALIZATION ***
1350 DIM S(NQ, NBW), F(NQ)
1360 FOR I = 1 TO NQ
1370 F(I) = 0
1380 FOR J = 1 TO NBW
1390 S(I, J) = 0: NEXT J: NEXT I
1400 ' *** GLOBAL STIFFNESS MATRIX **
1410 FOR N = 1 TO NE
```

```
1420 PRINT , "FORMING STIFFNESS MATRIX OF ELEMENT "; N
1430 GOSUB 1600
1440 PRINT "PLACING STIFFNESS IN GLOBAL LOCATIONS"
1450 FOR II = 1 TO 3
1460 NR = NOC(N, II): F(NR) = F(NR) + FE(II)
1470 FOR JJ = 1 TO 3
1480 NC = NOC(N, JJ) – NR + 1
1490 IF NC <= 0 GOTO 1510
1500 S(NR, NC) = S(NR, NC) + SE(II, JJ)
1510 NEXT JJ: NEXT II
1520 NEXT N
1530 ' *** EQUATION SOLVING ***
1540 GOSUB 5000
1550 PRINT "NODE NO. FUNCTION VALUE"
1560 FOR I = 1 TO NN
1570 PRINT #2, F(I): NEXT I
1580 CLOSE #2
1590 END
1600 'ELEMENT STIFFNESS FORMATION
1610 I1 = NOC(N, 1): I2 = NOC(N, 2): I3 = NOC(N, 3)
1620 X1 = X(I1, 1): Y1 = X(I1, 2)
1630 X2 = X(I2, 1): Y2 = X(I2, 2)
1640 X3 = X(I3, 1): Y3 = X(I3, 2)
1650 X21 = X2 – X1: X32 = X3 – X2: X13 = X1 – X3
1660 Y12 = Y1 – Y2: Y23 = Y2 – Y3: Y31 = Y3 – Y1
1670 DJ = X13 * Y23 – X32 * Y31      'DETERMINANT OF JACOBIAN
1680 AE = ABS(DJ) / 24
1690 SE(1, 1) = 2 * AE: SE(1, 2) = AE: SE(1, 3) = AE
1700 SE(2, 1) = AE: SE(2, 2) = 2 * AE: SE(2, 3) = AE
1710 SE(3, 1) = AE: SE(3, 2) = AE: SE(3, 3) = 2 * AE
1720 A1 = FS(N) * ABS(DJ) / 6
1730 FE(1) = A1: FE(2) = A1: FE(3) = A1
1740 RETURN
5000 N1 = NQ – 1
5010 REM *** FORWARD ELIMINATION ***
5020 FOR K = 1 TO N!
5030 NK = NQ – K + 1
5040 IF NK > NBW THEN NK = NBW
5050 FOR I = 2 TO NK
5060 C1 = S(K, I) / S(K, 1)
5070 I1 = K + I – 1
5080 FOR J = I TO NK
5090 J1 = J – I + 1
5100 S(I1, J1) = S(I1, J1) – C1 * S(K, J): NEXT J
5110 F(I1) = F(I1) – C1 * F(K): NEXT I: NEXT K
5120 REM *** BACK SUBSTITUTION ***
5130 F(NQ) = F(NQ) / S(NQ, 1)
5140 FOR KK = 1 TO N1
5150 K = NQ – KK
5160 C1 = 1 / S(K, 1)
5170 F(K) = C1 * F(K)
5180 NK = NQ – K + 1
5190 IF NK > NBW THEN NK = NBW
5200 FOR J = 2 TO NK
5210 F(K) = F(K) – C1 * S(K, J) * F(K + J – 1)
5220 NEXT J
5230 NEXT KK
5240 RETURN
```

PROBLEMS

12.1. Use program MESHGEN to generate finite element meshes for the regions in Figs. P12.1a and b. Generate meshes using both triangular and quadrilateral elements. For the fillet in P12.1a, use $y = 42.5 - 0.5x + x^2/360$.

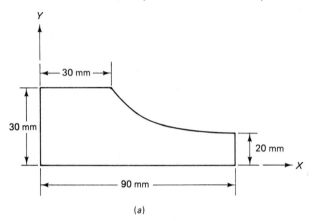

(a)

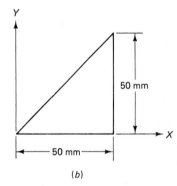

(b)   **Figure P12.1**

12.2. Generate a "graded" mesh for the region in Fig. P12.1a so that there are more elements near the left edge of the region. That is, the mesh density reduces along the $+x$ direction. Use MESHGEN with displaced midside nodes.

12.3. Use program CONTOUR to draw isotherms for the temperature distribution obtained in Example 10.4.

12.4. After solving Problem P5.6 using program FE2CST:
   (a) Use program PLOT2D to plot the original and deformed shape. The deformed shape requires selecting a scaling factor and using Eq. 12.3.
   (b) Use programs BESTFIT and CONTOUR and plot contours of maximum principal stress.

*12.5. Analyze the machine tool component shown in Fig. P12.5.
   (a) Use MESHGEN to generate the finite element mesh with triangles.
   (b) Use program FE2CST to analyze the problem.

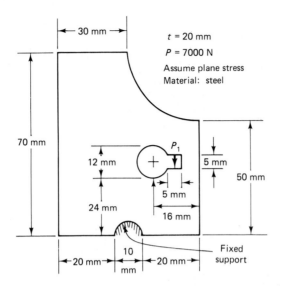

30 mm

t = 20 mm
P = 7000 N
Assume plane stress
Material: steel

70 mm

12 mm

P_1

5 mm

50 mm

5 mm

24 mm

16 mm

10
mm

20 mm   20 mm

Fixed
support

Figure P12.5

(c) Use program PLOT2D to draw both the original shape and the deformed shape.

(d) Use program BESTFIT to obtain nodal Von Mises stress values and then use program CONTOUR to draw contour lines.

12.6. Plot the mode shapes of the beam in Problem P11.4. For this you will need to modify PLOT2D and interface with INVITR.

*12.7. This problem illustrates the concept of a **dedicated** finite element program. Only design related parameters are input to the program, while mesh generation, boundary conditions and loading definition, finite element analysis, and postprocessing are automatically performed.

Consider the flywheel in Fig. P12.7. By modifying and interfacing programs MESHGEN, PLOT2D, AXYSYM, BESTFIT, and CONTOUR, develop a dedicated

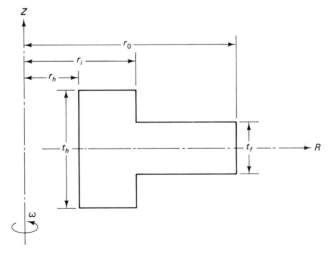

Figure P12.7

program that requires the user to input only the overall dimensions r_h, r_i, r_o, t_h, t_f and the values of E, ν, ρ, and ω. Your program may consist of independent programs executed through a batch or command file or can consist of one single program. Include the following features:

(a) A printout of all input data and output displacements and stresses.

(b) A plot of original and deformed shapes.

(c) Solve Problem P6.6. Provide contour plots of stress components.

12.8. Plot shearing stress contours for the torsion problem P10.11.

appendix

Proof of $dA = \det \mathbf{J} \, d\xi \, d\eta$

Consider a mapping of variables from x, y to u_1, u_2, given as

$$x = x(u_1, u_2) \qquad y = y(u_1, u_2) \tag{A1.1}$$

We assume that the above equations can be reversed to express u_1, u_2 in terms of x, y, and that the correspondence is unique.

If a particle moves from a point P in such a way that u_2 is held constant and only u_1 varies, then a curve in the plane is generated. We call this the u_1 curve (Fig. A1.1). Similarly, the u_2 curve is generated by keeping u_1 constant and letting u_2 vary. Let $\mathbf{r}$ represent the vector of a point P,

$$\mathbf{r} = x\mathbf{i} + y\mathbf{j} \tag{A1.2}$$

where $\mathbf{i}$, $\mathbf{j}$ are unit vectors along x, y, respectively. Consider the vectors

$$\mathbf{T}_1 = \frac{\partial \mathbf{r}}{\partial u_1} \qquad \mathbf{T}_2 = \frac{\partial \mathbf{r}}{\partial u_2} \tag{A1.3}$$

or, in view of Eq. A1.2,

$$\mathbf{T}_1 = \frac{\partial x}{\partial u_1}\mathbf{i} + \frac{\partial y}{\partial u_1}\mathbf{j} \qquad \mathbf{T}_2 = \frac{\partial x}{\partial u_2}\mathbf{i} + \frac{\partial y}{\partial u_2}\mathbf{j} \tag{A1.4}$$

We can show that $\mathbf{T}_1$ is a vector tangent to the u_1 curve and $\mathbf{T}_2$ is tangent to the u_2 curve (Fig. A1.1). To see this, we use the definition

$$\frac{\partial \mathbf{r}}{\partial u_1} = \lim_{\Delta u_1 \to 0} \frac{\Delta \mathbf{r}}{\Delta u_1} \tag{A1.5}$$

403

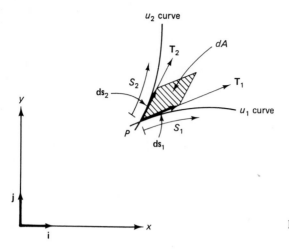

Figure A1.1

where $\Delta \mathbf{r} = \mathbf{r}(u_1 + \Delta u_1) - \mathbf{r}(u_1)$. In the limit, the chord $\Delta \mathbf{r}$ becomes the tangent to the u_1 curve (Fig. A1.2). However, $\partial \mathbf{r}/\partial u_1$ (or $\partial \mathbf{r}/\partial u_2$) is not a unit vector. To determine its magnitude (length), we write

$$\frac{\partial \mathbf{r}}{\partial u_1} = \frac{\partial \mathbf{r}}{\partial s_1} \frac{ds_1}{du_1} \tag{A1.6}$$

where s_1 is the arc length along the u_1 curve and ds_1 is the differential arc length. The magnitude of the vector

$$\frac{\partial \mathbf{r}}{\partial s_1} = \lim_{\Delta s_1 \to 0} \frac{\Delta \mathbf{r}}{\Delta s_1}$$

is the limiting ratio of the chord length to the arc length, which equals unity. Thus,

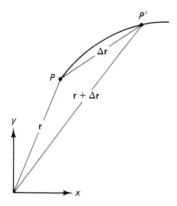

Figure A1.2

we conclude that the magnitude of the vector $\partial \mathbf{r}/\partial u_1$ is ds_1/du_1. We have

$$\mathbf{T}_1 = \left(\frac{ds_1}{du_1}\right)\mathbf{t}_1$$

$$\mathbf{T}_2 = \left(\frac{ds_2}{du_2}\right)\mathbf{t}_1$$

(A1.7)

where $\mathbf{t}_1$ and $\mathbf{t}_2$ are unit vectors tangent to the u_1 and u_2 curves, respectively. Using Eq. A1.7, we have the representation of the vectors $\mathbf{ds}_1$ and $\mathbf{ds}_2$ whose lengths are ds_1 and ds_2 (see Fig. A1.1):

$$\mathbf{ds}_1 = \mathbf{t}_1 \, ds_1 = \mathbf{T}_1 \, du_1$$

$$\mathbf{ds}_2 = \mathbf{t}_2 \, ds_2 = \mathbf{T}_2 \, du_2$$

(A1.8)

The differential area $\mathbf{dA}$ is a vector with magnitude dA and direction normal to the element area, which in this case is $\mathbf{k}$. The vector $\mathbf{dA}$ in view of Eqs. A1.4, and A1.8 is given by the determinant rule

$$\begin{aligned}
\mathbf{dA} &= \mathbf{ds}_1 \times \mathbf{ds}_2 \\
&= \mathbf{T}_1 \times \mathbf{T}_2 \, du_1 \, du_2 \\
&= \begin{vmatrix} \mathbf{i} & \mathbf{j} & \mathbf{k} \\ \dfrac{\partial x}{\partial u_1} & \dfrac{\partial y}{\partial u_1} & 0 \\ \dfrac{\partial x}{\partial u_2} & \dfrac{\partial y}{\partial u_2} & 0 \end{vmatrix} du_1, \, du_2 \\
&= \left(\frac{\partial x}{\partial u_1}\frac{\partial y}{\partial u_2} - \frac{\partial x}{\partial u_2}\frac{\partial y}{\partial u_1}\right) du_1 \, du_2 \, \mathbf{k}
\end{aligned}$$

(A1.9)

We denote the Jacobian matrix as

$$\mathbf{J} = \begin{bmatrix} \dfrac{\partial x}{\partial u_1} & \dfrac{\partial y}{\partial u_1} \\ \dfrac{\partial x}{\partial u_2} & \dfrac{\partial y}{\partial u_2} \end{bmatrix}$$

(A1.10)

The magnitude dA can now be written as

$$dA = \det \mathbf{J} \, du_1 \, du_2$$

(A1.11)

which is the desired result. Note that if we work with ξ, η coordinates instead of u_1, u_2 coordinates, as in the text, then

$$dA = \det \mathbf{J}\, d\xi\, d\eta$$

The relation derived above generalizes to three dimensions as

$$dV = \det \mathbf{J}\, d\xi\, d\eta\, d\zeta$$

where the Jacobian determinant $\det \mathbf{J}$ expresses the ratio of the volume element $dx\, dy\, dz$ to $d\xi\, d\eta\, d\zeta$.

Bibliography

There are many excellent published books and articles in various journals on the subject of the finite element method and its applications. A list of books published in the English language is given below. The list includes books dealing with both fundamental and advanced topics. Several of these have been used in the present text without explicit citation. On completing the material from various chapters from this book, the students and users should benefit from referring to other books and articles.

AKIN, J. E., *Application and Implementation of Finite Element Methods*. London: Academic, 1982.

AKIN, J. E., *Finite Element Analysis for Undergraduates*. London: Academic, 1986.

ALLAIRE, P. E., *Basics of the Finite Element Method—Solid Mechanics, Heat Transfer, and Fluid Mechanics*. Dubuque, IA: W. C. Brown, 1985.

AXELSSON, O., and V. A. BARKER, *Finite Element Solution of Boundary Value Problems*. Orlando, FL: Academic, 1984.

BAKER, A. J., *Finite Element Computational Fluid Mechanics*. New York: McGraw-Hill, 1983.

BARAN, N. M., *Finite Element Analysis on Microcomputers*. New York: McGraw-Hill, 1988.

BATHE, K. J., *Finite Element Procedures in Engineering Analysis*. Englewood Cliffs, NJ: Prentice-Hall, 1981.

BATHE, K. J., and E. L. WILSON, *Numerical Methods in Finite Element Analysis*. Englewood Cliffs, NJ: Prentice-Hall, 1976.

BECKER, E. B., G. F. CAREY, and J. T. ODEN, *Finite Elements—An Introduction*, Vol. 1. Englewood Cliffs, NJ: Prentice-Hall, 1981.

BICKFORD, W. M., *A First Course in the Finite Element Method*. Homewood, IL: Richard D. Irwin, 1990.

Bowes, W. H., and L. T. Russel, *Stress Analysis by the Finite Element Method for Practicing Engineers*. Lexington, MA: Lexington Books, 1975.

Brebbia, C. A., and J. J. Connor, *Fundamentals of Finite Element Techniques for Structural Engineers*. London: Butterworths, 1975.

Burnett, D. S., *Finite Element Analysis from Concepts to Applications*. Reading, MA: Addison-Wesley, 1987.

Carey, G. F., and J. T. Oden, *Finite Elements—A Second Course,* Vol. 2. Englewood Cliffs, NJ: Prentice-Hall, 1983.

Carey, G. F., and J. T. Oden, *Finite Elements—Computational Aspects,* Vol. 3. Englewood Cliffs, NJ: Prentice-Hall, 1984.

Chari, M. V. K., and P. P. Silvester, *Finite Elements in Electrical and Magnetic Field Problems*. New York: Wiley, 1981.

Cheung, Y. K., and M. F. Yeo, *A Practical Introduction to Finite Element Analysis*. London: Pitman, 1979.

Chung, T. J., *Finite Element Analysis in Fluid Dynamics*. New York: McGraw-Hill, 1978.

Ciarlet, P. G., *The Finite Element Method for Elliptic Problems*. Amsterdam: North-Holland, 1978.

Connor, J. C., and C. A. Brebbia, *Finite Element Techniques for Fluid Flow*. London: Butterworths, 1976.

Cook, R. D., *Concepts and Applications of Finite Element Analysis,* 2d ed. New York: Wiley, 1981.

Davies, A. J., *The Finite Element Method: A First Approach*. Oxford: Clarendon, 1980.

Desai, C. S., *Elementary Finite Element Method*. Englewood Cliffs, NJ: Prentice-Hall, 1979.

Desai, C. S., and J. F. Abel, *Introduction to the Finite Element Method*. New York: Van Nostrand Reinhold, 1972.

Fairweather, G., *Finite Element Galerkin Methods for Differential Equations*. New York: Dekker, 1978.

Fenner, R. T., *Finite Element Methods for Engineers*. London: Macmillan, 1975.

Gallagher, R. H., *Finite Element Analysis—Fundamentals*. Englewood Cliffs, NJ: Prentice-Hall, 1975.

Grandin, H., Jr., *Fundamentals of the Finite Element Method*. New York: Macmillan, 1986.

Hinton, E., and D. R. J. Owen, *Finite Element Programming*. London: Academic, 1977.

Hinton, E., and D. R. J. Owen, *An Introduction to Finite Element Computations*. Swansea, Great Britain: Pineridge Press, 1979.

Huebner, K. H., and E. A. Thornton, *The Finite Element Method for Engineers,* 2d ed. New York: Wiley-Interscience, 1982.

Hughes, T. J. R., *The Finite Element Method—Linear Static and Dynamic Finite Element Analysis*. Englewood Cliffs, NJ: Prentice-Hall, 1987.

Irons, B., and S. Ahmad, *Techniques of Finite Elements*. New York: Wiley, 1980.

Irons, B., and N. Shrive, *Finite Element Primer*. New York: Wiley, 1983.

Kikuchi, N., *Finite Element Methods in Mechanics*. Cambridge, Great Britain: Cambridge University Press, 1986.

LIVESLEY, R. K., *Finite Elements: An Introduction for Engineers*. Cambridge, Great Britain: Cambridge University Press, 1983.

LOGAN, D. L., *A First Course in the Finite Element Method*. Boston: PWS, 1986.

MARTIN, H. C., and G. F. CAREY, *Introduction to Finite Element Analysis: Theory and Application*. New York: McGraw-Hill, 1972.

MITCHELL, A. R., and R. WAIT, *The Finite Element Method in Partial Differential Equations*. New York: Wiley, 1977.

NAKAZAWA, S., and D. W. KELLY, *Mathematics of Finite Elements—An Engineering Approach*. Swansea, Great Britain: Pineridge Press, 1983.

NATH, B., *Fundamentals of Finite Elements for Engineers*. London: Athlone, 1974.

NAYLOR, D. J., and G. N. PANDE, *Finite Elements in Geotechnical Engineering*. Swansea, Great Britain: Pineridge Press, 1980.

NORRIE, D. H., and G. DE VRIES, *The Finite Element Method: Fundamentals and Applications*. New York: Academic, 1973.

NORRIE, D. H., and G. DE VRIES, *An Introduction to Finite Element Analysis*. New York: Academic, 1978.

ODEN, J. T., *Finite Elements of Nonlinear Continua*. New York: McGraw-Hill, 1972.

ODEN, J. T., and G. F. CAREY, *Finite Elements: Mathematical Aspects,* Vol. 4. Englewood Cliffs, NJ: Prentice-Hall, 1982.

ODEN, J. T., and J. N. REDDY, *An Introduction to the Mathematical Theory of Finite Elements*. New York: Wiley, 1976.

OWEN, D. R. J., and E. HINTON, *A Simple Guide to Finite Elements*. Swansea, Great Britain: Pineridge Press, 1980.

PAO, Y. C., *A First Course in Finite Element Analysis*. Newton, MA: Allyn & Bacon, 1986.

PINDER, G. F., and W. G. GRAY, *Finite Element Simulation in Surface and Subsurface Hydrology*. New York: Academic, 1977.

POTTS, J. F., and J. W. OLER, *Finite Element Applications with Microcomputers*. Englewood Cliffs, NJ: Prentice-Hall, 1989.

PRZEMIENIECKI, J. S., *Theory of Matrix Structural Analysis*. New York: McGraw-Hill, 1968.

RAO, S. S., *The Finite Element Method in Engineering,* 2d ed. Elmsford, NY: Pergamon, 1989.

REDDY, J. N., *An Introduction to the Finite Element Method*. New York: McGraw-Hill, 1984.

REDDY, J. N., *Energy and Variational Methods in Applied Mechanics with an Introduction to the Finite Element Method*. New York: Wiley-Interscience, 1984.

ROBINSON, J., *An Integrated Theory of Finite Element Methods*. New York: Wiley-Interscience, 1973.

ROBINSON, J., *Understanding Finite Element Stress Analysis*. Wimborne, Great Britain: Robinson and Associates, 1981.

ROCKEY, K. C., H. R. EVANS, D. W. GRIFFITHS, and D. A. NETHERCOT, *The Finite Element Method—A Basic Introduction*, 2d ed. New York: Halsted (Wiley), 1980.

ROSS, C. T. F., *Finite Element Programs for Axisymmetric Problems in Engineering*. Chichester, Great Britain: Ellis Horwood, 1984.

SEGERLIND, L. J., *Applied Finite Element Analysis,* 2d ed. New York: Wiley, 1984.

SHAMES, I. H., and C. L. DYM, *Energy and Finite Element Methods in Structural Mechanics.* New York: McGraw-Hill, 1985.

SILVESTER, P. P., and R. L. FERRARI, *Finite Elements for Electrical Engineers.* Cambridge, Great Britain: Cambridge University Press, 1983.

SMITH, I. M., *Programming the Finite Element Method.* New York: Wiley, 1982.

STASA, F. L., *Applied Finite Element Analysis for Engineers.* New York: Holt, Rinehart & Winston, 1985.

STRANG, G., and G. FIX, *An Analysis of the Finite Element Method.* Englewood Cliffs, NJ: Prentice-Hall, 1973.

TONG, P., and J. N. ROSSETOS, *Finite Element Method—Basic Techniques and Implementation.* Cambridge, MA: MIT Press, 1977.

URAL, O., *Finite Element Method: Basic Concepts and Applications.* New York: Intext Educational Publishers, 1973.

WACHSPRESS, E. L., *A Rational Finite Element Basis.* New York: Academic, 1975.

WAIT, R., and A. R. MITCHELL, *Finite Element Analysis and Applications.* New York: Wiley, 1985.

WHITE, R. E., *An Introduction to the Finite Element Method with Applications to Non-Linear Problems.* New York: Wiley-Interscience, 1985.

WILLIAMS, M. M. R., *Finite Element Methods in Radiation Physics.* Elmsford, NY: Pergamon, 1982.

YANG, T. Y., *Finite Element Structural Analysis,* Englewood Cliffs, NJ: Prentice-Hall, 1986.

ZIENKIEWICZ, O. C., *The Finite Element Method,* 3d ed. New York: McGraw-Hill, 1977.

ZIENKIEWICZ, O. C., and K. MORGAN, *Finite Elements and Approximation.* New York: Wiley-Interscience, 1982.

Index

AXYSYM computer program, 185–89
Area coordinates, 135
Aspect ratio, 152
Assembly
  of global stiffness **K** and load vector **F**, 58, 68, 107
  of global stiffness for banded solution, 115
  of global stiffness for skyline solution, 116
Axisymmetric elements, 165–93, 221, 329

B matrix, 51, 81, 139, 171, 199, 255
BEAM computer program, 241–43
BEAM_KM computer program, 361–63
BESTFIT computer program, 398–99
Bandwidth, 34, 62, 115–16, 215, 243, 266–67, 320, 374
Banded matrix, 34, 61, 115–16
Beams, 223
Beams on elastic supports, 234
Bending moment, 223, 232
Bibliography, 18, 407

Body forces 2–3, 54–55, 59–61, 82–85, 141–43, 172–73, 183–84, 200, 256
Boundary Conditions
  continuum, 4–5
  elimination approach, 64–69
  heat conduction, 277–78, 280, 282–85, 290–91, 294–95
  multipoint constraints, 63, 75–79, 151, 181, 263–64
  natural, 283–84
  penalty approach, 69–79
  scalar field problems, 302, 309–15
  specified displacement, 63–75, 150, 180, 183, 231, 263

CONTOUR computer program, 396–98
CST element. *See* Constant strain triangle.
CST_KM computer program, 363–67
Characteristic equation, 27, 344–47
Computer programs. *See* acronyms in this index.
Concentrated forces, 3, 45–46, 52, 59, 61, 140, 144, 165, 225

Conduction shape factor, 328
Conical (Belleville) Spring, 181–82
Constant strain triangle (CST), 130–64, 168–93, 291–308, 318–33
Contact, 74–75
Contour plotting, 382–83

D matrix, 7–9, 168
DATAFEM computer program, 392–96
Data handling, 381

Dedicated finite element program, 401
Degenerate quadrilaterals, 207
Degrees of freedom, 46, 101, 131–32, 169, 226–27, 235–36, 252
Displacement vector
  global **Q**, 46–47, 101, 131, 226, 252
  local/element **q**, 47, 101–02, 132–33, 168–69, 194–95, 226–27, 236, 252
Distributed stiffness, 234–35
Ducts, flow in, 314–15
Dynamic analysis, 334

411

Eigenvalue-eigenvectors, 27–28, 337, 343–55
Eight-node quadrilateral element, 209–11
Electric field problems, 311–12
Element
    beam, 223–34
    beam on elastic support, 234–35
    frame, 235–40
    hexahedral, 261–62
    one-dimensional linear, 45–79, 278–89
    one-dimensional quadratic, 79–86
    quadrilateral
        four-node, 194–207
        eight-node, 209–11
        nine-node, 207–09
    tetrahedral, 252–60
    triangle, three-node, 130–64, 168–93, 291–308, 318–33
    triangle, six-node, 211–13
    truss, 100–19
Element connectivity, 47, 79, 131–33, 194–95, 252–53, 261
Element mass matrices
    axisymmetric triangle, 340–41
    beam element, 341–43
    constant strain triangle (CST), 339–40
    frame element, 342
    one-dimensional bar, 338
    quadrilateral, four-node, 341
    tetrahedral element, 342–43
    truss, 338, 343
Element matrices for heat conduction, 280, 287, 293–95
Element stiffness matrices
    axisymmetric solids, 172
    beam, 230
    beam on elastic supports, 235
    constant strain triangle, 141
    frame, 237
    hexahedron, 262
    isoparametric, higher order, 207, 212
    one-dimensional, linear, 53
    one-dimensional, quadratic, 82
    quadrilateral, four-node, 200, 204–05. See also Element matrices for heat conduction.

tetrahedron, 256
trusses, 104
Elimination approach for handling boundary conditions, 64–69
Equation solving, 29–39
Equilibrium equations, 4, 65–66, 70–71, 225
Examples of scalar field problems, table, 274

FE2CST computer program, 154–58
FE2QUAD computer program, 215–19
FE3TETRA computer program, 266–69
FEM1D computer program, 91–93
FRAME computer program, 243–45
Fanning friction factor, 314
Fins, one-dimensional, 285–89
Fixed-end reactions, 232
Flow, 309–11, 314–15
Flywheel, 173, 183–85, 401–02
Forces. See Body Forces, Surface traction, Concentrated forces.
Fourier's law, 275
Frames, planar, 235–40
Functional approach, 279, 293, 312

GAUSS_R computer program, 40
Galerkin approach, 14–17
    in elasticity, 56–58, 104, 144–46, 174–75, 225–26, 256
    for handling boundary conditions, 65–66, 70–71
    in scalar field problems, 280–82, 286–88, 296–98, 305–306, 312
Gaussian elimination, 29–36
    with column reduction, 37–39. See also Skyline.
Gauss points and weights, tables, 203, 213
Gaussian quadrature, 201–205
Generalized eigenvalue problem, 337
Global stiffness matrix
    assembly, 58, 68, 107, 115–16

definition, 55, 58
    properties of, 61–62
Grid point stresses, 383–85

HEAT1D computer program, 317–18
HEAT2D computer program, 318–23
Half-bandwidth, 34, 62, 115–16, 215, 243, 266–67, 320, 374
Hamilton's principle, 335
Heat transfer, 275
    one-dimensional fins, 285–89
    one-dimensional heat conduction, 276–85
    two-dimensional heat conduction, 289–302
Helmholz, equation, 273
Hermite shape functions, 227–29
Hexahedron element, 261–62
Higher-order elements, 207–13, 261–62
Historical background, 1–2, 18
Hydraulic potential, 310–11

INVITR computer program, 356–59
Inclined rollers, 151. See also Multipoint constraints.
Initial strain. See Temperature effects.
Integration formula, tetrahedron, 255
Inverse iteration, 347–49
Isoparametric representation
    one-dimensional, 49–50
    quadrilateral, 196, 207–12
    three-dimensional 253–54, 262
    triangle, 135–36, 168–70, 291, 303
Isotherm, 275

JACOBI computer program, 359–61
Jacobi method, 350–54
Jacobian, 137, 254, 403

Kinetic energy, 336–37

Lagrangean, 334–35
Lagrange shape function, 49, 80, 195, 209, 261

Lame's constants, 19
Least square fit, 383–85
Loads. *See also* forces, Surface Fractions, Concentrated forces.
Lumped (diagonal) mass matrices, 343

MESHGEN computer program, 388–91
Magnetic field problems, 313–14
Mass matrix, derivation, 336–37
  matrices. *See* Element mass matrices.
Matrices
  eigenvalue-eigenvectors, 27–28, 337, 343–55
  diagonalization, 350–53
Matrix
  B, 51, 81, 139, 171, 199, 255
  D, 7–9, 168
  J, 137, 254, 403
  k. *See* Element stiffness.
  m. *See* Element mass.
  N. *See* Shape functions.
  adjoint, 26
  cofactor, 26
  determinant, 26, 43, 137, 254, 403
  positive definite, 28–29
  singular, 26
  upper triangular, 25, 32
Matrix algebra, 22–27
Matrix form, quadratic, 27
Mesh generation, 373–80
Mesh preparation with tetrahedra, 258–60, 262–66
Midside node, 213
Modeling. *See also* Boundary conditions.
  axisymmetric solids, 168–69, 179–84
  one-dimensional problems in elasticity, 45–48
  three-dimensional problems, 258–65
  two-dimensional problems in elasticity, 131–33, 150–52
Mode shapes. *See* Eigenvalue-eigenvectors.
Multipoint constraints, 63, 75–79, 151, 181, 263–64

Natural coordinates, 48, 79, 133–34, 195, 207–12, 227, 253, 261
Natural frequencies. *See* Eigenvalue-eigenvectors.
Nine-node quadrilateral element, 207–09
Nodal values from element values, 383–85
Node numbers
  global, 47, 101, 131–32, 226–27, 238, 252
  local, 47, 79, 102, 132, 195, 208, 210, 212, 227, 235, 252, 261
Nonlinearity, 181–82
Numerical integration, 172, 201–205, 212–13

One-dimensional problem
  elasticity, 44
  fin, 285
  heat conduction, 276
Orthogonal space, 348

PLOT2D computer program, 391
Penalty approach, 69–79, 151, 181, 263–64
  summary, 71
Plane strain, 9
Plane stress, 7–8
Plotting, 380–83
Potential energy, 10, 52, 55, 58, 63, 140, 165, 225. *See also* Functional approach.
  complementary, 304
Potential flow, 309–10
Preprocessing and Postprocessing, 372
Press fit, 179–81
Principal stresses, 146, 257
Principles
  Galerkin, 14–15
  Hamilton's, 335
  minimum potential energy, 10, 63
  Saint Venant, 18
  virtual work, 17

Quadratic shape functions, 79
Quadratic triangle, 211
Quadrilateral elements. *See* Element.

Rayleigh Ritz method, 12–13
Rayleigh quotient, 347
Reaction force, 65–66, 70–71
Reynolds number, 314

SKYLINE computer program, 40–42
Saint Venant's principle, 18
Scalar field problems
  electric and magnetic fields, 311–14
  flow in ducts, 314–15
  heat conduction, 275–302
  potential flow, 309–10
  seepage, 310–11
  torsion, 302–08
Seepage, 310–11
Serendipity elements, 209–11, 376
Shape functions
  axisymmetric, 168
  constant strain triangle (CST), 133–35
  Hermite, 227
  hexahedron, 261
  linear one-dimensional, 79–81
  quadratic one-dimensional 79–81
  quadrilateral
    four-node, 195–96
    eight-node, 209–11, 376
    nine-node, 207–09
  tetrahedron, 253, 341
  triangle, six-node, 211–12
  truss, 338
Shear force, 225, 232
Shifting, 348
Shrink fit, 179–81
Simultaneous equations, 29
  Gaussian elimination, 29
  skyline solution, 39
Six node triangular element, 211–13
Skyline
  stiffness assembly for, 116–17, 122–25
  theory and computer program, 37–42
Specified displacements, 63–75, 150, 180, 183, 231, 263
Stiffness matrix. *See* Element stiffness matrices, Global stiffness matrix.

Strain-displacement relations 6, 44, 51, 81, 130, 139, 167, 171, 199, 223, 255
Stream function, 309
Stress computations, 7–9, 51, 82, 86, 146, 149, 177–78, 205–206, 223, 257
Stress extrapolation, 184, 214, 383–85
Stress function, 302
Stress-strain relation, 7–9, 44, 131, 167. *See also* **D** matrix.
Stress tensor, component representation, 3
Sub-parametric element, 209
Summary of the finite element method, 66, 71
Surface Traction, 3, 44–45, 55, 143–44, 173–74, 201, 230, 237, 256–57

TORSION computer program, 323
TRUSS computer program, 119
TRUSSKY computer program, 122
Temperature effects
  axisymmetric solids, 178, 182–83
  combined heat transfer, stress analysis, 327
  constant strain triangle, plane stress and plane strain, 149–50
  initial strain, 9
  one-dimensional problems, 86–90
  trusses, 109–10
Tetrahedral element, 252
Three-dimensional problems, 251
Torsion, 302–08
Transformation matrix, 102, 114, 236

Triangular element
  linear, 130–64, 168–93, 291–308
  quadratic, 211–13
Trusses, 100
Two-dimensional problems
  elasticity, 130, 165, 194
  electric and magnetic fields, 311–14
  flow in ducts, 314–15
  heat conduction, 289
  potential flow, 309–10
  seepage, 310–11
  torsion, 302–08

Vibration, 334
Virtual work, principle, 17
Von Mises stress, 162

Winkler foundations, 234